Proceedings
of the
Steklov Institute of Mathematics

1980, Issue 1

Analytic Number Theory, Mathematical Analysis and their Applications

Collection of Papers
Dedicated to Academician
IVAN MATVEEVIČ VINOGRADOV
on his Eighty-fifth Birthday

edited by

S. M. Nikol′skiĭ

Translation of

ТРУДЫ
ордена Ленина
МАТЕМАТИЧЕСКОГО ИНСТИТУТА
имени В. А. СТЕКЛОВА

Том 143 (1977)

АКАДЕМИЯ НАУК

СОЮЗА СОВЕТСКИХ СОЦИАЛИСТИЧЕСКИХ РЕСПУБЛИК

ТРУДЫ

ордена Ленина

МАТЕМАТИЧЕСКОГО ИНСТИТУТА

имени В. А. СТЕКЛОВА

CXLIII

АНАЛИТИЧЕСКАЯ ТЕОРИЯ ЧИСЕЛ, МАТЕМАТИЧЕСКИЙ АНАЛИЗ И ИХ ПРИЛОЖЕНИЯ

сборник статей

Посвящается академику
Ивану Матвеевичу Виноградову
к его восьмидесятипятилетию

ИЗДАТЕЛЬСТВО "НАУКА" МОСКВА 1977

1980 *Mathematics Subject Classification*. Primary 06A10, 10A15, 10G10, 12A50, 41A55, 42A16, 42A20, 42A45, 42A55, 42C15, 46B15, 46E35, 49E15, 60E10, 60G50, 90D10, 90D26; Secondary 10B10, 10B15, 42A16, 42A20, 42A24.

ABSTRACT. The present collection of papers, dedicated to Academician Ivan Matveevič Vinogradov on his eighty-fifth birthday, is a continuation of volume 142 in this series. The papers—original work on various chapters of number theory, analysis and also their applications—are of interest to specialists and graduate students in mathematics.

Library of Congress Cataloging in Publication Data

Main entry under title:
Analytic number theory, mathematical analysis and their applications.
(Proceedings of the Steklov Institute of Mathematics; t. 143, 1980, issue 1 ISSN 0081-5438)
Translation of Analiticheskaîa teoriîa chisel, matematicheskiĭ analiz i ikh prilozheniîa.
Number in Russian series statement no. 143 (1977)
Includes bibliographies.
CONTENTS: Arhipov, G. I., Karacuba, A. A. and Cubarikov, V. N. An upper bound for the modulus of a multiple trigonometric sum.—Balašov, L. A. and Teljakovskiĭ, S. A. Some properties of lacunary series and the integrability of trigonometric series.—[etc.]
1. Numbers, Theory of—Addresses, essays, lectures. 2. Mathematical analysis—Addresses, essays, lectures. 3. Vinogradov, Ivan Matveevich, 1891– I. Vinogradov, Ivan Matveevich, 1891– II. Nikol′skiĭ, Sergeĭ Mikhaĭlovich. III. Series: Akademiîa nauk SSSR. Matematicheskiĭ institut. Proceedings of the Steklov Institute of Mathematics; 1980, issue 1.
QA1.A413 t. 143 [QA241] 510s [512′.7] 80-14363 ISBN 0-8218-3044-9

Translation authorized by the All-Union Agency for Author's Rights, Moscow

PROCEEDINGS OF THE STEKLOV INSTITUTE OF MATHEMATICS IN THE ACADEMY OF SCIENCES OF THE USSR

TABLE OF CONTENTS

RUSSIAN TABLE OF CONTENTS*

*The American Mathematical Society scheme for transliteration of Cyrillic may be found at the end of index issues of *Mathematical Reviews.*

Труды Мат. Инст. Стеклов.
143 (1977)

Proceedings of the Steklov
Institute of Mathematics
1980, Issue 1

AN UPPER BOUND FOR THE MODULUS OF A MULTIPLE TRIGONOMETRIC SUM

UDC 511

G. I. ARHIPOV, A. A. KARACUBA AND V. N. ČUBARIKOV

ABSTRACT. A general estimate is obtained for a multiple trigonometric sum with a polynomial in the exponent.

Bibliography: 11 titles.

This paper essentially completes the construction of a theory of multiple trigonometric sums with a polynomial in the exponent, which, in analogy with simple sums, will be called Weyl sums. This theory, in the main, repeats the method of Ivan Matveevič Vinogradov for estimates for Weyl sums (see [1] and [2]), which consists in reducing the estimation of an individual sum, by means of a shift of the interval of summation and raising to a power, to an estimation of the mean value of similar sums. The estimate of the mean value of a multiple trigonometric sum, like the corresponding theorem in the method of I. M. Vinogradov, carries the name of a mean-value theorem, and is central in the theory being developed. This theorem is proved by the p-adic method (see [3]–[5]), with the help of which the estimation of the number of solutions of a system of Diophantine equations is reduced to an estimation of the number of solutions of a system of congruences in which the unknowns run through a complete system of residues (see [3]–[8]). The theorem on the mean value of a double trigonometric sum was published in [6] and [7]; the theorem on the mean value of a multiple trigonometric sum is a straightforward generalization of it and was published in [8]; for completeness of exposition, we give here the full proof of the latter theorem (Theorem 1). The principal content of the present paper is constituted by two theorems, which are analogous to the theorems of I. M. Vinogradov on estimates for trigonometric sums with a polynomial of one variable [1], [2]. Theorem 2 gives an upper bound depending on the approximation of a coefficient in the polynomial by a rational fraction. In Theorem 3, a general upper bound is given for the modulus of a multiple Weyl trigonometric sum. For this, we need lemmas "on the intersection multiplicity of domains" (Lemmas 16 and 18; see [6] and [7]), upper bounds for rational multiple trigonometric sums and multiple trigonometric integrals (Lemmas 8, 12 and 13; see [9] and [10], and cf. also the lemmas of I. M. Vinogradov ([1], Chapter II, Lemma 4; Chapter III; Chapter IV, Lemma 7)).

We may remark that we consider the simplest and, at the same time, the most important cases of the mean-value theorem as well as multiple Weyl sums, namely the

1980 *Mathematics Subject Classification.* Primary 10G10.

cases when the unknowns or variables of summation are equivalent. By means of a slight modification of the proofs, the theorems mentioned above admit of a different kind of generalization and refinement.

NOTATION. A and Λ denote sets of real numbers $\alpha_{t_1,\ldots,t_r}$ and $\lambda_{t_1,\ldots,t_r}$, respectively, with the condition $0 \leqslant t_1, \ldots, t_r \leqslant n$, where $r \geqslant 1$ and $n \geqslant 2$ are integers;

$$A \times \Lambda = \sum_{t_1=0}^{n} \ldots \sum_{t_r=0}^{n} \alpha_{t_1,\ldots,t_r} \lambda_{t_1,\ldots,t_r};$$

$$f_A(x_1, \ldots, x_r) = \sum_{t_1=0}^{n} \ldots \sum_{t_r=0}^{n} \alpha_{t_1,\ldots,t_r} x_1^{t_1} \ldots x_r^{t_r};$$

$M = (n+1)^r$; Ω denotes the unit M-dimensional cube; $\exp(x) = e^x$; $J_n(k, P, \Lambda)$ denotes the number of solutions of the system of equations of the form

$$\sum_{j=1}^{2k} (-1)^j x_{1,j}^{t_1} \ldots x_{r,j}^{t_r} = \lambda_{t_1,\ldots,t_r}, \qquad 0 \leqslant t_1, \ldots, t_r \leqslant n, \tag{1}$$

where $\lambda_{t_1,\ldots,t_r}$ are fixed integers, and $x_{1,j}, \ldots, x_{r,j}$, $j = 1, \ldots, 2k$, are unknowns taking integral values from 1 to P. When Λ is the null set, the number of solutions of the system (1) will be denoted by $J_n(k, P)$. For positive integral m, by $S_m(A)$ we denote the trigonometric sum

$$S_m(A) = \sum_{x_1=1}^{P} \ldots \sum_{x_r=1}^{P} \exp(2\pi i m f_A(x_1, \ldots, x_r));$$

moreover, let $S_1(A) = S(A)$; the symbol

$$\left(\sum_{t_1=0}^{n} \ldots \sum_{t_r=0}^{n} \sum_{a_{t_1,\ldots,t_r}=0}^{p-1} \right)$$

signifies that the summation is carried out over sets of integers $a_{t_1,\ldots,t_r}$, where $0 \leqslant a_{t_1,\ldots,t_r} \leqslant p-1$, for all $t_1, \ldots, t_r$, $0 \leqslant t_1, \ldots, t_r \leqslant n$.

Chapter I

A THEOREM ON THE MEAN VALUE OF THE MODULUS OF A MULTIPLE TRIGONOMETRIC SUM

In this chapter, the fundamental theorem of the theory of multiple Weyl sums is proved. As stated above, the proof is carried out by the p-adic method, reducing the problem for an incomplete system of residues to one for a complete system of residues (see [3]–[8]). I. M. Vinogradov, having originated a new method for Waring's problem (see [2]), adapted it for estimating Weyl sums (see [1] and [2]), where essentially new results are obtained which are nearly the best possible. The connection between Vinogradov's method of solving Waring's problem–an estimation of the function $G(n)$ (see [2])–and his method of estimating Weyl sums is traced especially clearly in the monograph [1], where the most complete proof of the mean value theorem is given. In exactly the same way, the p-adic analogue of Waring's problem [3] allowed one to give a new p-adic proof of Vinogradov's mean-value theorem (see, for example, [4] or [5]), which, in turn, opened a way to the construction of a theory of multiple trigonometric sums (see [6]–[8]).

§1. Statement of the theorem and the simplest auxiliary assertions

THEOREM 1. *Let $\tau \geqslant 0$ and $P \geqslant 1$ be integers, $k(\tau, n) = M(\tau + r)$, and $u = u(\tau, n) = \min(\tau, rn)$. Then, for any integral $k \geqslant k(\tau, n)$,*

$$J_n(k, P) \leqslant k^{rnM+2M\tau} 3^{4r^2Mn\tau} (nr)^{rnMu} P^{2kr-\frac{rnM}{2}+\delta},$$

where

$$\delta = \delta(\tau, n) = \frac{rnM}{2}\left(1-\frac{1}{rn}\right)^{\tau}.$$

For proving Theorem 1, it is necessary to prove a number of auxiliary lemmas. The lemmas of this section are fairly simple and are similar to Lemma 1 of [5].

LEMMA 1. *The following relations hold*:

a) $J_n(k, P, \Lambda) = \int\limits_{\Omega} |S(A)|^{2k} \exp(-2\pi i A \times \Lambda)\, dA;$

b) $J_n(k, P, \Lambda) \leqslant J_n(k, P);$

c) $|S_m(A)|^{2k} = \sum\limits_{\Lambda} J_n(k, P, \Lambda) \exp(2\pi i m(A \times \Lambda));$

d) $\sum\limits_{\Lambda} J_n(k, P, \Lambda) = P^{2kr};$

e) $J_n(k, P) \geqslant (2k)^{-M} P^{2kr-\frac{rnM}{2}}.$

The proof is analogous to the proof of Lemma 1 of [5].

LEMMA 2. a) *If $x_{1,j}, \ldots, x_{r,j}$, $j = 1, \ldots, 2k$, are solutions of* (1) *for the null set Λ, then $x_{1,j} + a_1, \ldots, x_{r,j} + a_r$, $j = 1, \ldots, 2k$, is a solution of the system of equations* (1) *for a null set Λ.*

b) *If the solutions $x_{s,j}$, $s = 1, \ldots, r$, $j = 1, \ldots, 2k$, satisfy the condition of linear independence* mod *p, i.e. for every set of integers $c_{t_1,\ldots,t_r}$, not simultaneously congruent to* 0 (mod *p*) *there exists a k_0 such that*

$$\sum_{t_1=0}^{n} \cdots \sum_{t_r=0}^{n} c_{t_1,\ldots,t_r} x_{1,k_0}^{t_1} \cdots x_{r,k_0}^{t_r} \not\equiv 0 \pmod p, \tag{2}$$

then the condition of linear independence mod *p is satisfied by $x_{s,j} + a_s$, $s = 1, \ldots, r$, $j = 1, \ldots, 2k$.*

PROOF. a) By removing the parentheses in the expression

$$\sum_{j=1}^{2k} (-1)^j (x_{1,j} + a_1)^{t_1} \cdots (x_{r,j} + a_r)^{t_r}$$

and collecting together all the terms in $a_1^{v_1} \cdots a_r^{v_r}$, we can see that $x_{s,j} + a_s$, $s = 1, \ldots, r$, $j = 1, \ldots, 2k$, is also a solution of (1) for a null Λ.

b) We consider the set of numbers $c_{t_1,\ldots,t_r}$ not simultaneously congruent to 0 (mod *p*). Then

$$\sum_{t_1=0}^{n}\cdots\sum_{t_r=0}^{n} c_{t_1,\ldots,t_r}(x_{1,j}+a_1)^{t_1}\ldots(x_{r,j}+a_r)^{t_r}$$
$$=\sum_{t_1=0}^{n}\cdots\sum_{t_r=0}^{n} b_{t_1,\ldots,t_r}x_{1,j}^{t_1}\ldots x_{r,j}^{t_r}.$$

Let

$$u=\max_v\{v=t_1+\ldots+t_r\,|\,c_{t_1,\ldots,t_r}\not\equiv 0 \pmod p\}.$$

Then $b_{t_1,\ldots,t_r}=c_{t_1,\ldots,t_r}\not\equiv 0 \pmod p$ for $t_1+\ldots+t_r=u$. From the condition of linear independence of the $x_{s,j}$ it follows that for some j_0 the expression

$$\sum_{t_1=0}^{n}\cdots\sum_{t_r=0}^{n} b_{t_1,\ldots,t_r}x_{1,j_0}^{t_1}\ldots x_{r,j_0}^{t_r}$$

is not congruent to 0 (mod p). Part b) is thus proved.

§2. The lemma on a complete system

In this section, one of the fundamental lemmas is proved. The first statements and proofs of this lemma and its variants can be seen in [6] and [7]; the present proof is a repetition of that in [8].

LEMMA 3. *Let p be a prime and m a positive integer, and let T denote the number of solutions of the system of congruences*

$$\sum_{j=1}^{2M}(-1)^j x_{1,j}^{t_1}\ldots x_{r,j}^{t_r}\equiv\lambda_{t_1,\ldots,t_r} \pmod{p^{t_1+\ldots+t_r}} \tag{3}$$
$$(0\leqslant t_1,\ldots,t_r\leqslant n),$$

where $\lambda_{t_1,\ldots,t_r}$ are fixed integers, and $B\leqslant x_{s,j}<mp^{nr}+B$; $s=1,\ldots,r$; $j=1,\ldots,2M$. In addition, let the unknowns $x_{s,j}$ satisfy the condition (2) *of linear independence* mod p. *Then*

$$T\leqslant M!\,m^{2rM}p^{2Mr^2n-\frac{rnM}{2}}.$$

PROOF. Without any loss of generality we can assume that $B=0$ and $p>n$ (in the case $p\leqslant n$ there are no solutions of (3) satisfying the condition (2) of linear independence mod p). We represent each unknown in the form

$$x_{s,j}=x_{s,j,0}+px_{s,j,1}+\ldots+p^{rn-1}x_{s,j,rn-1}+p^{rn}x_{s,j,rn},$$
$$0\leqslant x_{s,j,0},\ldots,x_{s,j,rn-1}\leqslant p-1,\qquad 0\leqslant x_{s,j,rn}\leqslant m-1;$$
$$s=1,\ldots,r,\qquad j=1,\ldots,2M.$$

From (3) it follows that the following system of congruences must be satisfied:

$$\sum_{j=1}^{2M}(-1)^j x_{1,j,0}^{t_1}\ldots x_{r,j,0}^{t_r}\equiv\lambda_{t_1,\ldots,t_r}\pmod p \tag{4}$$
$$(0\leqslant t_1,\ldots,t_r\leqslant n).$$

Let us denote by T_0 the number of solutions of this system, and by $R_r(\nu)$ the number of solutions of the inequality $t_1+\ldots t_r\geqslant\nu$ in integers $t_1,\ldots,t_r$ with $0\leqslant t_1,\ldots,t_r\leqslant n$. We have

$$T_0 \leqslant p^{-M}\left(\sum_{t_1=0}^{n} \cdots \sum_{t_r=0}^{n} \sum_{a_{t_1,\ldots,t_r}=0}^{p-1}\right)\left|\sum_{x_1=0}^{p-1}\sum_{x_2=0}^{p-1} \exp\left(2\pi i f_A(x_1,\ldots,x_r)\right)\right|^{2M} = V,$$

where

$$a_{t_1,\ldots,t_r} = \frac{1}{p}\, a_{t_1,\ldots,t_r}.$$

We make the following change of variables:

$$\begin{aligned} x_1' &= x_1, \\ x_2' &= x_2 + x_1^{n+1}, \\ &\cdots\cdots\cdots \\ x_s' &= x_s + x_1^{(n+1)^{s-1}}, \\ &\cdots\cdots\cdots \\ x_r' &= x_r + x_1^{(n+1)^{r-1}}. \end{aligned}$$

We then get

$$V = p^{-M}\left(\sum_{t_1=0}^{n} \cdots \sum_{t_r=0}^{n} \sum_{a_{t_1,\ldots,t_r}=0}^{p-1}\right)\left|\sum_{x_1=0}^{p-1} \cdots \sum_{x_r=0}^{p-1} \exp\left(2\pi i f_A\left(x_1,\ x_2 + x_1^{n+1},\ \ldots,\ x_r + x_1^{(n+1)^{r-1}}\right)\right)\right|^{2M}.$$

Applying Hölder's inequality to V, we get

$$V \leqslant p^{(r-1)(2M-1)} \sum_{x_2,\ldots,x_r} V_1,$$

$$V_1 = p^{-M}\left(\sum_{t_1=0}^{n} \cdots \sum_{t_r=0}^{n} \sum_{a_{t_1,\ldots,t_r}=0}^{p-1}\right)\left|\sum_{x_1=0}^{p-1} \exp\left(2\pi i f_A\left(x_1,\ldots,x_r + x_1^{(n+1)^{r-1}}\right)\right)\right|^{2M},$$

where V_1 is the number of solutions, for fixed $x_2, \ldots, x_r$, of the following system of congruences:

$$\sum_{j=1}^{2M} (-1)^j x_{1,j}^{t_1}\left(x_2 + x_{1,j}^{n+1}\right)^{t_2} \cdots \left(x_r + x_{1,j}^{(n+1)^{r-1}}\right)^{t_r} \equiv 0 \pmod p \tag{5}$$

$$(0 \leqslant t_1, \ldots, t_r \leqslant n).$$

By removing the parentheses in the expression $x_{1,j}^{t_1} \ldots (x_r + x_{1,j}^{(n+1)^{r-1}})^{t_r}$, and collecting together all the terms in $x_2^{s_2} \ldots x_r^{s_r}$, we discover that (5) is equivalent to the system

$$\begin{gathered} x_{1,1} + \ldots + x_{1,M} \equiv x_{1,M+1} + \ldots + x_{1,2M} \pmod p, \\ \cdots\cdots\cdots\cdots\cdots\cdots \\ x_{1,1}^{M-1} + \ldots + x_{1,M}^{M-1} \equiv x_{1,M+1}^{M-1} + \ldots + x_{1,2M}^{M-1} \pmod p. \end{gathered}$$

The number of solutions of this system does not exceed $(M-1)!p^{M+1}$. Hence

$$T_0 \leqslant M!\, p^{(r-1)(2M-1)} p^{M+1} p^{r-1} = M!\, p^{2Mr-M+1}.$$

We fix $x_{s,j,0}, \ldots, x_{s,j,\nu-1}$; $s = 1, \ldots, r$; $j = 1, \ldots, 2M$. Let us consider the system

$$\sum_{j=1}^{2M} (-1)^j x_{1,j}^{t_1} \cdots x_{r,j}^{t_r} \equiv \lambda_{t_1,\ldots,t_r} \pmod{p^{\nu+1}},$$

$$t_1+\dots+t_r\geqslant\nu+1,\qquad 0\leqslant t_1,\dots,t_r\leqslant n,$$

and let

$$u_{s,j,\nu}=\sum_{\mu=0}^{\nu}p^{\mu}x_{s,j,\mu}.$$

Then

$$u_{s,j,\nu}=u_{s,j,\nu-1}+p^{\nu}x_{s,j,\nu},\qquad u_{1,j,\nu}^{t_1}\dots u_{r,j,\nu}^{t_r}$$

$$\equiv u_{1,j,\nu-1}^{t_1}\dots u_{r,j,\nu-1}^{t_r}+p^{\nu}\sum_{s=1}^{r}t_s\left(\prod_{q=1}^{r}u_{q,j,\nu-1}^{t_q}\right)u_{s,j,\nu-1}^{-1}x_{s,j,\nu}\pmod{p^{\nu+1}},$$

$$\sum_{j=1}^{2M}(-1)^j u_{1,j,\nu-1}^{t_1}\dots u_{r,j,\nu-1}^{t_r}\equiv\lambda_{t_1,\dots,t_r}\pmod{p^{\nu}}.$$

Therefore the $x_{s,j,\nu}$ are found, for some $\lambda'_{t_1,\dots,t_r}$, from the following system:

$$\begin{gathered}\sum_{j=1}^{2M}(-1)^j\sum_{s=1}^{r}t_s\left(\prod_{q=1}^{r}u_{q,j,\nu-1}^{t_q}\right)u_{s,j,\nu-1}^{-1}x_{s,j,\nu}\equiv\lambda'_{t_1,\dots,t_r}\pmod{p},\\ t_1+\dots+t_r\geqslant\nu+1,\qquad 0\leqslant t_1,\dots,t_r\leqslant n.\end{gathered}\tag{6}$$

Let T_ν be the number of solutions of this system. From (6) we form r subsystems. In the first subsystem, we put those congruences for which $t_1\geqslant\nu+1$, $t_2=\dots=t_r=0$; in the $(r-1)$th subsystem, those for which $t_1+\dots+t_{r-1}\geqslant\nu+1$, $t_{r-1}\neq 0$, $t_t=0$; and in the rth subsystem, those for which $t_1+\dots+t_r\geqslant\nu+1$, $t_r\neq 0$.

Let us consider the first subsystem of congruences:

$$\begin{gathered}\lambda''_{t_1,0,\dots,0}=\sum_{j=1}^{2M}(-1)^j u_{1,j,\nu-1}^{t_1-1}x_{1,j,\nu}\pmod{p},\\ n\geqslant t_1\geqslant\nu+1.\end{gathered}$$

By virtue of the condition (2), the congruences of this subsystem form a system of linearly independent congruences mod p. Hence the number of solutions of the first subsystem is at most $p^{2M-R_1(\nu+1)}$. Suppose that $x_{1,j,\nu},\dots,x_{s-1,j,\nu}$, $j=1,\dots,2M$, have been found. We estimate the number of solutions $x_{s,j,\nu}$ satisfying (6) for fixed $x_{1,j,\nu},\dots,x_{s-1,j,\nu}$, $j=1,\dots,2M$.

We consider the sth subsystem:

$$\begin{gathered}\lambda'_{t_1,\dots,t_s,0,\dots,0}\equiv\sum_{j=1}^{2M}(-1)^j\sum_{d=1}^{s}t_d\left(\prod_{q=1}^{s}u_{q,j,\nu-1}^{t_q}\right)u_{d,j,\nu-1}^{-1}x_{d,j,\nu}\pmod{p},\\ t_1+\dots+t_s\geqslant\nu+1,\qquad t_s\neq 0,\quad t_{s+1}=\dots=t_r=0.\end{gathered}$$

For some $\lambda''_{t_1,\dots,t_s,0,\dots,0}$ it is equivalent to the system

$$\begin{gathered}\lambda''_{t_1,\dots,t_s,0,\dots,0}\equiv\sum_{j=1}^{2M}(-1)^j\left(\prod_{q=1}^{s}u_{q,j,\nu-1}^{t_q}\right)u_{s,j,\nu-1}^{-1}x_{s,j,\nu}\pmod{p},\\ t_1+\dots+t_s\geqslant\nu+1,\qquad t_s\neq 0,\qquad t_{s+1}=\dots=t_r=0.\end{gathered}$$

The congruences of this system are linearly independent mod p (because of (2)); therefore, the number of the $x_{s,j,\nu}$, $j = 1, \ldots, 2M$, does not exceed $p^{2M-R_s(\nu+1)+R_{s-1}(\nu+1)}$.

Thus, for $\nu \leqslant rn-1$,

$$T_\nu \leqslant p^{2rM-R_r(\nu+1)}.$$

Trivially, we have $T_{rn} \leqslant m^{2rM}$. Hence

$$T_{rn} \leqslant m^{2rM}.$$

$$T \leqslant T_0 \ldots T_{rn-1} T_{rn} \leqslant M!\, m^{2rM} p^{2r^2nM - \sum_{\nu=1}^{rn} R_r(\nu)}. \tag{7}$$

Let $R_r^*(k)$ denote the number of solutions in integers $0 \leqslant t_1, \ldots, t_r \leqslant n$ of the equation $t_1 + \ldots + t_r = k$; then

$$\sum_{\nu=1}^{rn} R_r(\nu) = \sum_{\nu=1}^{rn}\sum_{k=\nu}^{rn} R_r^*(k) = \sum_{k=1}^{rn} kR_r^*(k) = \sum_{h_1=0}^{n} \ldots \sum_{h_r=0}^{n} (h_1 + \ldots + h_r) = \frac{rnM}{2}.$$

Substituting the quantity thus obtained in (7), we have the estimate stated in the lemma.

§3. A fundamental recurrence inequality

In this section we prove the second fundamental lemma—the analogue of the recurrence inequality of Lemma 3 of [5] (see [6]–[8]).

LEMMA 4. *Let $P \geqslant 1$, p a prime, $\frac{1}{2}P^{1/nr} < p \leqslant P^{1/nr}$, and let $P_1 = [Pp^{-1}] + 1$. Then*

$$J_n(k, P) \leqslant k^{2M} 3^{3Mr^2n-2} p^{2kr-2Mr+2Mr^2n-\frac{rnM}{2}} J_n(k-M, P_1)$$
$$+ 4(M-1)!\,(nr)^k p^{2kr-k} J_n(k, P_1).$$

PROOF. We separate all the solutions of the system (1) into two classes. In the first class, we put all those solutions for which the unknowns $x_{1,j}, \ldots, x_{r,j}$, both for $j = 1, 3, \ldots, 2k-1$ and for $j = 2, 4, \ldots, 2k$, satisfy the condition (2) of linear independence mod p. The remaining solutions are put in the second class.

Let J_1 denote the number of solutions of the first class, and J_2 those of the second class. Then $J_n(k, P) = J_1 + J_2$.

We introduce the notation

$$S(x_1, \ldots, x_r) = \sum_{z_1=0}^{P_1-1} \ldots \sum_{z_r=0}^{P_1-1} e^{2\pi i f_A(x_1+pz_1, \ldots, x_r+pz_r)}.$$

Let us estimate J_1. For every solution of the first class, one can indicate M even indices j such that the unknowns $x_{s,j}$, $s = 1, \ldots, r$, as a set satisfy the condition (2) of linear independence. The same holds for the odd indices too. Let J_1^* be the number of solutions of the first class for which the above-mentioned even indices are $2, 4, \ldots, 2M$, and the odd ones are $1, 3, \ldots, 2M-1$. Then, evidently,

$$J_1 \leqslant (C_k^M)^2 J_1^*,$$

where

$$J_1^* = \int_\Omega \left| \sum_{x_1, j} \cdots \sum_{x_{r,j}}' \prod_{j=1}^{M} S(x_{1,j}, \ldots, x_{r,j}) \right|^2 \left| \sum_{y_1=1}^{p} \cdots \sum_{y_r=1}^{p} S(y_1, \ldots, y_r) \right|^{2(k-M)} dA,$$

and the prime on the summation sign signifies that the sum is taken over all $x_{1,j}$, $\ldots, x_{r,j}$, $j = 1, \ldots, M$, satisfying the condition (2) of linear independence. Applying Hölder's inequality, we find that $J_1 \leqslant (C_k^M)^2 p^{r(2k-2M-1)} J_3$, where

$$J_3 = \int_\Omega \left| \sum_{x_{1,j}} \cdots \sum_{x_{r,j}}' \prod_{j=1}^{M} S(x_{1,j}, \ldots, x_{r,j}) \right|^2 \sum_{y_1} \cdots \sum_{y_r} | S(y_1, \ldots, y_r) |^{2(k-M)} dA.$$

It follows from Lemma 1a) that J_3 is the number of those solutions of the system

$$\sum_{j=1}^{2M} (-1)^j (x_{1,j} - y_1)^{t_1} \ldots (x_{r,j} - y_r)^{t_r} + \sum_{j=2M+1}^{2k} (-1)^j p^{t_1 + \ldots + t_r} z_{1,j}^{t_1} \ldots z_{r,j}^{t_r} = 0,$$
$$0 \leqslant t_1, \ldots, t_r \leqslant n, \quad t_1 + \ldots + t_r \geqslant 1,$$

for which $x_{1,j}, \ldots, x_{r,j}$, $j = 1, 3, \ldots, 2M - 1$ and $j = 2, 4, \ldots, 2M$, satisfy (2). By Lemma 1b), the condition (2) will also be satisfied by $x_{1,j} - y_1, \ldots, x_{r,j} - y_r$, $j = 1, \ldots, 2M$. Hence $J_3 \leqslant p^r T J_n(k - M, P_1)$, where T is the number of solutions of the system (3).

For estimating T, we utilize the lemma on a complete system. We get

$$J_1 \leqslant (C_k^M)^2 M!\,([P_1 p^{-nr+1}] + 1)^{2Mr} p^{2r^2 nM - \frac{rnM}{2}} p^{2kr - 2rM} J_n(k - M, P_1)$$
$$\leqslant k^{2M} 3^{3Mr^2 n - 2} p^{2kr - 2rM + 2r^2 nM - \frac{rnM}{2}} J_n(k - M, P_1).$$

Let us estimate J_2. For solutions of the second class, $x_{1,j}, \ldots, x_{r,j}$ do not satisfy, for $j = 1, 3, \ldots, 2k - 1$ or for $j = 2, 4, \ldots, 2k$, the condition (2) of linear independence mod p. Without any loss of generality, one can assume that for each $j = 1, 3, \ldots, 2k - 1$ there is a set of numbers $c_{t_1, \ldots, t_r}$, not all congruent to 0 (mod p), such that

$$\sum_{t_1=0}^{n} \cdots \sum_{t_r=0}^{n} c_{t_1, \ldots, t_r} x_{1,j}^{t_1} \ldots x_{r,j}^{t_r} = F(x_{1,j}, \ldots, x_{r,j}) \equiv 0 \pmod p. \qquad (8)$$

Thus

$$J_2 \leqslant 2J_4,$$
$$J_4 = \left(\sum_{t_1=0}^{n} \cdots \sum_{t_r=0}^{n} \sum_{c_{t_1, \ldots, t_r}=0}^{p-1}{}' \right) \int_\Omega S_1 \, dA,$$

$$S_1 = \sum_{\substack{x_{1,2j-1}=1 \\ F(x_{1,2j-1}, \ldots, x_{r,2j-1}) \equiv 0 \, (p) \\ j=1, \ldots, k}}^{p} \cdots \sum_{x_{r,2j-1}=1}^{p} \; \sum_{\substack{x_{1,2m}=1 \\ m=1, \ldots, k}}^{p} \cdots \sum_{x_{r,2m}=1}^{p}{}'' \prod_{j=1}^{k} S(x_{1,2j-1}, \ldots, x_{r,2j-1})$$
$$\times \prod_{m=1}^{k} \overline{S(x_{1,2m}, \ldots, x_{r,2m})};$$

the prime signifies that the summation is carried out over sets of numbers $c_{t_1, \ldots, t_r}$, not all congruent to 0 (mod p), where $c_{0,\dots,0}$ can take only two values 0 or 1, and Σ'' signifies that the summation is carried out over those $x_{1,j}, \dots, x_{r,j}$, $j = 1, \dots, 2k$, for which the following system is satisfied:

$$\sum_{j=1}^{k} x_{1,2j-1}^{t_1} \dots x_{r,2j-1}^{t_r} \equiv \sum_{j=1}^{k} x_{1,2j}^{t_1} \dots x_{r,2j}^{t_r} \pmod p,$$

$$0 \leqslant t_1, \dots, t_r \leqslant n.$$

Let V be the number of solutions of this system. For a polynomial $F(x_1, \dots, x_r)$, not all of whose coefficients are congruent to 0 (mod p) and the full degree of which does not exceed nr, the number of solutions of $F(x_1, \dots, x_r) \equiv 0 \pmod p$, $1 \leqslant x_s \leqslant p$, $s = 1, \dots, r$, does not exceed nrp^{r-1}.

Since $x_{1,j}, \dots, x_{r,j}$, $j = 1, 3, \dots, 2k-1$, satisfy (8), their number is at most $(nr)^k p^{k(r-1)}$. Consequently,

$$V \leqslant (nr)^k p^{k(r-1)} p^{kr-2rM} T_0,$$

where T_0 is the number of solutions of (4). For T_0 we have established the estimate

$$T_0 \leqslant (M-1)!\, p^{2M_r - M+1}.$$

Therefore

$$V \leqslant (M-1)!\,(nr)^k p^{2kr-k-M+1}.$$

For estimating $|S_1|$, we utilize the inequality of the arithmetic and geometric means. We get

$$|S_1| \leqslant \sum_{\substack{x_{1,j}=1 \\ F(x_{1,j}, \dots, x_{r,j}) \equiv 0\,(p) \\ j=1,3,\dots,2k-1}}^{p} \dots \sum_{x_{r,j}=1}^{p} \sum_{\substack{x_{1,j}=1 \\ j=2,\dots,2k}}^{p} \dots \sum_{x_{r,j}=1}^{p} \frac{1}{2k} \sum_{j=1}^{2k} |S(x_{1,j}, \dots, x_{r,j})|^{2k}.$$

Hence, from Lemma 2a), we conclude that

$$J_2 \leqslant 2\left(\sum_{t_1=0}^{n} \dots \sum_{t_r=0}^{n} \sum_{c_{t_1,\dots,t_r}=0}^{p-1}{}'\right)(M-1)!\,(nr)^k p^{2kr-k-M+1} J_n(k, P_1)$$
$$\leqslant 4(M-1)!\,(nr)^k p^{2kr-k} J_n(k, P_1)$$

This proves Lemma 4.

§4. Proof of Theorem 1

Evidently it is sufficient to prove the theorem for the case $k = k(\tau, n)$. For $\tau = 0$, the assertion of the theorem is valid for all $P \geqslant 1$. For $P \leqslant (2nr)^{2nr}$ and for $P \leqslant (2kr)^2$ the theorem holds for all $\tau \geqslant 1$, since

$$k^{rnM+2M\tau} \cdot 3^{4r^2nM\tau} (nr)^{Mrnu(\tau,n)} P^{-\frac{rnM}{2}+\delta(\tau,n)} \geqslant 1.$$

Let P_0 be an integer. We assume that the theorem holds for $P < P_0$, $\tau \leqslant m+1$ and for $P \leqslant P_0$, $\tau \leqslant m$. We shall prove the assertion of the theorem for $P \leqslant P_0$, $\tau = m+1$.

For estimating $J_n(k(m+1, n), P)$, we use Lemma 4. We get

$$J_n(k(m+1, n), P) \leqslant (k(m+1, n))^{2M} \cdot 3^{3Mr^2n-2} p^{2rk(m+1,n)+2r^2nM-\frac{rnM}{2}-2rM}$$
$$\times J_n(k(m+1, n)-M, P_1) + 4(M-1)!\,(nr)^{k(m+1,n)} p^{2rk(m+1,n)-k(m+1,n)}$$
$$\times J_n(k(m+1, n), P_1). \tag{9}$$

Since

$$k - M \geqslant Mm + rM, \qquad P_1 < P,$$

by virtue of the induction hypothesis we have

$$J_n(k(m, n), P_1) \leqslant (k(m, n))^{2Mm+rnM} \cdot 3^{4r^2nMm} (nr)^{nrMu(m,n)} P_1^{2rk(m,n)-\frac{rnM}{2}+\delta(m,n)},$$
$$J_n(k(m+1, n), P_1) \leqslant (k(m+1, n))^{2M(m+1)+rnM} \cdot 3^{4r^2nM(m+1)}$$
$$\times (nr)^{nrMu(m+1,n)} P_1^{2rk(m+1,n)-\frac{rnM}{2}+\delta(m+1,n)}.$$

We substitute these estimates in (9), and use the fact that

$$P_1^{\delta(m,n)} \leqslant 3^{r^2nM} P^{\delta(m+1,n)}, \qquad pP_1 \geqslant P,$$

$$8(M-1)!\,(nr)^k p^{-k+\frac{rnM}{2}-\delta(m+1,n)} \left(1+\frac{p}{P}\right)^{2kr} < 1 \quad \text{for } p > \frac{1}{2}(2nr)^2,$$

$$\left(1+\frac{p}{P}\right)^{2kr} < 3 \quad \text{for } P > (2kr)^2.$$

We then obtain the estimate affirmed in the theorem.

Chapter II

THE SIMPLEST ESTIMATE FOR A MULTIPLE TRIGONOMETRIC SUM

In this chapter, we obtain an estimate for a multiple trigonometric sum depending on the approximation of a coefficient in the polynomial by a rational fraction. The auxiliary lemmas, stated in §1, are due to I. M. Vinogradov [1], [2] and are given here without proofs. A slightly more precise estimate for a multiple trigonometric sum depending on the approximation of the highest coefficient in the polynomial by a rational fraction is obtained in [8].

§1. Auxiliary assertions

LEMMA 5. *To each integral y corresponds its own point $(Y_{n-1}, \ldots, Y_1)$ of $(n-1)$-dimensional space, determined by the expansion*

$$f(x+y) - f(y) = \alpha_n x^n + Y_{n-1}x^{n-1} + \ldots + Y_1 x$$

of the difference $f(x+y) - f(y)$ in powers of x.

Let Y be a positive integer less than P. A necessary condition that the point corresponding to some number y of the sequence $0, \ldots, Y$ by adding terms not exceeding

$$L_{n-1} = YP^{-n+1-k}, \ldots, L_1 = YP^{-1-k},$$

respectively, to its coordinates may be made comparable with the point corresponding to some definite y_0 *of the same sequence is the satisfaction, for* $s = n, \ldots, 2$, *of the inequalities*

$$\left(\frac{n!}{(s-1)!}\alpha_s(y-y_0)\right) \leqslant \frac{n!}{s!}\left(\frac{3}{2}n\right)^{n-s} L_{s-1},$$

$k \geqslant 0$ *being integral.*

The proof can be seen in [1], Chapter IV, Lemma 7.

LEMMA 6. *Suppose that* m, p *and* r *are integral,* $m > 0, p > 1, r > 1$,

$$\alpha = \frac{am}{q} + \frac{\theta m}{q^2}; \qquad (a, q) = 1; \qquad 0 < q < p^r; \ |\theta| \leqslant 1,$$

and let y *run through* $\leqslant Y$ *consecutive integers. Then, for* $W \geqslant 1$, *the number of values of* y *with* $(\alpha y) < Wp^{-r+1}$ *will be at most*

$$(3m + 2Wqp^{-r+1})\left(\frac{Y}{q} + 1\right).$$

The proof can be seen in [2], Lemma 4, p. 381.*

§2. Statement and proof of Theorem 2

THEOREM 2. *Suppose that, for some* $t_1, \ldots, t_r$ *such that* $t_1 + \ldots + t_r \geqslant 2$, $0 \leqslant t_1, \ldots, t_r \leqslant n$, *the relation*

$$\alpha_{t_1, \ldots, t_r} = \frac{a}{q} + \frac{\theta}{q^2}, \qquad (a, q) = 1, \qquad |\theta| \leqslant 1,$$
$$1 < q < P^{t_1 + \ldots + t_r},$$

holds. Define the number ω *as follows*:

$$q = P^{\omega}, \qquad \text{if } 1 < q < P;$$
$$\omega = 1, \qquad \text{if } P \leqslant q \leqslant P^{t_1 + \ldots + t_r - 1};$$
$$q = P^{t_1 + \ldots + t_r - \omega}, \qquad \text{if } P^{t_1 + \ldots + t_r - 1} < q < P^{t_1 + \ldots + t_r}.$$

Then

$$|S_m(A)| < 3^{10r^2 n + 31 \log \frac{\omega}{\rho}} m^{\frac{3\rho}{\omega}} P^{r-\rho},$$

where

$$\rho = \frac{\omega}{5Mrn \log \frac{3rnM}{2\omega}}.$$

PROOF. Without any loss of generality, one can assume that $t_1 > 0$ for the coefficient $\alpha_{t_1, \ldots, t_r}$ under consideration. For the variable x_1, we shift the interval of summation by a quantity y. Moreover,

$$S_m(A) = \sum_{x_1=1}^{P} \cdots \sum_{x_r=1}^{P} \exp(2\pi i m f(x_1 + y, x_2, \ldots, x_r)) + 2\theta_1 y P^{r-1};$$
$$|\theta_1| \leqslant 1.$$

Editor's note.* The reference is to Lemma 4 of I. M. Vinogradov, *An upper bound for the modules of a trigonometric sum*, Izv. Akad. Nauk SSSR Ser. Mat. **14 (1950), 199–214; English transl. in Amer. Math. Soc. Transl. (1) **2** (1962).

We sum the inequality thus obtained over integral values of y from 1 to $[P^{1-\rho}]$. We have

$$|S_m(A)| \leqslant W + 2YP^{r-1};$$

$$W = Y^{-1}\sum_{y=1}^{Y}\left|\sum_{x_1=1}^{P}\dots\sum_{x_r=1}^{P}\exp(2\pi i m f_{A(y)}(x_1,\dots,x_r))\right|;$$

$$f_{A(y)}(x_1,\dots,x_r) = \sum_{t_2=0}^{n}\dots\sum_{t_r=0}^{n} x_2^{t_2}\dots x_r^{t_r}\sum_{t_1=0}^{n}\alpha_{t_1,\dots,t_r}(x_1+y)^{t_1} \tag{10}$$

$$= \sum_{u=0}^{n}\sum_{t_2=0}^{n}\dots\sum_{t_r=0}^{n} x_1^u x_2^{t_2}\dots x_r^{t_r}\sum_{v=u}^{n}\alpha_{v,t_2,\dots,t_r}C_v^u y^{v-u}$$

$$= \sum_{u=0}^{n}\sum_{t_2=0}^{n}\dots\sum_{t_r=0}^{n} x_1^u x_2^{t_2}\dots x_r^{t_r} a_{u,t_2,\dots,t_r}(y).$$

We raise (10) to the power $2t$ and use Hölder's inequality to get

$$|S_m(A)|^{2t} \leqslant 2^{2t}W^{2t} + 2^{4t}Y^{2t}P^{(r-1)\cdot 2t}$$

$$\leqslant 2^{2t}Y^{-1}\sum_{y=1}^{Y}\left|\sum_{x_1=1}^{P}\dots\sum_{x_r=1}^{P}\exp(2\pi i m f_{A(y)}(x_1,\dots,x_r))\right|^{2t} + 2^{4t}Y^{2t}P^{2t(r-1)}.$$

Let us consider the domain $\Omega(y)$ defined by the conditions

$$|\beta_{u,t_2,\dots,t_r}(y) - a_{u,t_2,\dots,t_r}(y)| < YP^{-(u+t_2+\dots+t_r)-1}$$
$$(0 \leqslant u,\ t_2,\dots,t_r \leqslant n;\ u+t_2+\dots+t_r \geqslant 1).$$

Then, for points $B(y)$ belonging to this domain,

$$|S_m(A(y)) - S(B(y))|$$
$$= \left|\sum_{x_1=1}^{P}\dots\sum_{x_r=1}^{P}(\exp(2\pi i m f_{A(y)}(x_1,\dots,x_r)) - \exp(2\pi i m f_{B(y)}(x_1,\dots,x_r))\right|$$
$$\leqslant 2\pi M P^{r-1}Y.$$

We apply the inequality $|a+b|^{2t} \leqslant 2^{2t}(|a|^{2t}+|b|^{2t})$ to the sums $S_m(B(y))$ and $S_m(A(y)) - S_m(B(y))$. We get

$$|S_m(A)|^{2t} \leqslant 2^{4t}Y^{-1}\sum_{y=1}^{Y}|S_m(B(y))|^{2t}$$
$$+ 2^{4t}Y^{-1}\sum_{y=1}^{Y}|S_m(A(y)) - S_m(B(y))|^{2t} + 2^{4t}Y^{2t}P^{2t(r-1)}$$
$$\leqslant 2^{4t}Y^{-1}\sum_{y=1}^{Y}|S_m(B(y))|^{2t} + 2^{4t}((2\pi M)^{2t}+1)Y^{2t}P^{2t(r-1)}.$$

We integrate both sides of the inequality with respect to $B(y)$ belonging to $\Omega(y)$, and get

$$|S_m(A)|^{2t} \leqslant 2^{4t}P^{\frac{rnM}{2}+M-1}Y^{-M}\sum_{y=1}^{Y}\int_{\Omega(y)}|S_m(B(y))|^{2t}\,dB(y) + 2(8\pi M)^{2t}Y^{2t}P^{2t(r-1)}.$$

Let G be the maximum number of domains $\Omega(y)$ which intersect with an arbitrary fixed domain $\Omega(z)$. We collect the integrals over nonintersecting domains $\Omega(z)$, and replace this sum by an integration over the unit cube Ω. Then the preceding inequality is converted into the following one:

$$|S_m(A)|^{2t} \leqslant 2^{4t} P^{\frac{rnM}{2}+M-1} Y^{-M} G \int_{\Omega} |S_m(\beta)|^{2t}\, d\beta + 2(8\pi M)^{2t} Y^{2t} P^{2t(r-1)}. \qquad (11)$$

By Lemma 5, the number of intersections of $\Omega(y)$ and $\Omega(z)$ does not exceed the number of solutions of the inequality

$$\left(\frac{mn \cdots t_1 a(y-z) + mn \cdots t_1 \cdot \theta q^{-1}(y-z)}{q}\right) \leqslant n \ldots (t_1+1)\left(\frac{3}{2}n\right)^{n-t_1} Y P^{-(t_1+\ldots+t_r)}.$$

It follows from Lemma 6 that

$$G < \left(m\frac{n!}{(t_1-1)!} + 2\frac{n!}{t_1!}\left(\frac{3}{2}n\right)^{n-t_1} YP^{-(t_1+\ldots+t_r)}q\right)(Yq^{-1}+1)$$

$$\leqslant 6mn!\left(\frac{3}{2}n\right)^{n} P^{1-\omega}.$$

For estimating $|S_m(A)|$, we apply Theorem 1 to inequality (11) for $\tau = [rn \log(3rnM/2\omega)] + 1$. We get

$$|S_m(A)|^{2t} \leqslant 2^{4t} P^{\frac{rnM}{2}+M-1} P^{-M(1-\rho)} P^{1-\omega} \cdot 6mn!\left(\frac{3}{2}n\right)^{n} t^{rnM+2M\tau} \cdot 3^{3r^2nM\tau}(nr)^{3rnMu}$$

$$\times P^{2tr-\frac{rnM}{2}+\delta_\tau} + 2(8\pi M)^{2t} P^{(r-\rho)2t}.$$

Since $M\rho + \delta_\tau - \omega \leqslant -2t\rho$, we have

$$|S_m(A)|^{2t} \leqslant m 2^{4(t+1)} n!\left(\frac{3}{2}n\right) t^{rnM+2M\tau} \cdot 3^{3r^2nM\tau}(nr)^{3rnMu} P^{2t(r-\rho)}.$$

From this inequality, the estimate stated above follows. Theorem 2 is thus proved.

Chapter III

THE GENERAL ESTIMATE FOR A MULTIPLE TRIGONOMETRIC SUM

In this chapter, a general estimate is obtained for the sum $S(A)$ (see the Notation subsection). All points $\alpha_{t_1,\ldots,t_r}$ with $t_1 + \ldots + t_r \geqslant 1$, $0 \leqslant \alpha_{t_1,\ldots,t_r} \leqslant 1$, are divided into two classes depending on their approximations by rational fractions. For points of the second class, which constitute an overwhelming majority, a single sufficiently precise estimate is obtained for $|S(A)|$; for points of the first class, an estimate is obtained which, in many cases, is unimprovable. To derive an estimate for $|S(A)|$ for points of the first class, we would require estimates for complete rational multiple trigonometric sums and multiple trigonometric integrals. These estimates were obtained in [9] and [10]; here we give a more complete version of the exposition of these results; in particular, we prove the unimprovability of the above-mentioned estimates for the class of polynomials considered (Lemma 8, 12 and 13). For deriving an estimate for $|S(A)|$ for points of the second class, we require the lemmas on the multiplicity of intersection of domains (Lemmas 16 and 18), which are analogous to the corresponding lemmas in [6] and [7], and are their straightforward generalizations.

§1. Lemmas on complete rational multiple trigonometric sums and multiple trigonometric integrals

LEMMA 7. *Let* $f(x) = \alpha_0 + \ldots + \alpha_n x^n$, $\alpha = \max_{0\leqslant j\leqslant n}|\alpha_j|$, $0 \leqslant \beta \leqslant 1$, $0 \leqslant x \leqslant 1$. *Then the measure of those points for which* $|f(x)| \leqslant \alpha^\beta$ *does not exceed* $32n\alpha^{-(1-\beta)/n}$.

PROOF. Let us denote the measure of this set by μ. Then one can indicate $n+1$ distinct points $x_0 < \ldots < x_n$ on the set under consideration such that the distance between any two of them will be at least μ/n. In fact, for $n = 1$ and arbitrary μ the distance between the right and left ends of this set is at least μ. Suppose that this assertion is true for n points and arbitrary μ. We shall prove it for $n+1$ points and arbitrary μ. We take, on the interval $0 \leqslant x \leqslant 1$, the point on the extreme left of this set, and consider the point of the set nearest to it at a distance $\geqslant \mu/n$. The part of the original set from the point thus obtained up to the point on the extreme right of this set will be denoted by E. Its measure is at least $\mu(n-1)/n$. By virtue of the induction hypothesis, one can indicate on it n points with the distance between any two of them being at least μ/n. Thus, we have proved that $|x_i - x_j| \geqslant |i-j|\mu/n$. From Taylor's formula, we have

$$f(x_j) - f(x_0) = \sum_{k=1}^{n} \frac{f^{(k)}(x_0)}{k!}(x_j - x_0)^k; \qquad j = 1, \ldots, n.$$

We regard these equalities as linear equations in $f^{(k)}(x_0)/k!$. We get

$$\left|\frac{f^{(k)}(x_0)}{k!}\right| \leqslant \sum_{m=1}^{n} \frac{|f(x_m) - f(x_0)|}{x_m - x_0}\frac{\Delta_m}{\Delta}, \tag{12}$$

where Δ_m and Δ are matrix determinants:

$$\Delta_m = |1, (x_\nu - x_0), \ldots, (x_\nu - x_0)^{k-2}, (x_\nu - x_0)^k, \ldots, (x_\nu - x_0)^{n-1}|$$
$$(\nu = 1, \ldots, m-1, m+1, \ldots, n);$$
$$\Delta = |1, (x_\nu - x_0), \ldots, (x_\nu - x_0)^{n-1}|$$
$$(\nu = 1, \ldots, n).$$

It is well known (see [11], Vol. 2, Part VII, §1, problem 10) that

$$\Delta_m = \frac{\Delta\sigma_{n-k}}{(x_n - x_m)\cdots(x_{m+1} - x_m)},$$

where σ_{n-k} is the $(n-k)$th elementary symmetric function of the numbers $x_\nu - x_0$ $(\nu = 1, \ldots, m-1, m+1, \ldots, n)$. Therefore, from (12) we have

$$\left|\frac{f^{(k)}(x_0)}{k!}\right| \leqslant 2\alpha^\beta\sigma_{n-k}\sum_{m=1}^{n}\frac{n^{n-m+1}}{m(n-m)!\,\mu^{n-m+1}}$$

$$\leqslant 2\alpha^\beta\mu^{-n}\sigma_{n-k}\sum_{m=1}^{n}\frac{n^{n-m+1}}{m(n-m)!} \leqslant 2^{2n+1}\alpha^\beta\mu^{-n}\sigma_{n-k}.$$

Let $|\alpha_k| = \alpha$. Then

$$k!\,\alpha = |f^{(k)}(0)| = \left|\sum_{r=k}^{n} \frac{f^{(r)}(x_0)}{(r-k)!}(-x_0)^{r-k}\right| \leqslant \sum_{r=k}^{n} \frac{2^{2n+1}r!\,\alpha^{\beta}\mu^{-n}\sigma_{n-r}}{(r-k)!}. \tag{13}$$

Since

$$\sigma_{n-r} \leqslant \frac{n!}{r!\,(n-r)!}\,\frac{n!}{r!},$$

it follows from (13) that

$$\mu^n \leqslant \alpha^{\beta-1}2^{2n+1}\sum_{r=k}^{n}\frac{r!}{k!\,(r-k)!}\,\frac{n!}{r!\,(n-r)!}\,\frac{n!}{r!} \leqslant \alpha^{\beta-1}2^{3n+2}\,n! \leqslant (n+1)^n\,2^{2(n+1)}\alpha^{\beta-1}.$$

Thus $\mu \leqslant 32n\alpha^{(\beta-1)/n}$. This proves Lemma 7.

LEMMA 8. a) *Let* $\alpha = \max_{0\leqslant t_1,\ldots,t_r\leqslant n}|\alpha_{t_1,\ldots,t_r}|$ *and*

$$I = \int_0^1 \cdots \int_0^1 \exp(2\pi i f_A(x_1,\ldots,x_r))\,dx_1\ldots dx_r.$$

Then

$$|I| \leqslant \min\left(1,\ (62n)^r\alpha^{-\frac{1}{n}}(\ln(\alpha+1)+2)^{r-1}\right)$$

b) *Let* $\alpha > 1$ *and*

$$I_r(\alpha) = \int_0^1 \cdots \int_0^1 \exp(2\pi i\alpha x_1^n\ldots x_r^n)\,dx_1\ldots dx_r.$$

Then

$$|I_r(\alpha)| \geqslant \frac{1}{2\pi n^r(r-1)!}\alpha^{-\frac{1}{n}}(\ln\alpha)^{r-1}.$$

PROOF. Assertion a) holds for $r = 1$ ([1], Chapter II, Lemma 4). We carry out an induction on the number of variables in the polynomial. Let us assume that the lemma holds for $r-1$ variables. We shall prove it for r variables. Without any loss of generality, we can assume that $s_1 > 0$ for $|\alpha_{s_1,\ldots,s_r}| = \alpha$. Let

$$t = [\ln(\alpha+1)]+1;$$

$$E_0 = \{x_r \mid |\varphi_{s_1,\ldots,s_{r-1}}(x_r)| \leqslant 1\};$$

$$E_k = \left\{x_r \mid \alpha^{\frac{k-1}{t}} < |\varphi_{s_1,\ldots,s_{r-1}}(x_r)| \leqslant \alpha^{\frac{k}{t}}\right\},\qquad k = 1,\ldots,t-1;$$

$$E_t = \left\{x_r \mid \alpha^{\frac{t-1}{t}} < |\varphi_{s_1,\ldots,s_{r-1}}(x_r)|\right\},$$

where $\varphi_{s_1,\ldots,s_{r-1}}(x_r)$ is defined by

$$f_A(x_1,\ldots,x_r) = \sum_{t_1=0}^{n}\cdots\sum_{t_{r-1}=0}^{n} x_1^{t_1}\ldots x_{r-1}^{t_{r-1}}\varphi_{t_1,\ldots,t_{r-1}}(x_r).$$

Then

$$|I| \leqslant \operatorname{mes} E_0 + \sum_{k=1}^{t-1} \operatorname{mes} E_k \cdot \max_{x_r \in E_k} \left| \int_0^1 \cdots \int_0^1 \exp(2\pi i f_A(x_1, \ldots, x_r))\, dx_1 \ldots dx_{r-1} \right|$$

$$+ \max_{x_r \in E_t} \left| \int_0^1 \cdots \int_0^1 \exp(2\pi i f_A(x_1, \ldots, x_r))\, dx_1 \ldots dx_{r-1} \right|.$$

By Lemma 7,

$$\operatorname{mes} E_k \leqslant 32 n \alpha^{-\frac{1}{n} + \frac{k}{nt}}, \qquad k = 0, \ldots, t-1,$$

and, by the induction hypothesis,

$$\max_{x_r \in E_k} \left| \int_0^1 \cdots \int_0^1 \exp(2\pi i f_A(x_1, \ldots, x_r))\, dx_1 \ldots dx_{r-1} \right|$$

$$\leqslant \min(1, (62n)^{r-1} \alpha^{-\frac{k-1}{tn}} (\ln(\alpha+1)+2)^{r-2}).$$

Hence

$$|I| \leqslant 32 n \alpha^{-\frac{1}{n}}$$

$$+ \sum_{k=1}^{t-1} 32 n \alpha^{-\frac{1}{n} + \frac{k}{tn}} \min\left(1, (62n)^{r-1} \alpha^{-\frac{k-1}{tn}} (\ln(\alpha+1)+2)^{r-2}\right)$$

$$+ \min\left(1, (62n)^{r-1} \alpha^{-\frac{t-1}{tn}} (\ln(\alpha+1)+2)^{r-2}\right)$$

$$\leqslant \min\left(1, (62n)^{r} \alpha^{-\frac{1}{n}} (\ln(\alpha+1)+2)^{r-1}\right).$$

Assertion a) is thus proved.

We pass on to the proof of assertion b). First of all, we have

$$I_r(\alpha) = \frac{(-1)^{r-1}}{(r-1)!} \int_0^1 \exp(2\pi i \alpha x^n) (\ln x)^{r-1}\, dx.$$

We verify this by induction. It holds for $r = 1$. We assume its validity for $r - 1$ variables, and prove it for r variables. By the induction hypothesis,

$$I_r(\alpha) = \int_0^1 I_{r-1}(\alpha x^n)\, dx = \int_0^1 \frac{(-1)^{r-2}}{(r-2)!} \left(\int_0^1 \exp(2\pi i \alpha x^n y^n) (\ln y)^{r-2}\, dy \right) dx.$$

After the change of variables $z = xy$, we get

$$I_r(\alpha) = \frac{(-1)^{r-2}}{(r-2)!} \int_0^1 \frac{dx}{x} \int_0^x \exp(2\pi i \alpha z^n) (\ln z - \ln x)^{r-2}\, dz.$$

We integrate the last integral by parts:

$$I_r(\alpha)=\frac{(-1)^{r-2}}{(r-2)!}\int_0^1 (d\ln x)\int_0^x \exp(2\pi i\alpha z^n)\left(\sum_{k=0}^{r-2}(-1)^k C_{r-2}^k(\ln x)^k(\ln z)^{r-2-k}\right)dz$$

$$=\frac{(-1)^{r-2}}{(r-2)!}\sum_{k=0}^{r-2}\frac{(-1)^k}{k+1}C_{r-2}^k\int_0^1 d(\ln x)^{k+1}\int_0^x \exp(2\pi i\alpha z^n)(\ln z)^{r-2-k}\,dz$$

$$=\frac{(-1)^{r-1}}{(r-2)!}\sum_{k=0}^{r-2}\frac{(-1)^k}{k+1}C_{r-2}^k\int_0^1 \exp(2\pi i\alpha z^n)(\ln z)^{r-1}\,dz.$$

Since

$$\frac{1}{r-1}\sum_{k=0}^{r-2}(-1)^k\frac{r-1}{k+1}C_{r-2}^k=\frac{1}{r-1},$$

the validity of the assertion for r variables follows. We determine the lower bound of $|I_r(\alpha)|$ for $\alpha>1$. Let $J=\operatorname{Im}(I_r(\alpha))$. We have

$$J=\frac{1}{(r-1)!}\int_0^1 \sin(2\pi\alpha y^n)\left(\ln\frac{1}{y}\right)^{r-1}dy$$

$$=\frac{1}{n^r(r-1)!}\int_0^1 \sin(2\pi\alpha z)\left(\ln\frac{1}{z}\right)^{r-1}z^{\frac{1}{n}-1}\,dz$$

$$=\frac{1}{\pi\alpha n^r(r-1)!}\int_0^1\left(\ln\frac{1}{z}\right)^{r-1}z^{\frac{1}{n}-1}\,d(\sin^2\pi\alpha z)$$

$$=-\frac{1}{\pi\alpha n^r(r-1)!}\int_0^1 \sin^2(\pi\alpha z)\,d\left(\left(\ln\frac{1}{z}\right)^{r-1}z^{\frac{1}{n}-1}\right).$$

Besides,

$$J=-\frac{1}{\pi\alpha n^r(r-1)!}\int_{1/2\alpha}^{1+1/2\alpha}\cos^2(\pi\alpha z)\,d\left(\left(\ln\frac{1}{z-1/2\alpha}\right)^{r-1}\left(z-\frac{1}{2\alpha}\right)^{\frac{1}{n}-1}\right).$$

Since

$$-\frac{d}{dz}\left(\left(\ln\frac{1}{z}\right)^{r-1}z^{\frac{1}{n}-1}\right)\geqslant 0\quad(0<z<1)$$

is a monotonically decreasing function, on adding the expressions for J we get

$$2J>-\frac{1}{\pi\alpha n^r(r-1)!}\int_{1/\alpha}^1 d\left(\left(\ln\frac{1}{z}\right)^{r-1}z^{\frac{1}{n}-1}\right)=\frac{1}{\pi n^r(r-1)!}\alpha^{-\frac{1}{n}}(\ln\alpha)^{r-1}.$$

Thus

$$|I_r(\alpha)|\geqslant J\geqslant\frac{1}{2\pi n^r(r-1)!}\alpha^{-\frac{1}{n}}(\ln\alpha)^{r-1}.$$

This completes the proof of the lemma.

LEMMA 9. *Let $q=q_1\ldots q_k$ be a product of relatively prime numbers, $q=q_sQ_s$. Then ($\alpha_{t_1,\ldots,t_r}$ being integers)*

$$S\left(\frac{f(x_1,\ldots,x_r)}{q}\right)=S\left(\frac{Q_1^{-1}f(Q_1x_1,\ldots,Q_1x_r)}{q_1}\right)\ldots S\left(\frac{Q_k^{-1}f(Q_kx_1,\ldots,Q_kx_r)}{q_k}\right).$$

PROOF. If y_{ij}, $i=1,\ldots,r; j=1,\ldots,k$, runs through a complete system of residues mod q_j, then $x_i=Q_1y_{i1}+\ldots+Q_ky_{ik}$, $i=1,\ldots,r$, runs through a complete system of residues mod q. Hence we get (with $a_{0,\ldots,0}=0$)

$$f(x_1,\ldots,x_r)\equiv\sum_{j=1}^{k} f(Q_jy_{1j},\ldots,Q_jy_{rj}) \pmod q.$$

The lemma follows trivially from this congruence.

LEMMA 10. *Let p be a prime and l a positive integer, and let*

$$\varphi(x)=a_1x+\ldots+a_nx^n;\qquad (a_1,\ldots,a_n,\ p)=1.$$

Then

$$\left|S\left(\frac{\varphi(x)}{p^l}\right)\right|<n^{2n}p^{l\left(1-\frac{1}{n}\right)}.$$

The proof can be seen in [1], Chapter III, Lemma 6.

LEMMA 11. *Let $f(x)=a_0+a_1x+\ldots+a_nx^n$, $(a_0,a_1,\ldots,a_n,p)=1$, and let $N_p(\alpha,\beta)$ denote the number of solutions of the congruence $f(x)\equiv 0 \pmod{p^\beta}$, $\alpha\geqslant\beta$, $1\leqslant n\leqslant p^\alpha$. Then*

$$N_p(\alpha,\ \beta)\leqslant 5n^{2n}p^{\alpha-\frac{\beta}{n}}.$$

PROOF. Without any loss of generality, it can be assumed that $(a_1,\ldots,a_n,p)=1$ (otherwise the congruence $f(x)\equiv 0 \pmod{p^\beta}$ has no solutions), and that $n\geqslant 2$ (for $n=1$, the estimate is trivial). Since $x\equiv x_1 \pmod{p^\beta}$ implies that $f(x)\equiv f(x_1) \pmod{p^\beta}$, therefore

$$N_p(\alpha,\ \beta)=p^{\alpha-2\beta}\sum_{a=1}^{p^\beta}\sum_{x=1}^{p^\beta}\exp\left(2\pi i\,\frac{af(x)}{p^\beta}\right).$$

We decompose the sum over a into $\beta+1$ sums, collecting together those terms for a for which $(a,p)=1$; $p\,|\,a$ but $p^2\nmid a$; and so on. We set

$$N_p(\alpha,\ \beta)=p^{\alpha-2\beta}\sum_{k=0}^{\beta}S_k;\qquad S_k=\sum_{\substack{a=1\\ p^k\,\|\,a}}^{p^\beta}\sum_{x=1}^{p^\beta}\exp\left(2\pi i\,\frac{af(x)}{p^\beta}\right).$$

By Lemma 10, we have ($a=a_1p^k$, $(a_1,p)=1$)

$$|S_k|\leqslant p^{\beta-k}\left|\sum_{x=1}^{p^\beta}\exp\left(2\pi i\,\frac{a_1f(x)}{p^{\beta-k}}\right)\right|\leqslant n^{2n}p^{2\beta-k-\frac{\beta-k}{n}}.$$

Hence

$$N_p(\alpha,\ \beta)\leqslant p^{\alpha-2\beta}\sum_{k=0}^{\beta}|S_k|\leqslant n^{2n}p^{\alpha-\frac{\beta}{n}}\sum_{k=0}^{\beta}p^{-k+\frac{k}{n}}\leqslant 5n^{2n}p^{\alpha-\frac{\beta}{n}}.$$

This proves the lemma.

LEMMA 12. *Let*

$$f(x_1, \dots, x_r) = \sum_{t_1=0}^{n} \dots \sum_{t_r=0}^{n} a_{t_1, \dots, t_r} x_1^{t_1} \dots x_r^{t_r}, \ (a_{0, \dots, 0, 1}, \dots, a_{n, \dots, n}, \ p) = 1.$$

Then

$$\left| S\left(\frac{f(x_1, \dots, x_r)}{p^\alpha}\right)\right| \leqslant (5n^{2n})^r (\alpha+1)^{r-1} p^{r\alpha - \frac{\alpha}{n}}.$$

We give the proof by induction on the number of variables in the polynomial. The lemma holds for $r = 1$ (Lemma 10). We assume that the lemma holds for $r - 1$ variables and for arbitrary α, and prove it for r variables.

Let $(a_{s_1, \dots, s_r}, p) = 1$, $s_1 > 0$, and

$$f(x_1, \dots, x_r) = \sum_{t_1=0}^{n} \dots \sum_{t_{r-1}=0}^{n} x_1^{t_1} \dots x_{r-1}^{t_{r-1}} \varphi_{t_1, \dots, t_{r-1}}(x_r).$$

Then

$$\left| S\left(\frac{f(x_1, \dots, x_r)}{p^\alpha}\right)\right| \leqslant \sum_{k=0}^{\alpha} \sum_{\substack{x_r=1 \\ p^k \| \varphi_{s_1, \dots, s_{r-1}}(x_r)}}^{p^\alpha} \left| \sum_{x_1=1}^{p^\alpha} \dots \sum_{x_{r-1}=1}^{p^\alpha} \exp\left(2\pi i \frac{f(x_1, \dots, x_r)}{p^\alpha}\right)\right|.$$

By the induction hypothesis and Lemma 11, we have

$$\left|\left(S \frac{f(x_1, \dots, x_r)}{p^\alpha}\right)\right| \leqslant \sum_{k=0}^{\alpha} (5n^{2n})^{r-1} (\alpha+1)^{r-2} p^{(r-1)\alpha - \frac{\alpha-k}{n}} 5n^{2n} p^{\alpha - \frac{k}{n}}$$

$$= (5n^{2n})^r (\alpha+1)^{r-1} p^{r\alpha - \frac{\alpha}{n}}.$$

The lemma is proved.

LEMMA 13. a) *Suppose that the coefficients of the polynomial $f(x_1, \dots, x_r)$, together with q, are relatively prime, i.e. $(a_{0, \dots, 1}, \dots, a_{n, \dots, n}, q) = 1$. Then*

$$\left| S\left(\frac{f(x_1, \dots, x_r)}{q}\right)\right| \leqslant (5n^{2n})^{r\nu(q)} (\tau(q))^{r-1} q^{r - \frac{1}{n}},$$

where $\nu(q)$ denotes the number of distinct prime divisors of q, and $\tau(q)$ the number of divisors of q.

b) *Let $p \geqslant 3$ be a prime, m, n positive integers, $n > 1$, $(n, p) = 1$, $\alpha = mn$, $(a, p) = 1$, and let*

$$S\left(\frac{ax_1^n \cdots x_r^n}{p^\alpha}\right) = \sum_{x_1=1}^{p^\alpha} \dots \sum_{x_r=1}^{p^\alpha} \exp\left(2\pi i \frac{ax_1^n \cdots x_r^n}{p^\alpha}\right).$$

Then

$$S\left(\frac{ax_1^n \cdots x_r^n}{p^\alpha}\right) \geqslant \frac{m^{r-1}}{(r-1)!} \left(1 - \frac{1}{p}\right)^{r-1} p^{r\alpha - m}.$$

PROOF. Let $q = p_1^{\alpha_1} \dots p_k^{\alpha_k}$ be the canonical factorization of q into prime factors. Then, from the multiplicativity of the sum $S(f(x_1, \dots, x_r)/q)$ (Lemma 9) and from the estimate for $|S(f(x_1, \dots, x_r)/p^\alpha)|$ (Lemma 12), we have

$$\left| S\left(\frac{f(x_1, \dots, x_r)}{q}\right)\right| \leqslant (5n^{2n})^{r\nu(q)} (\tau(q))^{r-1} q^{r - \frac{1}{n}}.$$

This proves assertion a).

We prove assertion b) by induction. Assertion b) holds for $r = 1$ (see [2], p. 270).* We assume its validity for $r-1$ variables, and prove it for r variables. We have

$$S\left(\frac{ax_1^n\cdots x_r^n}{p^\alpha}\right)=\sum_{k=0}^{m-1} T_k+p^{r\alpha-m},$$

$$T_k=\sum_{x_1=1}^{p^\alpha}\cdots\sum_{x_{r-1}=1}^{p^\alpha}\sum_{\substack{p^k\,\|\,x_r\\ x_r=1}}^{p^\alpha}\exp\left(2\pi i\,\frac{ax_1^n\cdots x_r^n}{p^\alpha}\right).$$

By the induction hypothesis,

$$T_k\geqslant\varphi(p^{\alpha-k})\,p^{(r-1)kn}\frac{(m-k)^{r-2}}{(r-2)\,!}\left(1-\frac{1}{p}\right)^{r-2}p^{(\alpha-kn)(r-1)-m+k}$$

$$=\frac{(m-k)^{r-2}}{(r-2)\,!}\left(1-\frac{1}{p}\right)^{r-1}p^{\alpha-k+(r-1)kn+(\alpha-kn)(r-1)-m+k}=\frac{(m-k)^{r-2}}{(r-2)\,!}\left(1-\frac{1}{p}\right)^{r-1}p^{r\alpha-m}.$$

Therefore

$$S\left(\frac{ax_1^n\ldots x_r^n}{p^\alpha}\right)\geqslant\left(1-\frac{1}{p}\right)^{r-1}p^{r\alpha-m}\sum_{k=0}^{m-1}\frac{(m-k)^{r-2}}{(r-2)\,!}\geqslant\frac{m^{r-1}}{(r-1)\,!}\left(1-\frac{1}{p}\right)^{r-1}p^{r\alpha-m}.$$

This completes the proof of the lemma.

LEMMA 14. *Let $f(x_1,\ldots,x_r)$ be a real differentiable function for $1\leqslant x_1,\ldots,x_r\leqslant P$, where, within the interval of variation of the x_i, the function $\partial f(x_1,\ldots,x_r)/\partial x_i$ is piecewise monotonic and of constant sign with respect to each variable x_i for any fixed values of the remaining variables. Further, suppose that for $0<\delta<1$*

$$\left|\frac{\partial f(x_1,\ldots,x_r)}{\partial x_i}\right|\leqslant\delta,\qquad i=1,\ldots,r.$$

Then

$$\sum_{x_1=1}^{P}\cdots\sum_{x_r=1}^{P}\exp(2\pi i f(x_1,\ldots,x_r))$$

$$=\int_1^P\cdots\int_1^P\exp(2\pi i f(x_1,\ldots,x_r))\,dx_1\ldots dx_r+O\left(\frac{P^{r-1}}{1-\delta}\right).$$

The proof of this lemma is obtained by a systematic application of van der Corput's lemma (see [1], Chapter II, Lemma 3).

§2. Lemmas on the intersection multiplicity of domains

For subsequent investigations, we define the functions $B_{u_1,\ldots,u_r}$ and $A^{(s)}_{u_1,\ldots,u_r}$ by the equalities

*Editors' note. The reference is to I. M. Vinogradov, *The method of trigonometric sums in the theory of numbers*, Trudy Mat. Inst. Steklov. 23 (1947), Chapter II, Lemmas 4 and 5; English transl., Interscience, New York, 1954.

$$f(x_1+y_1, \dots, x_r+y_r) - f(x_1+z_1, \dots, x_r+z_r)$$
$$= \sum_{t_1=0}^{n} \dots \sum_{t_r=0}^{n} \alpha_{t_1, \dots, t_r} ((x_1+y_1)^{t_1} \dots (x_r+y_r)^{t_r} - (x_1+z_1)^{t_1} \dots$$
$$\dots (x_r+z_r)^{t_r}) = \sum_{t_1=0}^{n} \dots \sum_{t_r=0}^{n} x_1^{t_1} \dots x_r^{t_r} B_{t_1, \dots, t_r};$$

$$B_{u_1, \dots, u_r} = \sum_{t_1=u_1}^{n} \dots \sum_{t_r=u_r}^{n} \alpha_{t_1, \dots, t_r} C_{t_1}^{u_1} \dots C_{t_r}^{u_r} (y_1^{t_1-u_1} \dots y_r^{t_r-u_r} - z_1^{t_1-u_1} \dots z_r^{t_r-u_r});$$

$$A_{u_1, \dots, u_r}^{(s)} = \sum_{\substack{t_1=u_1 \\ t_1+\dots+t_r=s+u_1+\dots+u_r}}^{n} \dots \sum_{t_r=u_r}^{n} \alpha_{t_1, \dots, t_r} C_{t_1}^{u_1} \dots C_{t_r}^{u_r} (y_1^{t_1-u_1} \dots y_r^{t_r-u_r} - z_1^{t_1-u_1} \dots z_r^{t_r-u_r}).$$

LEMMA 15. *The following equality holds*:

$$A_{u_1, \dots, u_r}^{(s)} = \frac{1}{u_1! \dots u_r!\, s!} \sum_{\substack{v_1=u_1 \\ v_1+\dots+v_r=s-1+u_1+\dots+u_r}}^{n} \dots \sum_{v_r=u_r}^{n} v_1! \dots v_r!\, H_{v_1, \dots, v_r}^{(u_1, \dots, u_r, s)} A_{v_1, \dots, v_r}^{(1)},$$

where the coefficients of the polynomial $H_{v_1, \dots, v_r}^{(u_1, \dots, u_r, s)}$ *are integral and nonnegative, and their sum does not exceed* sr^{s-1}.

PROOF. We define the generating function

$$g(W_1, \dots, W_r) = \sum_{\substack{v_1=u_1 \\ v_1+\dots+v_r=s-1+u_1+\dots+u_r}}^{n} \dots \sum_{v_r=u_r}^{n} H_{v_1, \dots, v_r}^{(u_1, \dots, u_r, s)} W_1^{v_1} \dots W_r^{v_r}$$
$$= W_1^{u_1} \dots W_r^{u_r} \frac{(y_1W_1+\dots+y_rW_r)^s - (z_1W_1+\dots+z_rW_r)^s}{(y_1W_1+\dots+y_rW_r) - (z_1W_1+\dots+z_rW_r)}.$$

Hence, on the one hand,

$$g(W_1, \dots, W_r)[(y_1-z_1)W_1+\dots+(y_r-z_r)W_r]$$
$$= \sum_{\substack{v_1=u_1 \\ v_1+\dots+v_r=s-1+u_1+\dots+u_r}} \dots \sum_{v_r=u_r} H_{v_1, \dots, v_r}^{(u_1, \dots, u_r, s)} W_1^{v_1} \dots W_r^{v_r} [(y_1-z_1)W_1+\dots+(y_r-z_r)W_r]$$
$$= \sum_{\substack{v_1=u_1 \\ v_1+\dots+v_r=s+u_1+\dots+u_r}}^{n} \dots \sum_{v_r=u_r}^{n} W_1^{v_1} \dots W_r^{v_r} \left(\sum_{j=1}^{r} (y_j-z_j) H_{v_1, \dots, v_j-1, \dots, v_r}^{(u_1, \dots, u_r, s)} \right).$$

On the other hand,

$$g(W_1, \dots, W_r)[(y_1-z_1)W_1+\dots+(y_r-z_r)W_r]$$
$$= W_1^{u_1} \dots W_r^{u_r} [(y_1W_1+\dots+y_rW_r)^s - (z_1W_1+\dots+z_rW_r)^s]$$
$$= \sum_{\substack{v_1=u_1 \\ v_1+\dots+v_r=s+u_1+\dots+u_r}}^{n} \dots \sum_{v_r=u_r}^{n} W_1^{v_1} \dots W_r^{v_r} (y_1^{v_1-u_1} \dots y_r^{v_r-u_r} - z_1^{v_1-u_1} \dots z_r^{v_r-u_r})$$
$$\times \frac{s!}{(v_1-u_1)! \dots (v_r-u_r)!}.$$

We equate the coefficients of identical power-products $W_1^{v_1} \cdots W_r^{v_r}$ in the identities thus obtained. We have

$$\frac{s!}{(v_1-u_1)!\dots(v_r-u_r)!}\left(y_1^{v_1-u_1}\dots y_r^{v_r-u_r}-z_1^{v_1-u_1}\dots z_r^{v_r-u_r}\right)$$
$$=\sum_{j=1}^{r}(y_j-z_j)\,H_{v_1,\dots,v_j-1,\dots,v_r}^{(u_1,\dots,u_r,s)}.$$

Using this equality we transform $A_{u_1,\dots,u_r}^{(s)}$ to the form

$$A_{u_1,\dots,u_r}^{(s)}=\sum_{\substack{v_1=u_1\\ v_1+\dots+v_r=s+u_1+\dots+u_r}}^{n}\dots\sum_{v_r=u_r}^{n}\frac{1}{s!}\frac{v_1!}{u_1!}\dots\frac{v_r!}{u_r!}\alpha_{v_1,\dots,v_r}\frac{s!}{(v_1-u_1)!\dots(v_r-u_r)!}$$
$$\times\left(y_1^{v_1-u_1}\dots y_r^{v_r-u_r}-z_1^{v_1-u_1}\dots z_r^{v_r-u_r}\right)$$
$$=\frac{1}{u_1!\dots u_r!\,s!}\sum_{\substack{v_1=u_1\\ v_1+\dots+v_r=s+u_1+\dots+u_r}}^{n}\dots\sum_{v_r=u_r}^{n}v_1!\dots v_r!\,\alpha_{v_1,\dots,v_r}$$
$$\times\left(\sum_{j=1}^{r}(y_j-z_j)\,H_{v_1,\dots,v_j-1,\dots,v_r}^{(u_1,\dots,u_r,s)}\right).$$

We find the coefficient of $H_{v_1,\dots,v_r}^{(u_1,\dots,u_r,s)}$ here. Then the expression for $A_{u_1,\dots,u_r}^{(s)}$ can be written as follows:

$$A_{u_1,\dots,u_r}^{(s)}=\sum_{\substack{v_1=u_1\\ v_1+\dots+v_r=s-1+u_1+\dots+u_r}}^{n}\dots\sum_{v_r=u_r}^{n}v_1!\dots v_r!\,H_{v_1,\dots,v_r}^{(u_1,\dots,u_r,s)}$$
$$\times\sum_{j=1}^{r}\alpha_{v_1,\dots,v_j+1,\dots,v_r}(v_j+1)(y_j-z_j)$$
$$=\frac{1}{u_1!\dots u_r!\,s!}\sum_{\substack{v_1=u_1\\ v_1+\dots+v_r=s-1+u_1+\dots+u_r}}\dots\sum_{v_r=u_r}v_1!\dots v_r!\,H_{v_1,\dots,v_r}^{(u_1,\dots,u_r,s)}A_{v_1,\dots,v_r}^{(1)}.$$

Thus the lemma is proved.

LEMMA 16. *Let*

$$|B_{u_1,\dots,u_r}|<P^{-(u_1+\dots+u_r)}\ (0\leqslant u_1,\dots,u_r\leqslant n,\ rn-1\geqslant u_1+\dots+u_r\geqslant 1)\tag{14}$$

and let the variables $y_1,\dots,y_r$ *and* $z_1,\dots,z_r$ *take all integral values from* 1 *to* $Y\leqslant P$. *Then the number of* $(y_1,\dots,y_r)$ *satisfying* (14) *for any fixed* $(z_1,\dots,z_r)$ *does not exceed the number of solutions of the system*

$$\left(\frac{(rn)!}{(u_1+\dots+u_r+1)!}\frac{(rn+1)!}{(u_1+\dots+u_r+2)!}A_{u_1,\dots,u_r}^{(1)}\right)$$
$$\leqslant\frac{(rn)!}{(u_1+\dots+u_r+1)!}\frac{(rn+1)!}{(u_1+\dots+u_r+2)!}(4r^2n^2)^{rn-u_1-\dots-u_r-1}P^{-(u_1+\dots+u_r)}$$
$$(0\leqslant u_1,\dots,u_r\leqslant n,\qquad rn-1\geqslant u_1+\dots+u_r\geqslant 1).$$

We carry out the proof by induction on the sum $\Sigma_1^r u_j$. The lemma holds for $\Sigma_1^r u_j=rn-1$, since

$$B_{n,\dots,n-1,\dots,n}=A_{n,\dots,n-1,\dots,n}^{(1)}=n\alpha_{n,\dots,n}(y_k-z_k),$$
$$|B_{n,\dots,n-1,\dots,n}|<P^{-rn+1},$$

where the $n-1$ in the expressions for $B_{n,\dots,n-1,\dots,n}$ and $A^{(1)}_{n,\dots,n-1,\dots,n}$ stands in the kth place. Let us assume that the lemma holds for all $u_1, \dots, u_r$ satisfying $\Sigma_1^r u_j \geqslant s+1$. We shall prove it for all $u_1, \dots, u_r$ for which $u_1 + \dots + u_r = s$. By Lemma 15,

$$B_{u_1,\dots,u_r} = \sum_{s=1}^{rn-u_1-\dots-u_r} A^{(s)}_{u_1,\dots,u_r}$$

$$= \sum_{s=1}^{rn-u_1-\dots-u_r} \sum_{\substack{v_1=u_1 \\ v_1+\dots+v_r=s-1+u_1+\dots+u_r}}^{n} \cdots \sum_{v_r=u_r}^{n} \frac{v_1!\dots v_r!}{u_1!\dots u_r!\,s!} H^{(u_1,\dots,u_r,s)}_{v_1,\dots,v_r} A^{(1)}_{v_1,\dots,v_r}.$$

We multiply both sides by

$$\frac{(rn)!}{(u_1+\dots+u_r+1)!}\,\frac{(rn+1)!}{(u_1+\dots+u_r+2)!}.$$

We get

$$\frac{(rn)!}{(u_1+\dots+u_r+1)!}\,\frac{(rn+1)!}{(u_1+\dots+u_r+2)!} A^{(1)}_{u_1,\dots,u_r}$$

$$= \frac{(rn)!}{(u_1+\dots+u_r+1)!}\,\frac{(rn+1)!}{(u_1+\dots+u_r+2)!} B_{u_1,\dots,u_r}$$

$$- \sum_{s=2}^{rn-u_1-\dots-u_r} \sum_{\substack{v_1=u_1 \\ v_1+\dots+v_r=s-1+u_1+\dots+u_r}}^{n} \cdots \sum_{v_r=u_r}^{n} \frac{v_1!\dots v_r!}{u_1!\dots u_r!}\,\frac{(v_1+\dots+v_r+2)!}{s!\,(u_1+\dots+u_r+1)!} \tag{15}$$

$$\times \frac{(v_1+\dots+v_r+1)!}{(u_1+\dots+u_r+2)!} H^{(u_1,\dots,u_r,s)}_{v_1,\dots,v_r} \frac{(rn)!}{(v_1+\dots+v_r+1)!}$$

$$\times \frac{(rn+1)!}{(v_1+\dots+v_r+2)!} A^{(1)}_{v_1,\dots,v_r}.$$

The coefficients which precede

$$\frac{(rn)!}{(v_1+\dots+v_r+1)!}\,\frac{(rn+1)!}{(v_1+\dots+v_r+2)!} A^{(1)}_{v_1,\dots,v_r} \tag{16}$$

will be integers, since the numbers

$$\frac{(v_1+\dots+v_r+2)!}{s!\,(u_1+\dots+u_r+1)!}, \qquad \frac{(v_1+\dots+v_r+1)!}{(u_1+\dots+u_r+2)!}$$

$$(v_1+\dots+v_r+2=s+u_1+\dots+u_r+1,\ \ s>1)$$

are integers.

Therefore (16) can be replaced by a number congruent to it mod 1. Taking into account this remark, the induction hypothesis and (15), we see that for the validity of (14) it is necessary that the following linear Diophantine inequalities be satisfied:

$$\left(\frac{(rn)!}{(u_1+\ldots+u_r+1)!}\,\frac{(rn+1)!}{(u_1+\ldots+u_r+2)!}A^{(1)}_{u_1,\ldots,u_r}\right)$$
$$\leqslant \frac{(rn)!}{(u_1+\ldots+u_r+1)!}\,\frac{(rn+1)!}{(u_1+\ldots+u_r+2)!}P^{-(u_1+\ldots+u_r)}$$
$$+\sum_{s=2}^{rn-u_1-\ldots-u_r}\ \sum_{\substack{v_1=u_1\\ v_1+\ldots+v_r=s-1+u_1+\ldots+u_r}}^{n}\cdots\sum_{v_r=u_r}^{n}\frac{v_1!\ldots v_r!}{u_1!\ldots u_r!}\,\frac{(v_1+\ldots+v_r+2)!}{s!\,(u_1+\ldots+u_r+1)!}$$
$$\times\frac{(v_1+\ldots+v_r+1)!}{(u_1+\ldots+u_r+2)!}sr^{s-1}P^{s-1}\frac{(rn)!}{(v_1+\ldots+v_r+1)!}\,\frac{(rn+1)!}{(v_1+\ldots+v_r+2)!}$$
$$\times(4r^2n^2)^{rn-v_1-\ldots-v_r-1}P^{-(v_1+\ldots+v_r)}.$$

We use the following results:

$$\frac{v_1!\ldots v_r!}{u_1!\ldots u_r!}\leqslant n^{s-1}\qquad(v_1+\ldots+v_r=s-1+u_1+\ldots+u_r,\qquad n\geqslant v_j\geqslant u_j,\ j=1,\ldots,r),$$

$$\sum_{\substack{v_1=u_1\\ v_1+\ldots+v_r=s-1+u_1+\ldots+u_r}}^{n}\cdots\sum_{v_r=u_r}^{n}1\leqslant C^{s-1}_{s+r-2}\leqslant(2rn)^{s-1}.$$

We get

$$\left(\frac{(rn)!}{(u_1+\ldots+u_r+1)!}\,\frac{(rn+1)!}{(u_1+\ldots+u_r+2)!}A^{(1)}_{u_1,\ldots,u_r}\right)$$
$$\leqslant\frac{(rn)!}{(u_1+\ldots+u_r+1)!}\,\frac{(rn+1)!}{(u_1+\ldots+u_r+2)!}(4r^2n^2)^{rn-u_1-\ldots-u_r-1}P^{-(u_1+\ldots+u_r)}$$
$$\times\left((4r^2n^2)^{-rn+u_1+\ldots+u_r+1}+\sum_{s=2}^{rn-u_1-\ldots-u_r}\frac{1}{(s-1)!}\left(\frac{1}{2}\right)^{s-1}\right)$$
$$\leqslant\frac{(rn)!}{(u_1+\ldots+u_r+1)!}\,\frac{(rn+1)!}{(u_1+\ldots+u_r+2)!}(4r^2n^2)^{rn-u_1-\ldots-u_r-1}P^{-(u_1+\ldots+u_r)}$$
$$\times\left(e^{1/2}-1+\frac{1}{16}\right)\leqslant\frac{(rn)!}{(u_1+\ldots+u_r+1)!}\,\frac{(rn+1)!}{(u_1+\ldots+u_r+2)!}$$
$$\times(4r^2n^2)^{rn-u_1-\ldots-u_r-1}P^{-(u_1+\ldots+u_r)}.$$

Thus the lemma is proved.

LEMMA 17. *Let m be a positive integer, and let λ and β be real,*

$$\Phi(y)=m\frac{ay+\lambda y}{q}+\beta,\qquad(a,q)=1,\qquad q>0,$$

where y runs through $\leqslant Y$ consecutive integers. Then, for $V\geqslant 0$, the number of values of y with $(\Phi(y))\leqslant Vq^{-1}$ does not exceed

$$(\lambda Ym+m+2V),\qquad \textit{if}\quad Y\leqslant q,$$
$$(\lambda Ym+m+2V)\frac{2Y}{q},\qquad \textit{if}\quad Y>q.$$

The proof can be seen in [1], Chapter IV, Lemma 8.

LEMMA 18. *Let* $P \geqslant (50r^4n^4)^{2rn}$ *and*

$$\tau_{t_1,\ldots,t_r} = P^{t_1+\ldots+t_r-\frac{1}{2r}};$$

$$\alpha_{t_1,\ldots,t_r} = \frac{a_{t_1,\ldots,t_r}}{q_{t_1,\ldots,t_r}} + \frac{\theta_{t_1,\ldots,t_r}}{q_{t_1,\ldots,t_r}\tau_{t_1,\ldots,t_r}}; \qquad (a_{t_1,\ldots,t_r}, q_{t_1,\ldots,t_r}) = 1,$$

$$0 < q_{t_1,\ldots,t_r} \leqslant \tau_{t_1,\ldots,t_r}; \qquad |\theta_{t_1,\ldots,t_r}| \leqslant 1$$

$$(0 \leqslant t_1, \ldots, t_r \leqslant n, \quad t_1 + \ldots + t_r \geqslant 2);$$

and let Q_0 *denote the least common multiple of* $q_{t_1,\ldots,t_r}$ $(0 \leqslant t_1, \ldots, t_r \leqslant n;$ $t_1 + \ldots + t_r \geqslant 2)$. *Let*

$$f(x_1 + y_1, \ldots, x_r + y_r) - f(x_1, \ldots, x_r) = \sum_{t_1=0}^{n} \cdots \sum_{t_r=0}^{n} B_{t_1,\ldots,t_r} x_1^{t_1} \ldots x_r^{t_r}.$$

Consider the domain $\Omega(y_1, \ldots, y_r)$ *defined by*

$$|\gamma_{t_1,\ldots,t_r} - B_{t_1,\ldots,t_r}| < P^{-(t_1+\ldots+t_r)}$$

$$(rn - 1 \geqslant t_1 + \ldots + t_r \geqslant 1, \quad 0 \leqslant t_1, \ldots, t_r \leqslant n).$$

Let G denote the number of domains $\Omega(y_1, \ldots, y_r)$ *intersecting any fixed domain* $\Omega(y_1^{(0)}, \ldots, y_r^{(0)})$. *Then, for* $Q_0 \geqslant P^{1/2r-1/10M}$,

$$G < (8r^4n^4)^{2rn} Y^{r-1} P^{1-\frac{1}{2r}+\frac{1}{10M}}.$$

PROOF. Two cases are possible:

a) For $0 \leqslant t_1, \ldots, t_r \leqslant n$, $t_1 + \ldots t_r \geqslant 2$, one can find $(t_1, \ldots, t_r)$ such that

$$q_{t_1,\ldots,t_r} \geqslant P^{\frac{1}{2r}-\frac{1}{10M}}$$

(one can assume, without loss of generality, that $t_1 > 0$).

b) For all $0 \leqslant t_1, \ldots, t_r \leqslant n$, $t_1 + \ldots + t_r \geqslant 2$, we have

$$q_{t_1,\ldots,t_r} < P^{\frac{1}{2r}-\frac{1}{10M}}.$$

We consider case a). It follows from Lemma 16 that G does not exceed the number of solutions of the inequality

$$\left(\frac{(rn)!}{(t_1+\ldots+t_r)!} \frac{(rn+1)!}{(t_1+\ldots+t_r+1)!} A^{(1)}_{t_1-1,\ldots,t_r}\right)$$
$$\leqslant \frac{(rn)!}{(t_1+\ldots+t_r)!} \frac{(rn+1)!}{(t_1+\ldots+t_r+1)!} (4r^2n^2)^{rn-t_1-\ldots-t_r} P^{-(t_1+\ldots+t_r-1)}.$$

We apply Lemma 17 for fixed $y_2, \ldots, y_r$. We get

$$G < \left(\frac{(rn)!}{(t_1+\ldots+t_r)!} \frac{(rn+1)!}{(t_1+\ldots+t_r+1)!} t_1 P^{-(t_1+\ldots+t_r)+\frac{1}{2r}} Y + 1\right.$$
$$\left. + 2q_{t_1,\ldots,t_r} \frac{(rn)!}{(t_1+\ldots+t_r)!} \frac{(rn+1)!}{(t_1+\ldots+t_r+1)!} (4r^2n^2)^{rn-t_1-\ldots-t_r} P^{-(t_1+\ldots+t_r-1)}\right)$$
$$\times (Yq^{-1}_{t_1,\ldots,t_r} + 1) Y^{r-1} \leqslant (8r^4n^4)^{2rn} Y^{r-1} P^{1-\frac{1}{2r}+\frac{1}{10M}}.$$

We consider case b). Let

$$Q_{u_1, \ldots, u_r} = \text{L.C.M.}\ (q_{u_1+1, u_2, \ldots, u_r}, q_{u_1, u_2+1, \ldots, u_r}, \ldots, q_{u_1, u_2, \ldots, u_r+1})$$
$$(0 \leqslant u_1, \ldots, u_r \leqslant n, \quad rn-1 \geqslant u_1+\ldots+u_r \geqslant 1).$$

Since

$$\frac{(rn)!}{(u_1+\ldots+u_r+1)!} \frac{(rn+1)!}{(u_1+\ldots+u_r+2)!} (4r^2n^2)^{rn-u_1-\ldots-u_r-1} \times P^{-(u_1+\ldots+u_r)} Q_{u_1, \ldots, u_r} < \frac{1}{10};$$

$$\frac{(rn)!}{(u_1+\ldots+u_r+1)!} \frac{(rn+1)!}{(u_1+\ldots+u_r+2)!} ((u_1+1) Y\tau^{-1}_{u_1+1, \ldots, u_r} Q_{u_1+1, \ldots, u_r} q^{-1}_{u_1+1, \ldots, u_r} + \ldots + (u_r+1) Y\tau^{-1}_{u_1, \ldots, u_r+1} Q_{u_1, \ldots, u_r+1} q^{-1}_{u_1, \ldots, u_r+1}) < \frac{1}{10},$$

it follows from Lemma 16 that the number

$$\frac{(rn)!}{(u_1+\ldots+u_r+1)!} \frac{(rn+1)!}{(u_1+\ldots+u_r+2)!} \left(\frac{(u_1+1)\, a_{u_1+1, \ldots, u_r} y_1}{q_{u_1+1, \ldots u_r}} \cdot \ldots + \frac{(u_r+1)\, a_{u_1, \ldots, u_r+1} y_r}{q_{u_1, \ldots, u_r+1}} \right) \tag{17}$$
$$(0 \leqslant u_1, \ldots, u_r \leqslant n, \quad 1 \leqslant u_1+\ldots+u_r \leqslant rn-1)$$

is an integer.

Let the canonical factorization of Q_0 into prime factors be

$$Q_0 = p_1^{\alpha_1} \ldots p_s^{\alpha_s} = Q_k p_k^{\alpha_k}.$$

Then one can find numbers $(u_1^{(k)}, \ldots, u_r^{(k)})$ such that

$$p_k^{\alpha_k} \| Q_{u_1^{(k)}, \ldots, u_r^{(k)}}, \qquad p_k^{\alpha_k} \| q_{u_1^{(k)}+1, \ldots, u_r^{(k)}}.$$

Since (17) is an integer, G does not exceed the number of solutions of the system

$$\frac{(rn)!}{(u_1^{(k)}+\ldots+u_r^{(k)}+1)!} \frac{(rn+1)!}{(u_1^{(k)}+\ldots+u_r^{(k)}+2)!} ((u_1^{(k)}+1) b_{u_1^{(k)}+1, \ldots, u_r^{(k)}} y_1 + \ldots + (u_r^{(k)}+1) b_{u_1^{(k)}, \ldots, u_r^{(k)}+1} y_r) \equiv 0 \pmod{Q_{u_1^{(k)}, \ldots, u_r^{(k)}}}$$
$$(k=1, \ldots, s; \quad 1 \leqslant y_1, \ldots, y_r \leqslant Y) \tag{18}$$

for some integers

$$b_{u_1^{(k)}+1, \ldots, u_r^{(k)}}, \ldots, b_{u_1^{(k)}, \ldots, u_r^{(k)}+1}; \ (b_{u_1^{(k)}+1, \ldots, u_r^{(k)}}, p_k) = 1.$$

If $Y \leqslant Q_0$, one can write y_1 uniquely in the form

$$y_1 = Q_1 y_{1,1} + \ldots + Q_s y_{1,s};$$
$$0 \leqslant y_{1,k} \leqslant p_k^{\alpha_k}, \qquad k=1, \ldots, s.$$

Then the number of solutions of (18) does not exceed the number of solutions of the system

$$\frac{(rn)!}{(u_1^{(k)}+\ldots+u_r^{(k)}+1)!}\,\frac{(rn+1)!}{(u_1^{(k)}+\ldots+u_r^{(k)}+2)!}\,((u_1^{(k)}+1)\,d_{u_1^{(k)}+1,\ldots,u_r^{(k)}}y_{1k}$$
$$\ldots+(u_r^{(k)}+1)\,d_{u_1^{(k)},\ldots,u_r^{(k)}+1}y_r)\equiv 0\ (\mathrm{mod}\ p_k^{\alpha_k})$$
$$(k=1,\ \ldots,\ s)$$

for some integers

$$d_{u_1^{(k)}+1,\ldots,u_r^{(k)}},\ \ldots,\ d_{u_1^{(k)},\ldots,u_r^{(k)}+1};\ (d_{u_1^{(k)}+1,\ldots,u_r^{(k)}},\ p_k)=1.$$

We determine the quantities β_k, $k = 1, \ldots, s$, from the conditions

$$p_k^{\beta_k}\,\Big\|\,\frac{(rn)!}{(u_1^{(k)}+\ldots+u_r^{(k)}+1)!}\cdot\frac{(rn+1)!}{(u_1^{(k)}+\ldots+u_r^{(k)}+2)!}\,(u_1^{(k)}+1).$$

For a fixed set $(y_2, \ldots, y_r)$, the number of y_{1k} satisfying the above system does not exceed $p_k^{\beta_k}$. Therefore for G we have the estimate

$$G\leqslant p_1^{\beta_1}\ldots p_s^{\beta_s}Y^{r-1}\leqslant (rn)!\,(rn+1)!\,n!\,Y^{r-1}.$$

In the case $Y > Q_0$ we have

$$G\leqslant (rn)!\,(rn+1)!\,n!\,Q_0^{r-1}\left(\frac{Y}{Q_0}+1\right)^r.$$

We thus obtain the estimate for G in case b). This completes the proof of the lemma.

§3. Statement and proof of Theorem 3

We now obtain the general estimate for a multiple trigonometric sum. For simplicity of presentation, we do not calculate the constants which depend on the parameters n and r, and so use Vinogradov's symbol $\ll$, which in our case stands for $\leqslant c(n, r)$. We may also note that the above-mentioned constants can be calculated, as the corresponding constants in the mean-value theorem (Theorem 1), in the lemmas on multiple trigonometric sums and integrals (§1) and in the lemmas on the multiplicity of intersection of domains (§2) have already been calculated.

THEOREM 3. *Let*

$$\alpha_{t_1,\ldots,t_r}=\frac{a_{t_1,\ldots,t_r}}{q_{t_1,\ldots,t_r}}+z_{t_1,\ldots,t_r},\ (a_{t_1,\ldots,t_r},\ q_{t_1,\ldots,t_r})=1\quad(0\leqslant t_1,\ldots,t_r\leqslant n)$$

and let Q denote the least common multiple of $q_{t_1,\ldots,t_r}$ with $t_1 + \ldots + t_r \geqslant 1$. Call $\alpha_{t_1,\ldots,t_r}$ a point of the first class if $Q < P^{1/M}$, $|z_{t_1,\ldots,t_r}| < P^{-(t_1+\ldots+t_r)+1/M}$; the remaining points will be called points of the second class. Then, for points of the second class,

$$|S|\ll P^{r-\rho},$$

where

$$\rho=\frac{1}{50Mr^2n\log(4r^2nM)},$$

while, for points of the first class,

$$|S| \ll P^r Q^{-\frac{1}{n}+\varepsilon},$$

and if

$$\delta_{t_1,\ldots,t_r} = z_{t_1,\ldots,t_r} P^{t_1+\ldots+t_r}, \qquad \delta = \max_{0 \leqslant t_1,\ldots,t_r \leqslant n} |\delta_{t_1,\ldots,t_r}|,$$

then

$$|S| \ll P^r Q^{-\frac{1}{n}+\varepsilon} \delta^{-\frac{1}{n}} (\ln(\delta+1)+1)^{r-1}, \qquad \delta \geqslant 1.$$

PROOF. We consider a point of the first class. Let

$$x_j = Q\xi_j + \eta_j, \qquad 1 \leqslant \eta_j \leqslant Q, \qquad -\frac{\eta_j}{Q} < \xi_j \leqslant \frac{P-\eta_j}{Q}; \qquad 1 \leqslant j \leqslant r;$$

then

$$S = \sum_{\eta_1=1}^{Q} \cdots \sum_{\eta_r=1}^{Q} \exp\left(2\pi i \sum_{t_1=0}^{n} \cdots \sum_{t_r=0}^{n} \frac{a_{t_1,\ldots,t_r}}{q_{t_1,\ldots,t_r}} \eta_1^{t_1} \cdots \eta_r^{t_r}\right) W_{\eta_1,\ldots,\eta_r};$$

$$W_{\eta_1,\ldots,\eta_r} = \sum_{\xi_1,\ldots,\xi_r} \exp\left(2\pi i \sum_{t_1=0}^{n} \cdots \sum_{t_r=0}^{n} z_{t_1,\ldots,t_r} (Q\xi_1+\eta_1)^{t_1} \cdots (Q\xi_r+\eta_r)^{t_r}\right).$$

One can assume, without loss of generality, that $P \geqslant (50r^4n^4)^{2mM}$.

Lemma 14 is applicable to the sum $W_{\eta_1,\ldots,\eta_r}$, since the derivative with respect to ξ_j, $1 \leqslant j \leqslant r$, of the polynomial in the exponent does not exceed in absolute value the quantity

$$nMQP^{-1+\frac{1}{M}} < 0.01.$$

Therefore

$$W_{\eta_1,\ldots,\eta_r} = P^r Q^{-r} \int_0^1 \cdots \int_0^1 \exp\left(2\pi i \sum_{t_1=0}^{n} \cdots \sum_{t_r=0}^{n} \delta_{t_1,\ldots,t_r} x_1^{t_1} \cdots x_r^{t_r}\right) dx_1 \ldots dx_r$$
$$+ O(P^{r-1}Q^{1-r}) + O(1);$$

$$S = UV + O(P^{r-1}Q) + O(Q^r);$$

$$U = Q^{-r} \sum_{\eta_1=1}^{Q} \cdots \sum_{\eta_r=1}^{Q} \exp\left(2\pi i \sum_{t_1=0}^{n} \cdots \sum_{t_r=0}^{n} \frac{a_{t_1,\ldots,t_r}}{q_{t_1,\ldots,t_r}} x_1^{t_1} \cdots x_r^{t_r}\right);$$

$$V = P^r \int_0^1 \cdots \int_0^1 \exp\left(2\pi i \sum_{t_1=0}^{n} \cdots \sum_{t_r=0}^{n} \delta_{t_1,\ldots,t_r} x_1^{t_1} \cdots x_r^{t_r}\right) dx_1 \ldots dx_r.$$

We estimate V by Lemma 8a), and U by Lemma 13a). We thus obtain the assertion in the theorem for points of the first class.

We consider the second class. Let

$$\tau_{t_1,\ldots,t_r} = P^{t_1+\ldots+t_r-\frac{1}{2r}} \qquad (0 \leqslant t_1,\ldots,t_r \leqslant n, \quad t_1+\ldots+t_r \geqslant 1).$$

Then

$$\alpha_{t_1,\ldots,t_r} = \frac{a_{t_1,\ldots,t_r}}{q_{t_1,\ldots,t_r}} + \frac{\theta_{t_1,\ldots,t_r}}{q_{t_1,\ldots,t_r}\tau_{t_1,\ldots,t_r}}; \qquad (a_{t_1,\ldots,t_r}, q_{t_1,\ldots,t_r}) = 1,$$
$$0 < q_{t_1,\ldots,t_r} \leqslant \tau_{t_1,\ldots,t_r}, \qquad |\theta_{t_1,\ldots,t_r}| \leqslant 1.$$

As before, let Q_0 denote the least common multiple of $q_{t_1,\ldots,t_r}$; $0 \leqslant t_1, \ldots, t_r \leqslant n$, $t_1 + \ldots + t_r \geqslant 2$.

We consider two possibilities:

$$\text{a) } Q_0 \geqslant P^{\frac{1}{2r}-\frac{1}{10M}}, \qquad \text{b) } Q_0 < P^{\frac{1}{2r}-\frac{1}{10M}}.$$

We prove the theorem for case a). Shifting the intervals of summation over $x_1, \ldots, x_r$ by $y_1, \ldots, y_r$, and then summing over $y_1, \ldots, y_r$ from 1 to $Y = [P^{1-\rho}]$, we get

$$|S| \leqslant Y^{-r} \sum_{y_1=1}^{Y} \cdots \sum_{y_r=1}^{Y} \left| \sum_{x_1=1}^{P} \cdots \sum_{x_r=1}^{P} \exp(2\pi i f_A(x_1+y_1, \ldots, x_r+y_r)) \right| + 2rP^{r-1}Y.$$

Let $\tau = [rn \log(4r^2 nM)] + 1$ and $t = M(\tau + r)$. We apply Hölder's inequality to $|S|^{2t}$, and get

$$|S|^{2t} \leqslant 2^{2t} Y^{-r} \sum_{y_1=1}^{Y} \cdots \sum_{y_r=1}^{Y} \left| \sum_{x_1=1}^{P} \cdots \sum_{x_r=1}^{P} \exp(2\pi i f_A(x_1+y_1, \ldots, x_r+y_r)) \right|^{2t} + (4r)^{2t} P^{2t(r-\rho)}.$$

Let

$$S(y_1, \ldots, y_r) = \sum_{x_1=1}^{P} \cdots \sum_{x_r=1}^{P} \exp(2\pi i f_A(x_1+y_1, \ldots, x_r+y_r));$$

$$S_\beta(y_1, \ldots, y_r) = \sum_{x_1=1}^{P} \cdots \sum_{x_r=1}^{P} \exp(2\pi i f_\beta(x_1, \ldots, x_r));$$

$$f_A(x_1+y_1, \ldots, x_r+y_r) = \sum_{t_1=0}^{n} \cdots \sum_{t_r=0}^{n} B_{t_1,\ldots,t_r} x_1^{t_1} \cdots x_r^{t_r};$$

$$|\beta_{t_1,\ldots,t_r} - B_{t_1,\ldots,t_r}| \leqslant P^{-(t_1+\ldots+t_r)-\rho}$$

$$(0 \leqslant t_1, \ldots, t_r \leqslant n, \quad 1 \leqslant t_1 + \ldots + t_r \leqslant rn - 1).$$

This set of $\beta_{t_1,\ldots,t_r}$ $(0 \leqslant t_1, \ldots, t_r \leqslant n,\ 1 \leqslant t_1 + \ldots + t_r \leqslant rn - 1)$ will be denoted by $\Omega(y_1, \ldots, y_r)$. By integrating over $\Omega(y_1, \ldots, y_r)$, we obtain

$$|S|^{2t} \leqslant 2^{4t} P^{\frac{rnM}{2}+M\rho} Y^{-r} \sum_{y_1=1}^{Y} \cdots \sum_{y_r=1}^{Y} \int\limits_{\Omega(y_1,\ldots,y_r)} |S_\beta(y_1, \ldots, y_r)|^{2t}\, d\beta + (4r)^{2t} P^{2t(r-\rho)}.$$

By Lemma 17 and Theorem 1, we get

$$|S|^{2t} \ll P^{\frac{rnM}{2}+M\rho} Y^{-r} Y^{r-1} P^{1-\frac{1}{2r}+\frac{1}{10M}} P^{2tr-\frac{rnM}{2}+\delta_\tau} + P^{2t(r-\rho)}.$$

Since $-M\rho + 1/2r - 1/M - \rho - \delta_\tau \geqslant 2t\rho$, the theorem is proved for case a).

In case b), we make the following change of the summation variable:

$$x_j = Q_0\xi_j + \eta_j, \qquad 1 \leqslant \eta_j \leqslant Q_0, \qquad -\frac{\eta_j}{Q_0} < \xi_j \leqslant \frac{P-\eta_j}{Q_0}, \qquad j = 1, \ldots, r.$$

We have

$$S=\sum_{\eta_1=1}^{Q_0}\cdots\sum_{\eta_r=1}^{Q_0}\exp\left(2\pi i\sum_{t_1=0}^{n}\cdots\sum_{t_r=0}^{n}\frac{a_{t_1,\ldots,t_r}}{q_{t_1,\ldots,t_r}}\eta_1^{t_1}\cdots\eta_r^{t_r}\right)W_{\eta_1,\ldots,\eta_r};$$

$$W_{\eta_1,\ldots,\eta_r}=\sum_{\xi_1,\ldots,\xi_r}\exp\left(2\pi i\sum_{\substack{t_1=0\\ t_1+\ldots+t_r\geqslant 2}}^{n}\cdots\sum_{t_r=0}^{n}z_{t_1,\ldots,t_r}(Q_0\xi_1+\eta_1)^{t_1}\cdots(Q_0\xi_r+\eta_r)^{t_r}\right.$$

$$\left.+2\pi i\sum_{j=1}^{r}\left(\xi_j\left(Q_0\frac{a_{0,\ldots,1,\ldots,0}}{q_{0,\ldots,1,\ldots,0}}+Q_0z_{0,\ldots,1,\ldots,0}\right)+\eta_jz_{0,\ldots,1,\ldots,0}\right)\right);$$

where the notation $a_{0,\ldots,1,\ldots,0}$; $q_{0,\ldots,1,\ldots,0}$; $z_{0,\ldots,1,\ldots,0}$ signifies that in the jth place we have 1, and in the remaining places we have 0.

Let

$$Q_0a_{0,\ldots,1,\ldots,0}\equiv a'_{0,\ldots,1,\ldots,0}\ (\mathrm{mod}\ q_{0,\ldots,1,\ldots,0}),\ |a'_{0,\ldots,1,\ldots,0}|<0.5q_{0,\ldots,1,\ldots,0}.$$

Then, applying van der Corput's lemma (see [1], Chapter II, Lemma 3) successively for each variable, we get

$$W_{\eta_1,\ldots,\eta_r}=P^rQ^{-r}\int_0^1\cdots\int_0^1\exp\left(2\pi i\sum_{\substack{t_1=0\\ t_1+\ldots+t_r\geqslant 2}}^{n}\cdots\sum_{t_r=0}^{n}z_{t_1,\ldots,t_r}P^{t_1+\ldots+t_r}x_1^{t_1}\cdots x_r^{t_r}\right.$$

$$+2\pi i\sum_{j=1}^{r}\left(\frac{Px_j-\eta_j}{Q_0}\left(\frac{a'_{0,\ldots,1,\ldots,0}}{q_{0,\ldots,1,\ldots,0}}+Q_0z_{0,\ldots,1,\ldots,0}\right)\right.$$

$$\left.\left.+\eta_jz_{0,\ldots,1,\ldots,0}\right)\right)dx_1\ldots dx_r+O(P^{r-1}Q_0^{1-r})+O(1).$$

If $Q_0<Q$, then for some j the coefficient of x_j is at least

$$\left(\frac{1}{q_{0,\ldots,1,\ldots,0}}-\frac{Q_0}{q_{0,\ldots,1,\ldots,0}P^{1-\frac{1}{2r}}}\right)\frac{P}{Q_0}>\frac{P}{2q_{0,\ldots,1,\ldots,0}Q_0}\geqslant\frac{P^{\frac{1}{10M}}}{2}.$$

By Lemma 8a)

$$|W_{\eta_1,\ldots,\eta_r}|\ll P^rQ_0^{-r}P^{-\frac{1}{10Mn}}(\log P)^{r-1}+P^{r-1}Q_0^{1-r}+1,$$

whence it follows that $|S|\ll P^{r-p}$. Let $Q_0=Q$. Then

$$S=UV+O(P^{r-1}Q);$$

$$U=Q^{-r}\sum_{\eta_1=1}^{Q}\cdots\sum_{\eta_r=1}^{Q}\exp\left(2\pi i\sum_{t_1=0}^{n}\cdots\sum_{t_r=0}^{n}\frac{a_{t_1,\ldots,t_r}}{q_{t_1,\ldots,t_r}}\eta_1^{t_1}\cdots\eta_r^{t_r}\right);$$

$$V=P^r\int_0^1\cdots\int_0^1\exp\left(2\pi i\sum_{t_1=0}^{n}\cdots\sum_{t_r=0}^{n}\delta_{t_1,\ldots,t_r}x_1^{t_1}\cdots x_r^{t_r}\right)dx_1\ldots dx_r.$$

We apply Lemma 13a) for estimating U, and Lemma 8a) for estimating V.

Since either Q or δ is greater than $P^{1/M}$, therefore, substituting in the expression for S either the estimate for U or the estimate for V, we obtain the assertion in Theorem 3. This completes the proof of the theorem.

BIBLIOGRAPHY

1. I. M. Vinogradov, *Trigonometrical sums in number theory*, rev. ed., "Nauka", Moscow, 1971; English transl., Statistical Publishing Soc., Calcutta, 1975.

2. ———, *Selected works*, Izdat. Akad. Nauk SSSR, Moscow, 1952. (Russian)

3. A. A. Karacuba, *Waring's problem for a congruence modulo the power of a prime*, Vestnik Moskov. Univ. Ser. I Mat. Meh. **1962**, no. 4, 28–38. (Russian)

4. ———, *Theorems on the mean and complete trigonometric sums*, Izv. Akad. Nauk SSSR Ser. Mat. **30** (1966), 183–206. (Russian)

5. ———, *The mean value of the modulus of a trigonometric sum*, Izv. Akad. Nauk SSSR Ser. Mat. **37** (1973), 1203–1227; English transl. in Math. USSR Izv. **7** (1973).

6. G. I. Arhipov, Candidate's Dissertation, Steklov Inst. Math. Acad. Sci. USSR, Moscow, 1975. (Russian)

7. ———, *A theorem on the mean value of the modulus of a multiple trigonometric sum*, Mat. Zametki **17** (1975), 143–153; English transl. in Math. Notes **17** (1975).

8. G. I. Arhipov and V. N. Čubarikov, *Multiple trigonometric sums*, Izv. Akad. Nauk SSSR Ser. Mat. **40** (1976), 209–220; English transl. in Math. USSR Izv. **10** (1976).

9. V. N. Čubarikov, *On a multiple trigonometric integral*, Dokl. Akad. Nauk SSSR **227** (1976), 1308–1310; English transl. in Soviet Math. Dokl. **17** (1976).

10. ———, *Multiple rational trigonometric sums and multiple integrals*, Mat. Zametki **20** (1976), 61–68; English transl. in Math. Notes **20** (1976).

11. G. Polya and G. Szegö, *Aufgaben und Lehrsatze aus der Analysis.* Vol. II, Springer-Verlag, Berlin, 1925; English transl. of 4th (1971) ed., 1975.

Translated by V. N. SINGH

Труды Мат. Инст. Стеклов.
143 (1977)

Proceedings of the Steklov
Institute of Mathematics
1980, Issue 1

SOME PROPERTIES OF LACUNARY SERIES AND THE INTEGRABILITY OF TRIGONOMETRIC SERIES

UDC 517.5

L. A. BALAŠOV AND S. A. TELJAKOVSKIĬ

ABSTRACT. A theorem is proved on the integrability, with a certain weight, of a function that is the sum of a lacunary trigonometric series. For trigonometric series for which the first differences of the coefficients form a lacunary sequence the theorem is used to obtain necessary and sufficient conditions for the series to be integrable, i.e., to be a Fourier series.

Bibliography: 8 titles.

The first section deals with the problem of integrability, with a certain weight, of a function F that is the sum of a lacunary trigonometric series. As a corollary we consider the question of continuity of F.

In the second section we obtain for a class of trigonometric series with infrequently varying coefficients necessary and sufficient conditions for the integrability of the series, i.e. for it to be a Fourier series.

§1. Let $n_1, n_2, \ldots$ be a lacunary sequence of positive (not necessarily integral) numbers, i.e.

$$\frac{n_{k+1}}{n_k} \geqslant q > 1, \qquad k=1,\ 2,\ \ldots, \tag{1.1}$$

and suppose that the numbers a_k and b_k satisfy the condition $\Sigma(a_k^2 + b_k^2) < \infty$. Then, as is well known (see, for example, [1], Chapter II, §§1.2 and 1.3), the series

$$a_0 + \sum_{k=1}^{\infty} (a_k \cos n_k x + b_k \sin n_k x) \tag{1.2}$$

converges almost everywhere, and its sum, denoted by $F(x)$, is integrable on each interval.

We assume that the function $\varphi(x)$ is defined for $x > 0$, is positive, and increases to infinity as $x \longrightarrow 0$. We consider the question of integrability on an interval of the form $[0, \epsilon]$ of the function

$$F\ (x)\ \varphi\ (x). \tag{1.3}$$

In the first place we are interested in the case $\varphi(x) = 1/x$. We obtain a necessary and sufficient condition for the integrability of the function (1.3) under the assumption that for some α and sufficiently small x

---

1980 *Mathematics Subject Classification.* Primary 42A55; Secondary 42A16, 42A20.

$$\varphi(x)\, x \log^{\alpha} \frac{1}{x} \downarrow 0 \qquad \text{as} \qquad x \to 0. \tag{1.4}$$

We write $\varphi \in \Phi$ if the function φ satisfies all the above conditions. We remark that it follows from (1.4) that as $x \longrightarrow 0$

$$\int_0^x t\varphi(t)\, dt = O(x^2 \varphi(x)) \tag{1.5}$$

and

$$\varphi(x) = O(\varphi(2x)). \tag{1.6}$$

For the sine series, i.e., for $a_k \equiv 0$, necessary and sufficient conditions for the integrability of the function $F(x)/x$ were obtained by Mary Weiss ([2]; see also [3], p. 32). Our arguments follow her proof in a certain sense.

In this paper we use A to denote positive quantities (generally different) that can depend only on q and the function φ. Here we have in mind that the corresponding estimates will hold if we choose A sufficiently large.

THEOREM 1. *Suppose that the sequence* $\{n_k\}$ *satisfies the condition* (1.1), *the series* $\Sigma(a_k^2 + b_k^2)$ *converges, and* $\varphi \in \Phi$. *Then for convergence of the integral*

$$\int_{+0} |F(x)|\, \varphi(x)\, dx \tag{1.7}$$

a necessary and sufficient condition is the convergence of the series

$$\sum_{p=0}^{\infty} \varphi_p \left\{ \left| \sum_{k=0}^{p} a_k \right| + \left[\sum_{k=p}^{\infty} (a_k^2 + b_k^2) \right]^{1/2} \right\}, \tag{1.8}$$

where

$$\varphi_p = \int_{1/n_{p+1}}^{1/n_p} \varphi(x)\, dx. \tag{1.9}$$

If (1.8) *converges and its sum equals* S, *then*

$$\int_0^{1/n_1} |F(x)|\, \varphi(x)\, dx \leqslant AS. \tag{1.10}$$

For $a_k \equiv 0$ and $\varphi(x) = 1/x$ Theorem 1 was obtained by Weiss [2].

We shall use the following known estimates(1) for integrals of sums of lacunary series.

LEMMA 1. *Let* $q > 1$. *There exist positive numbers* Λ *and* Q, *depending only on* q, *such that if* $0 \leqslant \beta \leqslant 1/q$, $\Sigma(c_k^2 + d_k^2) < \infty$, *and* $\{\lambda_m\}$ *is an arbitrary sequence of numbers satisfying the conditions* $\lambda_1 \geqslant \Lambda$ *and* $\lambda_{m+1}/\lambda_m \geqslant q$, $m = 1, 2, \ldots$, *then*

(1) They can be obtained from estimates given, for example, in [4], §11, by means of Hölder's inequality.

$$\frac{1}{Q}\left\{|c_0|+\left[\sum_{m=1}^{\infty}(c_m^2+d_m^2)\right]^{1/2}\right\}\leqslant\int_{\beta}^{1}\left|c_0+\sum_{m=1}^{\infty}(c_m\cos\lambda_m x+d_m\sin\lambda_m x)\right|dx$$
$$\leqslant Q\left\{|c_0|+\left[\sum_{m=1}^{\infty}(c_m^2+d_m^2)\right]^{1/2}\right\}. \tag{1.11}$$

Proof of Theorem 1. We assume that the sequence $\{n_k\}$ satisfies the conditions

$$1<q\leqslant\frac{n_{k+1}}{n_k}\leqslant q^2,\qquad k=1,\ 2,\ \ldots \tag{1.12}$$

It is possible to arrive at this case by adding to (1.2) terms with the coefficients a_k and b_k equal to zero. It is clear that this does not change (1.8), and it is thus enough to prove the theorem under the conditions (1.12). We observe that, by (1.12) and (1.16),

$$\frac{1}{A}\varphi_p\leqslant\frac{1}{n_p}\varphi\left(\frac{1}{n_p}\right)\leqslant A\varphi_p. \tag{1.13}$$

We estimate the integral $\int|F|\varphi\,dx$ over the interval $[1/n_{M+1}, 1/n_1]$. Let E_p denote the interval $[1/n_{p+1}, 1/n_p]$. We write the indicated integral as the sum

$$\int_{1/n_{M+1}}^{1/n_1}|F(x)|\,\varphi(x)\,dx=\sum_{p=1}^{M}\int_{E_p}|F(x)|\,\varphi(x)\,dx, \tag{1.14}$$

and for $x\in E_p$ we represent $F(x)$ in the form

$$F(x)=\sum_{k=1}^{p+r-1}[a_k(\cos n_k x-1)+b_k\sin n_k x]$$
$$+\sum_{k=0}^{p+r-1}a_k+\sum_{k=p+r}^{\infty}[a_k\cos n_k x+b_k\sin n_k x], \tag{1.15}$$

where r is the least natural number such that for the Λ in Lemma 1 we have

$$q^r\geqslant\Lambda. \tag{1.16}$$

We estimate the integrals of the first sum on the right-hand side of (1.15):

$$\sigma\equiv\sum_{p=1}^{M}\int_{E_p}\left|\sum_{k=1}^{p+r-1}[a_k(\cos n_k x-1)+b_k\sin n_k x]\right|\varphi(x)\,dx$$
$$\leqslant\sum_{p=1}^{M}\int_{E_p}\sum_{k=1}^{p+r-1}(|a_k|+|b_k|)\,n_k x\varphi(x)\,dx\leqslant\sum_{k=1}^{r-1}(|a_k|+|b_k|)\,n_k\int_0^{1/n_1}x\varphi(x)\,dx$$
$$+\sum_{k=r}^{M+r-1}(|a_k|+|b_k|)\,n_k\int_0^{1/n_{k+1-r}}x\varphi(x)\,dx.$$

From this, using (1.5) and (1.12), we get

$$\sigma\leqslant A\sum_{k=1}^{M+r-1}\varphi_k(|a_k|+|b_k|). \tag{1.17}$$

We now consider the integrals of the second and third sums on the right-hand side of (1.15). By (1.16), the sequence $\lambda_m = n_{m+p+r-1}/n_p$ satisfies for each p the conditions in Lemma 1. Therefore, we have the estimates from above

$$\sigma_p \equiv \int_{E_p} \left| \sum_{k=0}^{p+r-1} a_k + \sum_{k=p+r}^{\infty} (a_k \cos n_k x + b_k \sin n_k x) \right| \varphi(x)\, dx$$

$$\leqslant \frac{1}{n_p} \varphi\left(\frac{1}{n_{p+1}}\right) \int_{n_p/n_{p+1}}^{1} \left| \sum_{k=0}^{p+r-1} a_k + \sum_{k=p+r}^{\infty} \left(a_k \cos \frac{n_k}{n_p} x + b_k \sin \frac{n_k}{n_p} x\right) \right| dx$$

$$\leqslant \frac{A}{n_p} \varphi\left(\frac{1}{n_{p+1}}\right) \left\{ \left| \sum_{k=0}^{p+r-1} a_k \right| + \left[\sum_{k=p+r}^{\infty} (a_k^2 + b_k^2) \right]^{1/2} \right\} \tag{1.18}$$

and from below

$$\sigma_p \geqslant \frac{1}{An_p} \varphi\left(\frac{1}{n_p}\right) \left\{ \left| \sum_{k=0}^{p+r-1} a_k \right| + \left[\sum_{k=p+r}^{\infty} (a_k^2 + b_k^2) \right]^{1/2} \right\}. \tag{1.19}$$

Using (1.12) and (1.6), we see that

$$\frac{1}{n_p} \varphi\left(\frac{1}{n_{p+1}}\right) \leqslant A \varphi_{p+r} \quad \text{and} \quad \frac{1}{n_p} \varphi\left(\frac{1}{n_p}\right) \geqslant \frac{1}{A} \varphi_{p+r}. \tag{1.20}$$

We note also that in the first sums on the right-hand sides of (1.18) and (1.19) we can replace $p + r - 1$ by $p + r$.

From (1.17), (1.18), and (1.20) and the representations (1.14) and (1.15) we find that

$$\int_{1/n_{M+1}}^{1/n_1} |F(x)| \varphi(x)\, dx \leqslant A \sum_{k=1}^{M+r-1} \varphi_k (|a_k| + |b_k|) + A \sum_{p=r+1}^{M+r} \varphi_p \left\{ \left| \sum_{k=0}^{p} a_k \right| \right.$$

$$\left. + \left[\sum_{k=p}^{\infty} (a_k^2 + b_k^2) \right]^{1/2} \right\} \leqslant A \sum_{p=1}^{M+r} \varphi_p \left\{ \left| \sum_{k=0}^{p} a_k \right| + \left[\sum_{k=p}^{\infty} (a_k^2 + b_k^2) \right]^{1/2} \right\}. \tag{1.21}$$

From this we get the sufficiency of the conditions of the theorem and the estimate (1.10).

Similarly, from (1.17), (1.19), (1.20), (1.14) and (1.15) we get

$$\int_{1/n_{M+1}}^{1/n_1} |F(x)| \varphi(x)\, dx \geqslant \frac{1}{A} \sum_{p=r+1}^{M+r} \varphi_p \left\{ \left| \sum_{k=0}^{p} a_k \right| \right.$$

$$\left. + \left[\sum_{k=p}^{\infty} (a_k^2 + b_k^2) \right]^{1/2} \right\} - A \sum_{p=1}^{M+r-1} \varphi_p (|a_p| + |b_p|). \tag{1.22}$$

The necessity of the conditions of the theorem follows from (1.22). Indeed, if the series

$$\sum_{k=1}^{\infty} \varphi_k (|a_k| + |b_k|) \tag{1.23}$$

converges, it is immediately clear. But if (1.23) diverges, we first use the obvious estimate

$$\sum_{p=1}^{N} \varphi_p (|a_p| + |b_p|) = o\left(\sum_{p=1}^{N} \sqrt{p}\, \varphi_p (|a_p| + |b_p|) \right), \tag{1.24}$$

and then the next proposition.

LEMMA 2. *If ψ_p are positive and for some α the numbers $\psi_p p^\alpha$ decrease monotonically with increasing p, then*

$$\sum_{p=1}^{N} \sqrt{p}\, \psi_p (|a_p| + |b_p|) \leqslant B \sum_{p=1}^{N} \psi_p \left[\sum_{k=p}^{N} (a_k^2 + b_k^2) \right]^{1/2}, \tag{1.25}$$

where B depends only on α.

By (1.4), the ψ_p in Lemma 2 can be taken to be $\varphi(1/n_p)/n_p$. But since (1.13) holds, the necessity of the conditions of Theorem 1 when (1.23) diverges follows from (1.22), (1.24), and (1.25). Thus, to conclude the proof of Theorem 1 it suffices to prove Lemma 2.

PROOF OF LEMMA 2. The estimate (1.25) is the result of a chain of inequalities obtained by first using the fact that $\psi_k k^\alpha$ decreases, and then the Cauchy-Schwarz-Bunjakovskiĭ inequality. We set $\epsilon = 1 + |\alpha|$. Then

$$\sum_{k=1}^{N} \sqrt{k}\, \psi_k (|a_k| + |b_k|) \leqslant (\varepsilon + 1) \sum_{k=1}^{N} \sqrt{k}\, \psi_k (|a_k| + |b_k|) \sum_{p=1}^{k} \frac{p^\varepsilon}{k^{\varepsilon+1}}$$

$$= (\varepsilon + 1) \sum_{p=1}^{N} p^\varepsilon \sum_{k=p}^{N} \psi_k \frac{1}{k^{\varepsilon+1/2}} (|a_k| + |b_k|) \leqslant (\varepsilon + 1) \sum_{p=1}^{N} p^{\varepsilon+\alpha} \psi_p \sum_{k=p}^{N} \frac{|a_k| + |b_k|}{k^{\alpha+\varepsilon+1/2}}$$

$$\leqslant (\varepsilon + 1) \sum_{p=1}^{N} p^{\varepsilon+\alpha} \psi_p \left[\sum_{k=p}^{N} \frac{1}{k^{2\alpha+2\varepsilon+1}} \sum_{k=p}^{N} (|a_k| + |b_k|)^2 \right]^{1/2}$$

$$\leqslant 4(\varepsilon + 1) \sum_{p=1}^{N} \psi_p \left[\sum_{k=p}^{N} (a_k^2 + b_k^2) \right]^{1/2}.$$

The lemma is proved. We note that for ψ_p of the form p^c this lemma is a particular case of Theorem 346 in [5].

From Theorem 1 we can get sufficient conditions for the continuity of functions with lacunary Fourier series.

COROLLARY 1. *Assume that* (1.2) *is the Fourier series of the function F, and the sequence $\{n_k\}$ is lacunary, i.e.* (1.1) *holds. If*

$$\int_{+0} \frac{|F(x)|}{x\sqrt{\log(1/x)}}\, dx < \infty, \tag{1.26}$$

then F is continuous.

If for small x we set $\varphi(x) = (x(\log(1/x))^{1/2})^{-1}$, we get

$$\varphi_p = \int_{1/n_{p+1}}^{1/n_p} \varphi(x)\, dx \geqslant \int_{1/qn_p}^{1/n_p} \varphi(x)\, dx \geqslant \frac{1}{A\sqrt{p}}.$$

Therefore, by Theorem 1, (1.26) implies the convergence of the series

$$\sum_{p=1}^{\infty}\frac{1}{\sqrt{p}}\left[\sum_{k=p}^{\infty}(a_k^2+b_k^2)\right]^{1/2}. \tag{1.27}$$

By (1.25), we get from this the absolute convergence of the series (1.2), whence the function F is continuous.

We now show that *for each positive function*

$$\varphi(x)=o\left(\frac{1}{x\sqrt{\log(1/x)}}\right),\qquad x\to 0, \tag{1.28}$$

there exists a function F with Fourier series of the form

$$\sum_{k=1}^{\infty} b_k \sin 2^k x \tag{1.29}$$

which does not converge absolutely, although the integral (1.7) *converges.*

By (1.28), there are numbers $\epsilon_p \downarrow 0$ such that

$$\varphi_p=\int_{2^{-(p+1)}}^{2^{-p}}\varphi(x)\,dx\leqslant\frac{\varepsilon_p}{\sqrt{p}}.$$

We choose $\alpha_p \downarrow 0$ such that $\Sigma_1^\infty \alpha_p/p=\infty$, but $\Sigma_1^\infty \epsilon_p\alpha_p/p<\infty$. The desired example is given by the series (1.29) with $b_k=\alpha_k/k$. Indeed, this series does not converge absolutely. On the other hand,

$$\sum_{p=1}^{\infty}\varphi_p\left(\sum_{k=p}^{\infty}b_k^2\right)^{1/2}\leqslant 2\sum_{p=1}^{\infty}\varphi_p\frac{\alpha_p}{\sqrt{p}}\leqslant 2\sum_{p=1}^{\infty}\frac{\varepsilon_p\alpha_p}{p}<\infty.$$

Therefore, the integral (1.7) converges by Theorem 1.

Accordingly, in the class of all lacunary series the condition (1.26) in Corollary 1 is best possible in the sense indicated. But if we consider sparser sequences $\{n_k\}$, then it can be weakened.

We restrict ourselves to the following examples, which supplement Corollary 1.

1. If $\alpha>1$ and $n_k=2^{[k^\alpha]}$, i.e., $\log_2 n_k=[k^\alpha]$, then a sufficient condition for the continuity of F is the convergence of the integral (1.7) with

$$\varphi(x)=\frac{1}{x\log^{1-\frac{1}{2\alpha}}\frac{1}{x}}$$

for small x. Indeed, for this φ and for sufficiently large p

$$\varphi_p=2\alpha\left(\log^{\frac{1}{2\alpha}}n_{p+1}-\log^{\frac{1}{2\alpha}}n_p\right)\geqslant\frac{1}{A\sqrt{p}}.$$

Therefore, by Theorem 1, the convergence of (1.7) implies that of (1.27), which ensures the continuity of F.

2. Now let n_k be such that

$$\underbrace{\log_2\log_2\ldots\log_2}_{m} n_k=[k^\alpha],$$

where $m \geqslant 2$ and $\alpha > 0$. Then for a function $\varphi(x)$ given for small x by

$$\frac{1}{x\left(\log\frac{1}{x}\right)\left(\log\log\frac{1}{x}\right)\cdots\left(\underbrace{\log\log\ldots\log}_{m}\frac{1}{x}\right)^{1-\frac{1}{2\alpha}}},$$

we have $\varphi_p \geqslant 1/Ap^{1/2}$ for sufficiently large p, and convergence of (1.7) with this function φ is sufficient for the continuity of F.

It is not hard to show that the sufficient conditions for the continuity of the function given in both examples are best possible in the same sense as the conditions in Corollary 1. We shall not dwell on this.

§2. We consider the question of integrability of trigonometric series with "infrequently varying" coefficients–more precisely, series for which the first differences of the coefficients form a lacunary sequence.

THEOREM 2. *Let $n_0 = 0$ and let $n_1, n_2, \ldots$ be a lacunary sequence of natural numbers, i.e., the condition* (1.1) *holds. The series*

$$\frac{a_0}{2}+\sum_{m=1}^{\infty}(a_m\cos mx+b_m\sin mx), \tag{2.1}$$

with coefficients converging to zero and such that

$$a_m=a_{n_k},\quad b_m=b_{n_k}\quad \textit{for}\quad n_{k-1}<m\leqslant n_k,\quad k=1,\ 2,\ \ldots, \tag{2.2}$$

is a Fourier series if and only if

$$\sum_{m=1}^{\infty}\frac{|b_m|}{m}<\infty \tag{2.3}$$

and

$$\sum_{p=1}^{\infty}\left\{\sum_{k=p}^{\infty}[(\Delta a_{n_k})^2+(\Delta b_{n_k})^2]\right\}^{1/2}\log\frac{n_{p+1}}{n_p}<\infty. \tag{2.4}$$

PROOF. We assume first that (2.3) and (2.4) hold. It follows from (2.4) that

$$\sum_{m=1}^{\infty}(|\Delta a_m|+|\Delta b_m|)<\infty$$

whence the series (2.1) converges for all $x \in (0, 2\pi)$, and its sum $f(x)$ is continuous on $(0, 2\pi)$. Therefore, (2.1) is a Fourier series if the function f is integrable. Setting $b_0 = 0$, we find that

$$f(x)=\frac{1}{2\sin(x/2)}\sum_{k=0}^{\infty}\left[\Delta a_{n_k}\sin\left(n_k+\frac{1}{2}\right)x-\Delta b_{n_k}\cos\left(n_k+\frac{1}{2}\right)x\right]. \tag{2.5}$$

We make use of Theorem 1. For

$$\varphi_p = \int_{1/(n_{p+1}+1/2)}^{1/(n_p+1/2)} \frac{dx}{2\sin(x/2)} \tag{2.6}$$

we have the estimate $\varphi_p \leqslant (\pi/2)\log((2n_{p+1}+1)/(2n_p+1))$. Further,

$$\sum_{p=1}^{\infty} \varphi_p \left| \sum_{k=0}^{p} \Delta b_{n_k} \right| \leqslant \frac{\pi}{2} \sum_{p=1}^{\infty} |b_{n_{p+1}}| \log \frac{2n_{p+1}+1}{2n_p+1} \leqslant A \sum_{m=1}^{\infty} \frac{|b_m|}{m}.$$

Hence, by Theorem 1, the integrability of f follows from (2.3) and (2.4).

We now prove the converse assertion. Suppose (2.1) is the Fourier series of $f(x)$. Then

$$\sum_{k=0}^{\infty} [\Delta a_{n_k} \sin(2n_k+1)x - \Delta b_{n_k} \cos(2n_k+1)x]$$

is the Fourier series of $2f(2x)\sin x$. Since this series is lacunary, it converges almost everywhere; hence (2.5) holds for almost all x. Moreover,

$$\sum_{k=0}^{\infty} [(\Delta a_{n_k})^2 + (\Delta b_{n_k})^2] < \infty.$$

Thus Theorem 1 is applicable, and it gives us that the integrability of $f(x)$ on $[0, \pi]$ implies the convergence of the series

$$\sum_{p=1}^{\infty} \varphi_p \left\{ \left| \sum_{k=0}^{p} \Delta b_{n_k} \right| + \left[\sum_{k=p}^{\infty} ((\Delta a_{n_k})^2 + (\Delta b_{n_k})^2) \right]^{1/2} \right\},$$

where the φ_p are defined by (2.6). From this we get (2.4), since

$$\varphi_p \geqslant \log \frac{2n_{p+1}+1}{2n_p+1}.$$

But since

$$\sum_{p=1}^{\infty} \varphi_p \left| \sum_{k=0}^{p} \Delta b_{n_k} \right| \geqslant \sum_{p=1}^{\infty} |b_{n_{p+1}}| \log \frac{2n_{p+1}+1}{2n_p+1} \geqslant \frac{1}{A} \sum_{m=1}^{\infty} \frac{|b_m|}{m},$$

the condition (2.3) also holds, and the theorem is proved.

We derive some corollaries from Theorem 2. We first consider the case when the numbers n_k increase so rapidly that the sequence $\{\log n_k\}$ is lacunary.

COROLLARY 2. *Suppose that the coefficients of* (2.1) *converge to zero and satisfy* (2.2), *where the* n_k *are such that*

$$\frac{\log n_{k+1}}{\log n_k} \geqslant q > 1. \tag{2.7}$$

Then (2.1) *is a Fourier series if and only if*

$$\sum_{m=1}^{\infty} (|\Delta a_m| + |\Delta b_m|) \log m < \infty, \tag{2.8}$$

or, equivalently,

$$\sum_{m=1}^{\infty} \frac{|a_m| + |b_m|}{m} < \infty. \tag{2.9}$$

The sufficiency of (2.8) even for arbitrary series with coefficients converging to zero is well known. For cosine series this was established by S. Szidon [6], and for sine series by L. Tonelli ([7], p. 265).

The necessity of (2.8) in Corollary 2 follows from the necessity of (2.4) in Theorem 2. Indeed, by (2.7), $\log(n_{p+1}/n_p) \geqslant A^{-1}\log n_p$, and (2.4) implies the convergence of the series

$$\sum (|\Delta a_{n_p}| + |\Delta b_{n_p}|) \log n_p,$$

which is equivalent to condition (2.8).

It remains to prove the equivalence of (2.8) and (2.9). For brevity we assume $b_m \equiv 0$.

The estimate (2.9) follows from (2.8) for arbitrary a_m, since

$$\sum_{m=1}^{\infty} \frac{|a_m|}{m} = \sum_{m=1}^{\infty} \frac{1}{m} \left| \sum_{k=m}^{\infty} \Delta a_k \right| \leqslant \sum_{k=1}^{\infty} |\Delta a_k| \sum_{m=1}^{k} \frac{1}{m}.$$

But if (2.2) and (2.7) hold, we also have the converse assertion

$$\sum_{m=1}^{\infty} |\Delta a_m| \log m = \sum_{k=1}^{\infty} |\Delta a_{n_k}| \log n_k \leqslant 2 \sum_{k=1}^{\infty} |a_{n_k}| \log n_k$$

$$\leqslant 2 \sum_{k=1}^{\infty} \sum_{m=n_{k-1}+1}^{n_k} \frac{|a_{n_k}|}{m} \frac{\log n_k}{\log n_k - \log(n_{k-1}+1)} \leqslant A \sum_{m=1}^{\infty} \frac{|a_m|}{m}.$$

In conclusion we obtain a corollary for cosine series with monotonically decreasing coefficients.

COROLLARY 3. *Suppose that* $\{n_k\}$ *is a lacunary sequence, i.e., the condition* (1.1) *holds, and that the series*

$$\frac{a_0}{2} + \sum_{m=1}^{\infty} a_m \cos mx, \tag{2.10}$$

whose coefficients are monotonically decreasing and satisfy (2.2), *is a Fourier series. Then*

$$\sum_{m=2}^{\infty} \frac{a_m}{m\sqrt{\log m}} < \infty. \tag{2.11}$$

We add new terms to the sequence $\{n_k\}$ in such a way that the estimates (1.12) hold. The assertions of Theorem 2, on which we base our arguments, are equivalent for the original and the new sequences. Therefore, we can assume without loss of generality that the estimates (1.12) hold in the first place.

Since (2.10) is a Fourier series, Theorem 2 gives us (2.4). By Lemma 2, (2.4) implies the convergence of $\Sigma\, k^{1/2}|\Delta a_{n_k}|$, i.e., according to (1.12), the series

$\Sigma(\log m)^{1/2}|\Delta a_m|$ must converge, which is equivalent to the condition (2.11) for monotone a_m.

We show that *Corollary* 3 *cannot be strengthened, in the sense that there is no sequence of numbers* y_m *that increases to infinity for which the condition* (2.11) *can be replaced by*

$$\sum \frac{a_m y_m}{m\sqrt{\log m}} < \infty. \tag{2.12}$$

Indeed, assume that $y_m \uparrow \infty$ and $y_m/(m(\log m)^{1/2}) \downarrow 0$. We can always arrive at this case by replacing the y_m by smaller numbers. For brevity we introduce the notation $y_{2^m} = z_m$ and $z_{[m/2]} = t_m$. Since $a_m y_m/(m(\log m)^{1/2}) \downarrow 0$, we have, by Cauchy's theorem on series with monotone coefficients, that (2.12) is equivalent to the condition

$$\sum a_{2^m} \frac{z_m}{\sqrt{m}} < \infty. \tag{2.13}$$

We choose numbers $\alpha_k \downarrow 0$ so that $\Sigma\, \alpha_k k^{-1/2}$ converges but $\Sigma\, \alpha_k k^{-1/2} t_k$ diverges, and we set $a_{2^m} = \Sigma_{k=m}^{\infty}\, \alpha_k/k$. By Theorem 2, the series (2.10) in Corollary 3 with these a_{2^m} is a Fourier series, since we have (2.4):

$$\sum_{p=1}^{\infty}\left\{\sum_{k=p}^{\infty}(\Delta a_{2^k})^2\right\}^{1/2} = \sum_{p=1}^{\infty}\left(\sum_{k=p}^{\infty}\frac{\alpha_k^2}{k^2}\right)^{1/2} \leqslant 2\sum_{p=1}^{\infty}\frac{\alpha_p}{\sqrt{p}} < \infty.$$

But

$$\sum_{m=1}^{\infty} a_{2^m}\frac{z_m}{\sqrt{m}} = \sum_{k=1}^{\infty}\frac{\alpha_k}{k}\sum_{m=1}^{k}\frac{z_m}{\sqrt{m}} \geqslant \sum_{k=2}^{\infty}\frac{\alpha_k}{k}\sum_{m=[k/2]}^{k}\frac{t_k}{\sqrt{m}} \geqslant \frac{1}{2}\sum_{k=2}^{\infty}\frac{\alpha_k}{\sqrt{k}}t_k$$

and so (2.13) fails to hold, as well as (2.12). Thus, our statement is proved.

We mention a particular case of Corollary 2.

COROLLARY 4. *If the sequence* $\{\log n_k\}$ *is lacunary, i.e.* (2.7) *holds, and the coefficients of* (2.10) *are monotonically decreasing and satisfy* (2.2), *then this series is a Fourier series if and only if* $\Sigma a_m/m$ *converges.*

As shown by Szidon [6], the condition $\Sigma a_m/m < \infty$ is sufficient for an arbitrary series (2.10) with monotone coefficients to be a Fourier series. Corollary 4 determines a class of series (2.10) for which it is also necessary.

Under the additional condition $a_m = O(1/\log m)$ and for $n_k = 2^{2^k}$ Corollary 4 was established by L. A. Balašov [8]. We see that this additional condition on the coefficients a_m is superfluous. This enables us to simplify the derivation of the next assertion, which was obtained by Balašov in the same paper:

COROLLARY 5. *Suppose that the numbers* b_m *decrease monotonically, and that* $\Sigma\, b_m/m$ *diverges. Then there exist numbers* $a_m \downarrow 0$ *such that* $a_m \leqslant b_m$ *and the series* (2.10) *with these coefficients* a_m *is not a Fourier series.*

Indeed, we set $a_m = b_m$ for $m \leqslant 2$ and $a_m = b_{2^{2^k}}$ for $2^{2^{k-1}} < m \leqslant 2^{2^k}$, $k = 1, 2, \ldots$.

Applying Cauchy's theorem on series with monotone coefficients twice, we see that the conditions imposed on b_m imply that $\Sigma\, 2^k b_{2^{2^k}}$ diverges; hence, so does $\Sigma\, a_m/m$. By Corollary 4, the series (2.10) with these coefficients a_m cannot then be a Fourier series.

BIBLIOGRAPHY

1. V. F. Gapoškin, *Lacunary series and independent functions*, Uspehi Mat. Nauk **21** (1966), no. 6 (132), 3–82; English transl. in Russian Math. Surveys **21** (1966).

2. Mary Weiss, *A theorem on lacunary trigonometric series*, Orthogonal Expansions and their Continuous Analogues (Proc. Conf., Edwardsville, Ill., 1967), Southern Illinois Univ. Press, Carbondale, Ill. 1968.

3. Ralph P. Boas, Jr., *Integrability theorems for trigonometric transforms*, Springer-Verlag, Berlin and New York, 1967.

4. A. S. Belov, *On the sum of a lacunary series*, Trudy Moskov. Mat. Obšč. **33** (1976), 107–144; English transl. in Trans. Moscow Math. Soc. 1978, no. 1 (33).

5. G. H. Hardy, J. E. Littlewood and G. Pólya, *Inequalities,* Cambridge Univ. Press, New York, 1934.

6. S. Szidon, *Reihentheoretische Sätze and ihre Anwendungen in der Theorie der Fourierschen Reihen*, Math. Z. **10** (1921), 212–127.

7. L. Tonelli, *Serie trigonometriche*, Zanichelli, Bologna, 1928.

8. L. A. Balašov, *On trigonometric series with monotone coefficients*, Mat. Zametki **22** (1977), 77–84; English transl in Math. Notes **22** (1977).

Translated by H. H. McFADEN

Труды Мат. Инст. Стеклов.
143 (1977)

Proceedings of the Steklov
Institute of Mathematics
1980, Issue 1

INTERCELLULAR AVERAGES AND AN ERROR ESTIMATE FOR CUBATURE FORMULAS IN SOBOLEV SPACES AND THEIR GENERALIZATIONS

UDC 517.392

O. V. BESOV

ABSTRACT. We study some properties of intercellular averages of differentiable functions of several variables with their highest derivatives belonging to some symmetric space. The results are applied to the clarification of the asymptotic behavior of a sequence of norms of error functionals or certain more general functionals.

Bibliography: 11 titles.

In this paper will be studied some properties of intercellular averages of differentiable functions of several real variables in Sobolev spaces and their generalizations (spaces that are defined in terms of the metric of symmetric spaces, or Morrey spaces). The results are then used to clarify the asymptotic behavior of the norms of error functionals as the mesh width decreases, as well as that of the norms of some more general functionals. Functional methods, becoming widely used in the theory of quadrature formulas from the time Nikol'skii's work appeared and the first edition of his book [1] was published, let to the development of a number of different trends. In recent years the work of Sobolev has given rise to a theory of cubature formulas, which has been summed up in his monograph [2]. Many papers have been devoted to the study of quadrature and cubature formulas.

In the case of approximate (many-dimensional) integration, with a corresponding error functional

$$(l,\ f)=\int_{\Omega} f(x)\,dx-\sum_{k=1}^{N} c_k f(x^k),$$

the error can be estimated by an inequality

$$|(l,\ f)|\leqslant \|l\|_{B^*}\|f\|_B,$$

where B is a linear function space with a seminorm. The asymptotic behavior (as the mesh width $h \longrightarrow 0$) of a sequence of error functionals in the space $B = L_2^m(E^n)$ has been studied by Sobolev. His investigations in this direction were continued by Polovinkin [3], [4], who applied a new method for the study of analogous problems in the spaces $B = L_p^m(\Omega)$ with the seminorm

$$\|f\|_{L_p^m(\Omega)}=\left\{\int_{\Omega}\left[\sum_{|\alpha|=m}\frac{m\,!}{\alpha\,!}(D^\alpha f)^2\right]^{\frac{p}{2}}dx\right\}^{\frac{1}{p}},\qquad 1<p<\infty,$$

where Ω denotes a bounded domain in E^n.

1980 *Mathematics Subject Classification.* Primary 46E35, 41A55.

Here and everywhere below, $D^\alpha f$ stands for a generalized derivative in the sense of Sobolev.

The present paper arose in connection with Polovinkin's results. It proposes a different (and, as it seems, simpler and more natural) approach, tied to the study of intercellular averages of functions. This approach not only makes it possible to sharpen the corresponding theorem of Polovinkin for the Sovolev space $L_p^m(\Omega)$ (exchanging convergence with respect to sequences of the type $h = s^{-1}, s^{-2}, \ldots$, where $s \geqslant 2$ is an integer, for convergence as $h \longrightarrow 0$), but it also allows this result to be carried over to the more general space $B^m(\Omega)$ of functions with the seminorm $\|f\|_{B^m(\Omega)}$, defined when the derivatives $D^\alpha f$, $|\alpha| = m$, belong to some basic space $B(\Omega)$.

In order for cubature formulas to be applicable to functions in $B^m(\Omega)$ it is necessary that $B^m(\Omega)$ can be imbedded into the space of continuous functions: $B^m(\Omega) \subset C^m$.

Here it should be noted that it is sometimes possible, when asking for an estimate of the error of a cubature formula for a given function, to obtain a lowering of the order m through a judicious choice of the space $B^m(\Omega)$ containing the function. Thus, for instance, for a Morrey space $L_{p,a}^m(\Omega)$, $0 \leqslant a < 1$ (which for $a = 0$ coincides with $L_p^m(\Omega)$), it is true that $L_{p,a}^m(\Omega) \subset C(\Omega)$ when $m - n(1-a)/p > 0$, which means that one can lower the order m, for given p, in the transition from Sobolov spaces to Morrey spaces, which can be a distinct advantage in some cases.

Let us proceed to the description of the function spaces we are going to consider. Let $Q = [0, 1)^n$ be a half-open cube in the Euclidean space E^n, and let G be a set in E^n, h a positive number, and $\gamma = (\gamma_1, \ldots, \gamma_n)$ a vector with nonnegative components. Then

$$hG = \left\{x : \frac{x}{h} \in G\right\}, \quad h\gamma + G = \{x : x - h\gamma \in G\},$$

The number of different $\gamma \in G$ is denoted by $|G|$, and the number of elements in the set $\Gamma = \{\gamma\}$ by $|\Gamma|$. Summation over all $\gamma \in h^{-1}G$ is indicated by $\Sigma_{\gamma h \in G}$. Sets of the type $h\gamma + hQ$ will be called h-cells. Finally, we set

$$Q^{(N)} = \bigcup_{|\gamma_i| \leqslant N-1} (\gamma + Q) = [-N+1, N)^n, \qquad Q_\gamma^{(N)} = \gamma + Q^{(N)},$$

for every natural number N.

Throughout the following Ω will denote a Jordan measurable domain in E^n, even through some of the results will remain valid also for a domain Ω of finite Lebesgue measure. If Ω and the natural number N are given, then for every sufficiently small $h > 0$ we set

$$\Omega \# h = \{x : x \in h\gamma + \operatorname{int} hQ \subset \Omega\},$$

$$\Omega_h = \bigcup_{\gamma h \in \Omega_h} (h\gamma + hQ), \qquad \Omega_h + hQ^{(N)} \subset \Omega, \qquad \rho(\partial\Omega, \partial\Omega_h) \to 0 \quad (h \to 0),$$

$$\Omega^h = \{x : x \in h\gamma + hQ, \ \operatorname{mes}[\Omega \cap (h\gamma + hQ)] > 0\},$$

and

$$\mu = \operatorname{mes}^{\frac{1}{n}} \Omega, \qquad \mu_h = \operatorname{mes}^{\frac{1}{n}} \Omega_h, \qquad \mu^h = \operatorname{mes}^{\frac{1}{n}} \Omega^h,$$

so that

$$\Omega_h \subset \Omega \subset \Omega^h, \qquad \mu_h \leqslant \mu \leqslant \mu^h, \qquad \mu^h - \mu_h \to 0 \qquad (h \to 0).$$

By $B(\Omega)$ we denote a Banach space of measurable functions on Ω with a norm $\|f\|_{B(\Omega)} = \|f\|_\Omega$ and satisfying the following conditions I–IV:

I. If $f \in B(\Omega)$, if g is measurable on Ω, and if $|g(x)| \leqslant |f(x)|$ on Ω, then $g \in B(\Omega)$, and $\|g\|_\Omega \leqslant \|f\|_\Omega$ (i.e. $B(\Omega)$ is a Banach lattice).

II. Together with every function f in $B(\Omega)$, $B(\Omega)$ also contains every function g that is equimeasurable with f, and in addition $\|g\|_\Omega = \|f\|_\Omega$ (i.e. $B(\Omega)$ is a symmetric space). As usual, two functions f and g are called *equimeasurable* if

$$\lambda(f,\ t) = \lambda(g,\ t) \qquad \forall t > 0, \qquad \text{where } \lambda(f,\ t) = \operatorname{mes}\{x : |f(x)| > t\}.$$

III. If $f \in B(\Omega)$ and

$$f \in B(\Omega), \ \int_0^t g^*(u)\,du \leqslant \int_0^t f^*(u)\,du \qquad \forall t > 0,$$

then $g \in B(\Omega)$. Here $f^*(u)$ is the nonincreasing function on $[0, \operatorname{mes}\Omega]$ that is equimeasurable with $f(x)$. In the case that we are considering, with the Lebesgue measure, the last inequality is equivalent to the estimate

$$\sup_{\operatorname{mes} G \leqslant t} \int_G |g(x)|\,dx \leqslant \sup_{\operatorname{mes} G \leqslant t} \int_G |f(x)|\,dx.$$

Condition III is satisfied by L_p-spaces, Orlicz, Marcinkiewicz, and Lorentz spaces, as well as by many other spaces.

For a measurable set $G \subset \Omega$ we set $\|f\|_{B(G)} = \|f\|_G = \|\bar{f}\|_\Omega$, where $\bar{f}$ is the continuation by zero of f (or of the restriction of f to G) from G to Ω.

Furthermore, let $B(\mu Q)$ and $B_\mu(Q)$, with $\mu^n = \operatorname{mes}\Omega$, be symmetric spaces, with their respective norms defined so that

$$\|f(x)\|_{B_\mu(Q)} = \left\| f\left(\frac{x}{\mu}\right)\right\|_{B(\mu Q)} \quad \text{and} \quad \|g\|_{B(\mu Q)} = \|g\|_{\mu Q} = \|f\|_\Omega,$$

if f and g, defined on μQ and Ω, respectively, are equimeasurable.

It is easy to see that the norms $\|f(x/\nu)\|_{\nu Q}$ and $\|f(x/\mu)\|_{\mu Q}$ are equivalent for every fixed $\nu \in (0, \mu)$.

For all symmetric spaces it is true that

$$\sup_f \frac{\|f(x/\nu)\|_{\nu Q}}{\|f(x/\mu)\|_{\mu Q}} \to 1 \qquad \text{as} \quad \nu \uparrow \mu.$$

IV. There exists a function $\tau(t) \downarrow 0$ as $t \longrightarrow 0$ such that

$$\left\| f\left(\frac{x}{\mu}\right)\right\|_{\mu Q} \leqslant \tau(t) \left\| \frac{1}{t^n} f\left(\frac{x}{t}\right)\right\|_{tQ}.$$

for every $t \in (0, \epsilon)$ for some $\epsilon > 0$, and for every function f.

Let

$$G(h) = \bigcup_\Gamma (h\gamma + hQ) = \Omega^h \cap [(\Omega^h \setminus \Omega_h) + hQ^{(N)}],$$

so that mes $G(h) \longrightarrow 0$ as $h \longrightarrow 0$, and denote by $\sigma(h)$ a best possible coefficient (with regard to magnitude) in the inequality

$$\sum_{\gamma \in \Gamma} \|f\|_{\Omega \cap (h\gamma + hQ)} \leqslant \sigma(h) \|f\|_{\Omega \cap G(h)}. \tag{1}$$

as $h \longrightarrow 0$. It will be shown in Lemma 2 that

$$\sigma(h) \leqslant \tau(\nu) \tau_1(h) h^{-n}, \qquad \nu = \nu(h) = \text{mes}^{\frac{1}{n}} G(h), \tag{2}$$

where $\tau_1(h)$ is the coefficient in the inequality(1)

$$\left\| f\left(\frac{x}{h}\right) \right\|_{hQ} \leqslant \tau_1(h) \left\| f\left(\frac{x}{\mu}\right) \right\|_{\mu Q}, \qquad h \in (0, \varepsilon), \qquad \tau_1(h) \downarrow 0 \qquad (h \to 0). \tag{3}$$

The size of $\sigma(h)$ depends as much on the choice of the concrete space $B(\Omega)$ as on the rate of decrease of mes $G(h)$, which is determined by the structure of Ω, and on the rate of decrease of $\rho(\partial\Omega, \partial\Omega_h)$. Thus, for instance, if $\partial\Omega$ is a Lipschitz manifold with $\rho(\partial\Omega, \partial\Omega_h) \leqslant Kh$, then mes $G(h) \leqslant K_1 h^{n-1}$.

Let us indicate some examples of spaces $B(\Omega)$ satisfying (for the corresponding functional parameters ψ and M) the conditions I–IV (cf. [6] and [7]):

a. The space $L_p(\Omega)$, $1 < p \leqslant \infty$, with

$$\|f\|_{L_p(\Omega)} = \left\{ \int_\Omega |f(x)|^p \, dx \right\}^{\frac{1}{p}},$$

and the functions $\tau(t) = ct^{n-n/p}$ and $\tau_1(t) = c_1 t^{n/p}$.

b. The Marcinkiewicz space $M_\psi(\Omega)$, with

$$\|f\|_{M_\psi(\Omega)} = \sup_{D \subset \Omega} \frac{1}{\psi(\text{mes } D)} \int_D |f(x)| \, dx,$$

where $\psi(t)$ is a continuous function on $[0, \infty)$, monotonically increasing together with $\psi^*(t) = t/\psi(t)$. For $M_\psi(\Omega)$ one has the functions

$$\tau(t) \leqslant c \sup_{0 < \nu \leqslant 1} \frac{\psi(t^n \nu)}{\psi(\nu)}, \qquad \tau_1(t) \leqslant c_1 \sup_{0 < \nu \leqslant 1} \frac{t^n \psi(\nu)}{\psi(t^n \nu)}.$$

c. The Lorentz space $\Lambda_\psi(\Omega)$, with

$$\|f\|_{\Lambda_\psi(\Omega)} = \int_0^\infty \psi(\lambda(f, t)) \, dt,$$

where $\psi(t)$ is convex from below, continuous on $[0, \infty)$, and monotonically increasing together with $\psi_*(t) = t/\psi(t)$. Here

$$\tau(t) \leqslant c \sup_{0 < \nu \leqslant 1} \frac{t^n \psi(\nu)}{\psi(t^n \nu)}, \quad \tau_1(t) \leqslant c_1 \sup_{0 < \nu \leqslant 1} \frac{\psi(t^n \nu)}{\psi(\nu)}.$$

d. The Orlicz space $L_M^*(\Omega)$, with

---

(1) The properties IV and (3) can be given an equivalent formulation in terms of the dilatation operators σ_s: $\sigma_s f^*(t) = f^*(t/s)$ (cf. [5]). In fact, IV is equivalent to the estimate $\|\sigma_s\| = o(s)$ as $s \to +\infty$, and (3) is equivalent to the estimate $\|\sigma_s\| = o(1)$ as $s \to 0$. In the case of Orlicz spaces $B(\Omega) = L_M^*(\Omega)$ the second of these estimates is equivalent to the existence of a Δ_2-condition for the function $M(u)$, while the first estimate is equivalent to a Δ_2-condition for the complementary N-function $M^*(u)$. The author is grateful to E. M. Semenov for consultations concerning the properties of symmetric spaces.

$$\|f\|_{M(\Omega)}=\inf\left\{r>0:\int_\Omega M\left[\frac{f(x)}{r}\right]dx<1\right\},$$

where the N-function $M(u)$ for some $0<v_1<v_2$ and all $u>0$ satisfies

$$vM(u)\leqslant M[\psi(v)u] \quad \text{for} \quad v_0<v<\infty,$$
$$vM(u)\leqslant M[\psi_1(v)u] \quad \text{for} \quad 0<v<v_1,$$

which in the case of strictly monotonically increasing ψ and ψ_1 are equivalent, respectively, with the estimates

$$M(vu)\leqslant \Psi(v)M(u) \quad \text{for} \quad 0<v<v_0^{-1},$$
$$M(vu)\leqslant \Psi_1(v)M(u) \quad \text{for} \quad v_1^{-1}<v<\infty,$$

where

$$\Psi(v)=\frac{1}{\psi^{-1}(1/v)},\qquad \Psi_1(v)=\frac{1}{\psi_1^{-1}(1/v)} \text{ and } \tau(t)\leqslant c\psi\left(\frac{1}{t^n}\right)t^n,\qquad \tau_1(t)\leqslant c_1\psi_1(t^n).$$

Examples of N-functions with the above-mentioned properties are

$$M_1(u)=u^p+u^q,\qquad 1<p<q<\infty,\qquad \Psi(v)=v^p,\qquad \Psi_1(v)=v^q,$$
$$M_2(u)=u^p(1+|\ln|u||),\qquad 1<p<\infty,\qquad \Psi(v)=\Psi_1(v)=M_2(v),$$
$$M_3(u)=(1+|u|)\ln(1+|u|)-|u|,\qquad \Psi(v)=v,\ \Psi_1(v)=v^2.$$

It is well known [6] that functions belonging to an arbitrary symmetric space $B(\Omega)$ are summable on Ω.

The symbol $D^\alpha f$ stands for the generalized derivative in the sense of Sobolev of the function f of order $\alpha=(\alpha_1,\dots,\alpha_n)$ for integral $\alpha_i\geqslant 0$, $|\alpha|=\Sigma_1^n\alpha_i$. In view of its definition, $D^\alpha f$ is a locally summable function.

Let us now set for natural numbers m

$$D^m f(x)=f^{(m)}(x)=\sqrt{\sum_{|\alpha|=m}\frac{m!}{\alpha!}(D^\alpha f(x))^2},$$

and let us denote by $B^m(\Omega)$ the space of functions with generalized derivatives $D^\alpha f$ of all orders $|\alpha|\leqslant m$ with the seminorm

$$\|f\|_{B^m(\Omega)}=\|D^m f\|_{B(\Omega)}=\|f^{(m)}\|_\Omega.$$

In a similar way we define the spaces $B^m(\Omega\#h)$, $B^m(\Omega_h)$, etc.

The seminorms $\|f\|_{B^m(\Omega)}$ and $\|f\|_{B^m(\Omega\#h)}$ annihilate polynomials of degree $m-1$, and h-cellularly-polynomial functions of degree $m-1$, respectively.

We denote by $\widetilde{B_\mu^m}(E^n\#1)=\widetilde{B_\mu^m}(\#)$ the space of functions defined on $E^n\#1$ with 1-periodic derivatives $D^\alpha f$, $|\alpha|=m$, on $E^n\#1$, and with

$$\|f\|_{\widetilde{B_\mu^m}(\#)}=\|f^{(m)}\|_{B_\mu(Q)}.$$

By $\widetilde{B_\mu^m}(E^n)=\widetilde{B_\mu^m}$ we denote the space $\widetilde{B_\mu^m}(\#)$ of functions f of the form

$$f(x)=\tilde{f}(x)+P_m(x)=\tilde{f}(x)+\sum_{|\alpha|\leqslant m}c_\alpha x^\alpha,$$

where $\widetilde{f}(x)$ is a 1-periodic functions with a full set of derivatives of order m on E^n. Thus

$$\|f\|_{\widetilde{B_\mu^m}} = \|f\|_{\widetilde{B_\mu^m(\#)}} = \|f^{(m)}\|_{B_\mu(Q)}.$$

Let us finally consider the function spaces $B_\mu^m = B_{\mu,N}^m$ and $B_\mu^m(\#) = B_{\mu,\underline{N}}^m(\#)$, defined respectively on $Q^{(N)}$ and $Q^{(N)}\#1$, with the seminorms

$$\|f\|_{B_\mu^m} = \sup_{y+Q\subset Q^{(N)}} \|f^{(m)}(x+y)\|_{B_\mu(Q)},$$

$$\|f\|_{B_\mu^m(\#)} = \sup_{y+Q\subset Q^{(N)}} \|\overline{f^{(m)}}(x+y)\|_{B_\mu(Q)},$$

where $\overline{f^{(m)}}$ is the continuation with zero of $f^{(m)}$ from $Q^{(N)}\#1$ to $Q^{(N)}$.

In the following we study the asymptotic behavior as $h \longrightarrow 0$ of the norms of functionals l^h on the spaces $B^m(\Omega)$. The functionals l^h will be constructed by means of a "partition" of a fundamental functional l. Let us now proceed to the description of this construction.

Suppose that l is a continuous linear functional on the space $B_\mu^m = B_{\mu,N}^m$. It then clearly follows from the continuity of l that $(l, f) = 0$ if $f^{(m)} = 0$ on $Q^{(N)}$. The value of l on a function f with domain of definition $D \supset Q^{(N)}$ is defined as its value on the restriction of f to $Q^{(N)}$. An analogous convention will be used also in other similar cases.

We shall use the notation

$$(l(x),\ f(x)) = (l,\ f) \quad \left(l\left(\frac{x}{h}-\gamma\right),\ f(x)\right) = h^n(l(x),\ f(hx+h\gamma)),$$

$$(l^h,\ f) = \left(\sum_{\gamma h\in\Omega_h} l\left(\frac{x}{h}-\gamma\right),\ f(x)\right) = \left(l(x),\ h^n\sum_{\gamma h\in\Omega_h} f(hx+h\gamma)\right). \tag{4}$$

The functional l^h, defined on $B^m(\Omega)$ by (4), will be called a *partition* of the functional l. A partition l^h of a functional l is again a linear continuous functional on $B_m(\Omega)$, and its values are defined in terms of restrictions of functions to $\Omega_h + hQ^{(N)} \subset \Omega$.

In the following we shall have to do with sequences $\{l^h\}$ of functionals that are not necessarily partitions of some functional. Here h denotes a positive parameter, tending toward zero. In problems connected with cubature formulas, it will often be understood that h runs through a sequence of the type 1, 1/2, 1/3, 1/4, ...

The following definition was introduced for the spaces $L_p^m(\Omega)$ by Polovinkin; the term "partition" is proposed here.

Definition. Let l be a linear continuous functional on a space B_μ^m. Then a set $\{l^h\}$ of linear functionals will be called a *partition of l with boundary layer in* $B^m(\Omega)$ if the functionals l^h are of the form

$$l^h = \sum_{\gamma h\in\Omega} l_\gamma^h = \sum_{\gamma h\in\Omega_h} l\left(\frac{x}{h}-\gamma\right) + \sum_{\gamma h\in\Omega^h\setminus\Omega_h} l_\gamma^h$$

and if for some constant K

$$\| l_\gamma^h \|_{B^{m*}(\Omega \cap (h\gamma + hQ^{(N)}))} \leqslant K h^m \sigma_1(h) \quad \forall \gamma h \in \Omega^h \setminus \Omega_h, \tag{5}$$

where $\sigma_1(h) = (1/\sigma(h))$ as $h \longrightarrow 0$, and where the function $\sigma(h)$ is the same as in (1). In particular, it is possible in view of (2) to choose $\sigma_1(h) = h^n/\tau_1(h)$. In that case $\{l^h\}$ is called a *set of partitions of an error functional l with boundary layer in $B^m(\Omega)$* if $B^m(\Omega) \subset C(\Omega)$ and

$$(l, f) = \int_Q f(x)\, dx - \sum_{|\gamma_i| \leqslant N-1} c_\gamma f(\gamma),$$

$$(l_\gamma^h, f) = \int_{\Omega \cap hQ_\gamma^{(N)}} f(x)\, dx - \sum_{\beta h \in \Omega} c_{\gamma,\beta}^h f(h\beta),$$

where the values $f(h\beta)$ for $h\beta \in \partial\Omega$ are to be understood as limit values of $f(x)$ with respect to a set that depends on γ.

In view of this definition a partition of a functional l is a particular case of a partition of l with boundary layer in $B^m(\Omega)$.

REMARK. It follows from results of Sobolev (cf. [2], pp. 704 and 705) and Polovinkin [3], [4] that for $m > n/p$, and under certain regularity conditions on Ω, there exists a set of partitions with boundary layer in $L_p^m(\Omega)$ for every error functional l with integer-component nodes that annihilates every monomial x^α with $|\alpha| < m$.

The main result of the present paper is contained in the following theorem, dealing with the asymptotic behavior of the norms $\|l^h\|_{B^{m*}(\Omega)}$ as $h \longrightarrow 0$.

THEOREM. *Let Ω be a domain of cube type, and let $\{l^h\}$ be a set of partitions of a functional $l \in B_\mu^{m*}$ with boundary layer in $B^m(\Omega)$. Then, as $h \longrightarrow 0$,*

$$\| l^h \|_{B^{m*}(\Omega)} = \mu^n \| l \|_{\widetilde{B}_\mu^{m*}} h^m + o(h^m). \tag{6}$$

In the case $B^m(\Omega) = L_p^m(\Omega)$ the principal term of the asymptotic expansion was found by Polovinkin [3], [4] for h running through any sequence of the type s^{-1}, s^{-2}, s^{-3}, ... with $s \geqslant 2$ a natural number. Polovinkin's proof was based on the uniform convexity of $L_p^m(\Omega)$, $1 < p < \infty$, and on the existence of extremal functions. As we mentioned above, the method discussed in the present paper, which is founded only on the properties of intercellular averages, makes it possible in the very same case of Sobolev spaces to study the asymptotics of $\|l^h\|_{L_p^{m*}(\Omega)}$ for a step h that tends to zero in an arbitrary way. It is also possible to extend this result to more general spaces.

Let us first establish some auxiliary results.

A function $\varphi(x)$ will be called *h-periodic on G* if

$$\varphi(x + h\gamma) = \varphi(x) \qquad \text{when} \quad x \in G, \quad x + h\gamma \in G.$$

A function that is h-periodic on E^n will be called h-periodic.

LEMMA 1. *Let $G = \bigcup_\Gamma (h\gamma + hQ) \subset \Omega$ and $\nu^n = \operatorname{mes} G$. Then*

$$\left\| h^n \sum_{\gamma \in \Gamma} f(hx + h\gamma) \right\|_{B_\mu(Q)} \leqslant \tau(\nu) \| f \|_{G \# h},$$

with $\tau(t)$ defined as in condition IV.

PROOF. Since both sides of the proposed inequality are invariant for cellular rearrangements on G of the function f, it is sufficient to prove the inequality for some such rearrangement. Now, the norm of the arithmetic mean over all cellular rearrangements of f on G (i.e. the norm of the intercellular average of f) does not exceed the norm of f, as one can verify by Minkowski's inequality. Therefore it is sufficient to prove the desired inequality only for h-periodic functions $\varphi(x)$. For such functions the inequality takes on the form

$$\| \varphi (hx)\|_{B_{\mu}(Q)} \leqslant \frac{\tau (\nu)}{\nu} \| \varphi \|_{B(G)}$$

and is an obvious corollary of (and even equivalent to) the inequality of condition IV.

LEMMA 2. *Suppose that* $G = \bigcup_{\Gamma}(h\gamma + hQ) \subset \Omega^h$ *and* $\nu^n = G \leqslant \mu^n$. *Then*

$$\sum_{\gamma\in\Gamma} \| f \|_{\Omega\cap(h\gamma+hQ)} \leqslant \tau(\nu)\, \tau_1(h)\, h^{-n} \| f \|_{\Omega\cap G}, \tag{7}$$

with the same $\tau_1(t)$ *as in* (3).

PROOF. Let us redefine f, setting $f(x) = 0$ for $x \in G\backslash\Omega$. Then (7) can be written in the form

$$\sum_{\gamma\in\Gamma} \| f \|_{h\gamma+hQ} \leqslant \tau(\nu)\, \tau_1(h)\, h^{-n} \| f \|_{G}. \tag{8}$$

We shall prove (8) for an arbitrary function f with $\|f\|_G < \infty$. Clearly both sides of (8) are invariant for intercellular rearrangements of f on G. Therefore it is sufficient to establish the correctness of (8) for the intercellular average of f on G, i.e. for an arbitrary h-periodic function φ on G. But for an h-periodic function φ, (8) follows from (3):

$$\sum_{\gamma\in\Gamma} \| \varphi \|_{h\gamma+hQ} = \nu^n h^{-n} \| \varphi \|_{hQ} \leqslant \nu^n h^{-n} \tau_1(h) \left\| \varphi\left(\frac{hx}{\mu}\right)\right\|_{\mu Q}$$
$$\leqslant h^{-n} \tau_1(h)\, \tau(\nu) \left\| \varphi\left(\frac{hx}{\nu}\right)\right\|_{\nu Q} = \tau(\nu)\, \tau_1(h)\, h^{-n} \| \varphi \|_{G}.$$

LEMMA 3. *Suppose that* $\varphi \in B^m_\mu(Q)$, *and* $\int_Q D^\alpha \varphi dx = 0$ *for all* $|\alpha| \leqslant m - 1$. *Then*

$$\sum_{|\alpha|\leqslant m-1} \| D^\alpha \varphi \|_{B_\mu(Q)} \leqslant C \| D^m \varphi \|_{B_\mu(Q)}, \tag{9}$$

where C *does not depend on* φ.

PROOF. It is enough to consider the case $m = 1$, because the general case is an obvious corollary. Suppose first that φ is a continuously differentiable function on μQ. Then

$$\varphi(y) - \varphi(x) = \int_0^1 \frac{\partial}{\partial t} \varphi(x + t(y - x))\, dt,$$

for $x, y \in \mu Q$, so that

$$\varphi(x) - \mu^{-n} \int_{\mu Q} \varphi(y)\, dy = \mu^{-n} \int_{\mu Q}\int_0^1 \frac{\partial}{\partial t} \varphi(x + t(y - x))\, dt dy, \tag{10}$$
$$\left| \varphi(x) - \mu^{n} \int_{\mu Q} \varphi(y)\, dy \right| \leqslant C_1 \int_{\mu Q} \frac{\varphi^{(1)}(y)\, dy}{|x - y|^{n-1}}.$$

If φ is summable over μQ, (10) holds for almost every $x \in \mu Q$ (see, for example, [8]). The integral of potential type on the right side of (10) is a bounded operator on L_1 as well as on L_∞, and hence is also bounded on every symmetric space satisfying condition III (see [9] or [10]). This implies that (10) is a consequence of (9) for $m = 1$, which is just what we wanted to prove. We remark that for L_p^m Lemma 3 was established by Polovinkin, who used certain results of Sobolev concerning equivalent normings of L_p^m.

PROOF OF THE THEOREM. Let l^h be represented in the form

$$l^h = \overset{0}{l}{}^h + \lambda^h, \qquad \overset{0}{l}{}^h(x) = \sum_{\gamma h \in \Omega_h} l\left(\frac{x}{h} - \gamma\right),$$

so that $\overset{0}{l}{}^h$ is a partition of the functional l. We can then estimate $\|\lambda^h\|_{B^{m*}(\Omega)}$ by means of (5):

$$|(\lambda^h, f)| = \Big|\Big(\sum_{\gamma h \in \Omega^h \setminus \Omega_h} l_\gamma^h, f\Big)\Big|$$
$$\leqslant \sum_{\gamma h \in \Omega^h \setminus \Omega_h} \|l_\gamma^h\|_{B^{m*}(\Omega \cap hQ_\gamma^{(N)})} \|f^{(m)}\|_{\Omega \cap hQ_\gamma^{(N)}} \leqslant K h^m \sigma_1(h) C_N \sum_{\gamma h \in G} \|f^{(m)}\|_{\Omega \cap (h\gamma + hQ)},$$

where

$$\begin{aligned} C_N &= (2N-1)^n, \\ G &= \Omega^h \cap [(\Omega^h \setminus \Omega_h) + hQ^{(N)}]. \end{aligned} \tag{11}$$

If we then set $\nu^n = \operatorname{mes} G$ and apply (1) and (5), we get that

$$|(\lambda^h, f)| \leqslant K C_N h^m \sigma_1(h) \sigma(h) \|f^{(m)}\|_{\Omega \cap G} \leqslant K C_N h^m \sigma_1(h) \sigma(h) \|f^{(m)}\|_{\Omega}.$$

Consequently, $\|\lambda^h\|_{B^{m*}(\Omega)} = o(h^m)$ as $h \to 0$.

In this way the proof of the theorem is reduced to the proof of (6) for $\overset{0}{l}{}^h$ rather than for l^h; that is, to the special case of the theorem when $\{l^h\}$ is a partition of l (and not a partiton with boundary layer). Therefore we can assume in the remaining part of the proof that $l^h = \overset{0}{l}{}^h$ is a partition of l.

Let us first estimate the left side of (6) from below, under the assumption that $\|l\|_{\widetilde{B}_\mu^{m*}} > 0$. Thus, consider for arbitrary $\epsilon > 0$ the function

$$f(x) = \tilde{f}(x) + P_m^0(x) = \tilde{f}(x) + \sum_{|\alpha| = m} c_\alpha x^\alpha \in \widetilde{B}^m,$$

where $\widetilde{f}(x)$ is a 1-periodic function, and

$$(l, f) > (1 - \varepsilon)\|l\|_{\widetilde{B_\mu^{m*}}} \|f\|_{\widetilde{B_\mu^m}}.$$

For $f_h(x) = \widetilde{f}(x/h) + P_m^0(x/h)$ it follows from the definition of l^h that

$$(l^h, f_h) = \Big(l(x), h^n \sum_{\gamma h \in \Omega_h} f_h(hx + h\gamma)\Big) = \Big(l(x), h^n \sum_{\gamma h \in \Omega_h} \big(\tilde{f}(x + \gamma) + P_m^0(x + \gamma)\big)\Big)$$
$$= h^n \Big(l(x), \sum_{\gamma h \in \Omega_h} \Big(\tilde{f}(x) + P_m^0(x) + \sum_{|\alpha| < m} c_\alpha^{(\gamma)} x^\alpha\Big)\Big)$$
$$= h^n |h^{-1}\Omega_h| (l, f) = \operatorname{mes} \Omega_h (l, f) > (1 - \varepsilon) \operatorname{mes} \Omega_h \|l\|_{\widetilde{B_\mu^{m*}}} \|f\|_{\widetilde{B_\mu^m}}. \tag{12}$$

On the other hand, we can use the property

$$\lim_{\nu\uparrow\mu}\sup_f \frac{\| f(x/\nu)\|_{\nu Q}}{\| f(x/\mu)\|_{\mu Q}}=1 \tag{13}$$

of a symmetric space, and the h-periodicity of $D^m f_h$, to prove, for $0 < h \leqslant h_0 = h_0(\Omega)$ and some function

$$\delta(h)=\delta_1\left(\frac{\mu^h-\mu_h}{\mu_h}\right)\to 0\ (h\to 0),$$

that

$$\| f_h\|_{B^m(\Omega)}=\| D^m f_h\|_\Omega \leqslant \| D^m f_h\|_{\Omega_h}+\| D^m f_h\|_{\Omega\setminus\Omega_h}\leqslant (1+\delta(h))\| D^m f_h\|_{\Omega_h}$$
$$=(1+\delta(h))\,h^{-m}\left\| f^{(m)}\left(\frac{x}{\mu_h}\right)\right\|_{\mu_h Q}\leqslant (1+\delta(h))\,h^{-m}\| f\|_{\widetilde{B^m_\mu}}.$$

Together with (12) this shows that

$$(l^h,\ f_h)>\frac{(1-\varepsilon)\,\mu_h^n}{1+\delta(h)}\| l\|_{\widetilde{B^{m*}_\mu}}h^m\| f_h\|_{B^m(\Omega)},$$

which, if we recall that $\epsilon > 0$ and $\mu_k \longrightarrow \mu$, shows that the left side of (6) is not smaller than the right.

The second part of the proof of the theorem, consisting of a verification of an estimate from above of the norm of a partition functional $\|l^h\|_{B^{m*}(\Omega)}$, will be given later, after some preparatory remarks and lemmas.

We extend the functional l from $B^m_\mu = B^m_{\mu,N}$ to $B^m_\mu(\#) = B^m_{\mu,N}(\#)$ with preservation of the norm, using the Hahn-Banach theorem (see, for example, [11]).

We also extend the functional $l^h = \Sigma_{\gamma h\in\Omega_h} l(x/h-\gamma)$ by use of the following formula to vector-valued functions $\{f_\gamma\} = \{f_\gamma\}_{\gamma\in h^{-1}\Omega_h}$ with components $f_\gamma(x) \in B^m(\Omega\ \#\ h)$:

$$(l^h,\ \{f_\gamma\})=\Big(l(x),\ h^n\sum_{\gamma h\in\Omega_h} f_\gamma(hx+h\gamma)\Big), \tag{14}$$

so that, for $f_\gamma = f\ \forall\gamma\in h^{-1}\Omega_h$,

$$(l^h,\ \{f\})=(l^h,\ f).$$

Let the mapping $\gamma' = \pi(\gamma)$ define a rearrangement of the elements $\gamma \in h^{-1}\Omega_h$. Since

$$\Big(l^h(x),\ \{f(x+h\pi(\gamma)-h\gamma)\}=\Big(l(x),\ h^n\sum_{\gamma h\in\Omega_h} f(hx+h\pi(\gamma)\Big)=(l^h,\ f)$$

does not depend on the rearrangement π, it follows that

$$(l^h(x),\ \{S^h_\gamma f(x)\})=(l^h,\ f), \tag{15}$$

where $S^h_\gamma f(x)$ is the arithmetic mean of $f(x + h\pi(\gamma) - h\gamma)$ over all rearrangements π, or, which amounts to the same, the arithmetic mean of $f(x + h\gamma' - h\gamma)$ over all $\gamma' \in h^{-1}\Omega_h$; here $S^h_\gamma f(x)$ is defined on $h\gamma + h(Q^{(N)}\#1)$.

The function

$$S^h f(x)\,|_{\Omega_h}=\sum_{\gamma h\in\Omega_h}\chi_1\left(\frac{x}{h}-\gamma\right)S^h_\gamma f(x) \tag{16}$$

(where the $S_\gamma^h f$ are extended by zero, and χ_1 is the characteristic function of Q) is an h-periodic function on Ω_h. We shall denote by $S^h f(x)$ its h-periodic continuation outside Ω_h onto all of E^n, and, for the sake of simplicity, we let f stand for the intercellular average of the function.

LEMMA 4. *As* $h \to 0$,

$$\eta(h, f) = (l^h, f) - (l^h, S^h f) = \varepsilon(h, f) h^m \| f^{(m)} \|_{\Omega \# h},$$

where $\epsilon(h, f) \to 0$ *as* $h \to 0$ *uniformly with respect to* $f \in B^m(\Omega \# h)$.

PROOF. In view of (15), (14), and the fact that l has compact support we see that

$$\begin{aligned}
\eta(h, f) &= (l^h, \{S_\gamma^h f - S^h f\}) \\
&= \Big(l^h(x), \Big\{ \sum_{|\beta_i| \leqslant N-1} \chi_1\Big(\frac{x}{h} - \gamma - \beta\Big)[S_\gamma^h f(x) - S_\gamma^h f(x - h\beta)]\Big\}\Big) \\
&= \Big(l^h(x), \Big\{ \sum_{|\beta_i| \leqslant N-1} \chi_1\Big(\frac{x}{h} - \gamma - \beta\Big) |h^{-1}\Omega_h|^{-1} \sum_{\gamma' h \in \Omega_h} [f(x + h\gamma' - h\gamma) \\
&\qquad - f(x - h\beta + h\gamma' - h\gamma)]\Big\}\Big) = \Big(l(x),\ h^n \sum_{|\beta_i| \leqslant N-1} \chi_1(x - \beta) \\
&\qquad \times \sum_{\gamma' h \in \Omega_h} [f(hx + h\gamma') - f(hx - h\beta + h\gamma')]\Big) \\
&= \sum_{|\beta_i| \leqslant N-1} \Big(l(x),\ h^n \chi_1(x - \beta) \sum_{\gamma' h \in \Delta_\beta \Omega_h} (\pm 1) f(hx + h\gamma')\Big),
\end{aligned}$$

where $\Delta_\beta \Omega_h = \Omega_h \Delta(\beta h + \Omega_h)$ is the symmetric difference of the sets, and where the choice of the signs $\pm$ is determined by the vector γ'. Hence

$$|\eta(h, f)| \leqslant C \|l\|_{B_\mu^{m*}(\#)} \sum_{|\beta_i| \leqslant N-1} \Big\| h^n \sum_{\gamma h \in \Delta_\beta \Omega_h} (\pm 1) f(hx + h\gamma) \Big\|_{B_\mu^m(Q)}$$

and, in view of Lemma 1,

$$|\eta(h, f)| \leqslant C_1 h^m \sum_{|\beta_i| \leqslant N-1} \tau(\operatorname{mes} \Delta_\beta \Omega_h) \| f^{(m)} \|_{\Omega \# h},$$

which proves Lemma 4.

END OF THE PROOF OF THE THEOREM. We shall establish an estimate from above for the norm of the partition l^h of the functional l, i.e. of

$$\| l^h \|_{B^{m*}(\Omega)} = \sup_{\|f^{(m)}\|_\Omega = 1} (l^h, f) = \sup_{\|f^{(m)}\|_\Omega = 1} [(l^h, S^h f) + \eta(h, f)].$$

In view of Lemma 4, it is enough to prove that

$$B(h, f) = \frac{|(l^h, S^h f)|}{\| f^{(m)} \|_{\Omega \# h}} = \frac{\mu_h^n |(l(x), S^h f(hx))|}{\| f^{(m)} \|_{\Omega \# h}} \leqslant \mu^n \| l \|_{\widetilde{B}_\mu^{m*}} h^m + \varepsilon(h) h^m, \quad (17)$$

where $\epsilon(h) = o(1)$ does not depend on f.

Suppose that $b > 0$. In order to prove (17) we shall estimate $B(h, f)$ in one way in the case when

$$\| S^h f(hx) \|_{B^m_\mu(Q)} \leqslant b h^m \| f^{(m)} \|_{\Omega \# h}$$

and in another way in the opposite case.

In the first case

$$B(h, f) \leqslant \mu_h \| l \|_{B^{m*}_\mu(\#)} \| S^h f(hx) \|_{B^m_\mu(Q)} \| f^{(m)} \|^{-1}_{\Omega \# h} \leqslant \mu \| l \|_{B^{m*}_\mu} b h^m.$$

Choosing b arbitrarily small for $\|l\|_{\widetilde{B}^{m*}_\mu}$, and

$$b = \frac{\| l \|_{\widetilde{B^{m*}_\mu}}}{\| l \|_{B^{m*}_\mu} 2\mu} \quad \text{for } \| l \|_{\widetilde{B^{m*}_\mu}} > 0,$$

we obtain a lower estimate for $B(h, f)$.

In the second case, we can use (13) to show that for some $\epsilon(h) \longrightarrow 0$ (as $h \longrightarrow 0$) it is true that

$$\begin{aligned} \| f^{(m)} \|_{\Omega_h} \geqslant \| S^h f \|_{B^m(\Omega_h \# h)} &= h^{-m} \| (S^h f)^{(m)} (hx) \|_{B_{\mu_h}(Q)} \\ &\geqslant (1 - \varepsilon(h)) h^{-m} \| (S^h f)^{(m)} (hx) \|_{B_\mu(Q)} \end{aligned}$$

Hence

$$B(h, f) \leqslant \frac{\mu h^m}{(1 - \varepsilon(h))} \frac{| (l(x), S^h f(hx)) |}{\| S^h f(hx) \|_{B^m_\mu(Q)}}. \tag{18}$$

In order to complete the proof it remains to make use of the following fact.

LEMMA 5. *As* $h \longrightarrow 0$

$$\sup_{f \in B^m(\Omega)} \inf_{g \in \widetilde{B^m_\mu}} \frac{\| S^h f(hx) - g(x) \|_{B^m_\mu(Q)}}{\| f^{(m)} \|_{\Omega \# h}} = \varepsilon(h) h^m = o(h^m). \tag{19}$$

Before we proceed to the proof of the lemma, we shall show how it can be applied in order to obtain an estimate of $B(h, f)$ in the case when

$$\| S^h f(hx) \|_{B^m_\mu(Q)} \geqslant b h^m \| f^{(m)} \|_{\Omega \# h}.$$

We have that, if $f \in B^m(\Omega)$, and if $g(x) \in \widetilde{B}^m_\mu$ is sufficiently close to $S^h f(hx)$ in the sense of realization of (19), then

$$\| g \|_{B^m_\mu(Q)} \leqslant \| S^h f(hx) \|_{B^m_\mu(Q)} \left(1 + \frac{2\varepsilon(h)}{b}\right),$$

$$\begin{aligned} \frac{| (l(x), S^h f(hx) |}{\| S^h f(hx) \|_{B^m_\mu(Q)}} &\leqslant \frac{\| (l, g) | + \| l \|_{B^{m*}_\mu} \frac{2\varepsilon(h)}{b} \| S^h f(hx) \|_{B^m_\mu(Q)}}{\| S^h f(hx) \|_{B^m_\mu(Q)}} \\ &\leqslant \frac{| (l, g) |}{\| g^{(m)} \|_{B_\mu(Q)}} \left(1 + \frac{2\varepsilon(h)}{b}\right) + \frac{2\varepsilon(h)}{b} \| l \|_{B^{m*}_\mu} \leqslant \| l \|_{\widetilde{B^{m*}_\mu}} + \frac{4\varepsilon(h)}{b} \| l \|_{B^{m*}_\mu}. \end{aligned}$$

Together with (18), this shows that (17) is true. Therefore, in order to complete the proof of the theorem it only remains to provide the

PROOF OF THE LEMMA. Suppose that $f \in B^m(\Omega)$, $h\gamma^{(0)} + hQ \subset \Omega_h$, and (for the sake of simplicity) $\gamma^{(0)} = 0$.

Denote by $f_0(x)$ the restriction to $2Q$ of the function

$$\frac{h^n}{\operatorname{mes} \Omega_h} \sum_{\gamma h \in \Omega_h} f(hx + h\gamma),$$

so that $f_0(x) = S^h f(hx)$ for $x \in Q$. Further, let

$$\bar{f}_0(x) = f_0(x) - \sum_{|\alpha| \leqslant m} c_\alpha^{(0)} x^\alpha, \qquad x \in 2Q, \tag{20}$$

where the polynomial $\Sigma_{|\alpha| \leqslant m} c_\alpha^{(0)} x^\alpha$ is chosen in such a way that for the function $\Delta_1 \bar{f}_0(x) = \bar{f}_0(x + e_1) - \bar{f}_0(x)$ each one of the derivatives $D^\alpha[\Delta_1 \bar{f}_0(x)]$, $|\alpha| \leqslant m - 1$, has mean value zero over Q. The possibility of making such a choice is guaranteed by the fact that every polynomial of degree $m - 1$ (consequently also every polynomial of degree $m - 1$ with prescribed mean values for all its derivatives) can be expressed in the form

$$\Delta_1 \sum_{|\alpha| \leqslant m} c_\alpha^{(0)} x^\alpha = \sum_{|\alpha| \leqslant m} c_\alpha^{(0)} \Delta_1 x^\alpha = \sum_{k=1}^{m} \sum_{|\alpha| \leqslant m} c_\alpha^{(0)} x^{\alpha'} \Delta_1 x_1^k, \quad \alpha = (\alpha_1, \alpha'),$$

where $c_\alpha^{(0)} = 0$ for $|\alpha| = m$, $\alpha_1 < m$, a fact that can easily be established by equating coefficients for all the powers x_1^s $(s = m - 1, m - 2, \ldots, 0)$.

In view of Lemma 3, the function $\Delta_1 \bar{f}_0(x)$ satisfies

$$\sum_{|\alpha| \leqslant m-1} \| D^\alpha \Delta_1 \bar{f}_0 \|_{B_\mu(Q)} \leqslant C \| D^{(m)} \Delta_1 \bar{f}_0 \|_{B_\mu(Q)}. \tag{21}$$

Now consider on $[0,1) \times [0, 2)^{n-1}$ the function

$$f_1(x) = \bar{f}_0(x) + v(x_1) \Delta_1 \bar{f}_0(x),$$

where $v(t) \in C^\infty(E^1)$, $v(t) = 1$ for $t < 1/3$, and $v(t) = 0$ for $t > 2/3$. It is obvious that the 1-periodic continuation $\widetilde{f}_1(x)$ with respect to x_1 of the function $f_1(x)$ possesses derivatives $D^\alpha \widetilde{f}_1(x)$ of all orders $|\alpha| \leqslant m$ on $E^1 \times (0, 2)^{n-1}$.

In view of (21),

$$\| f_1 - \bar{f}_0 \|_{B_\mu^m(Q)} = \| v(x_1) \Delta_1 \bar{f}_0(x) \|_{B_\mu^m(Q)} \leqslant C_1 \| \Delta_1 \bar{f}_0 \|_{B_\mu^m(Q)} = C_1 \| \Delta_1 f_0 \|_{B_\mu^m(Q)}. \tag{22}$$

Moreover, let

$$\bar{f}_1(x) = f_1(x) - \sum_{\substack{|\alpha| \leqslant m \\ \alpha_1 = 0}} c_\alpha^{(1)} x^\alpha, \qquad x \in [0, 1) \times [0, 2)^{n-1},$$

where $c_\alpha^{(1)} = 0$ for $|\alpha| = m$, $\alpha_2 < m$, and where the polynomial is chosen in such a way that the function

$$\Delta_2 \bar{f}_1(x) = \bar{f}_1(x + e_2) - \bar{f}_1(x)$$

has derivatives $D^\alpha[\Delta_1 \bar{f}_1(x)]$ of mean value zero on Q for all $|\alpha| \leqslant m - 1$. As was stressed above, such a polynomial does indeed exist. It does not depend on x_1, because the mean values of the derivatives $D^\alpha f_1(x)$ and $D^\alpha f_1(x + e_2)$ for the functions $f_1(x)$ and $f_1(x + e_2)$ are already equal to zero for $|\alpha| \leqslant m$ and $\alpha_1 > 0$:

$$\int_Q D^{(\alpha_1,\alpha')} f_1 dx = \int_{(0,1)^{n-1}} \int_0^1 \frac{\partial}{\partial x_1} D^{(\alpha_1-1,\alpha')} f_1(x_1, x')\, dx_1 dx'$$

$$= \int_{(0,1)^{n-1}} D^{(\alpha_1-1,\alpha')} f_1(x_1, x') \big|_{x_1=0}^{x_1=1} dx' = \int_{(0,1)^{n-1}} 0 dx' = 0.$$

This chain of identities is valid in the case when $\widetilde{f}_1$ has all derivatives of order up to and including m continuous on $\overline{Q}$. The same result concerning the vanishing of the mean values is true also in the case when one only knows about the existence of generalized derivatives in the sense of Sobolev. One then has use of a standard limit procedure.

Consider the following function of $[0, 1)^2 \times [0, 2)^{n-2}$:

$$f_2(x) = f_1(x) + v(x_2)\Delta_2 f_1(x).$$

Its 1-periodic continuation $\widetilde{f}_2(x)$ with respect to x_1 and x_2 has derivatives $D^\alpha \widetilde{f}_2$, $|\alpha| \leqslant m$, on $E^2 \times (0, 2)^{n-2}$. In view of (21), and, due to the specific character of the added correcting polynomials (with the coefficients $c_\alpha^{(0)}$ and $c_\alpha^{(1)}$), for the norm $\|\cdot\| = \|\cdot\|_{B_\mu^m(Q)}$ we have

$$\|f_2 - f_1\| = \|v(x_2)\Delta_2 f_1(x)\| \leqslant C_1 \|\Delta_2 f_1\| = C_1\|\Delta_2 f_1\| \tag{23}$$
$$= C_1\|\Delta_2 f_0 + \Delta_2 v(x_1)\Delta_1 f_0\| \leqslant C_2\|\Delta_2 f_0\| + C_2\|\Delta_1 f_0(\cdot + e_2)\|.$$

The last of the inequalities in this sequence is obtained if one recalls that all derivatives of order up to and including $m - 1$ of $\Delta_1 f_0(x + e_2)$ have vanishing mean values on $Q - e_2$, because it then follows, precisely as in the proof of Lemma 3, that

$$\sum_{|\alpha| \leqslant m-1} \|D^\alpha \Delta_1 f_0(\cdot + e_2)\|_{B_\mu(Q)} \leqslant C(\|D^{(m)}\Delta_1 f_0\|_{B_\mu(Q)} + \|D^{(m)}\Delta_1 f_0(\cdot + e_2)\|_{B_\mu(Q)}.$$

By means of analogous constructions with respect to the variables $x_3, \ldots, x_n$, we finally end up with a function $f_n(x)$, $x \in Q$, which is 1-periodic with respect to each variable $x_1, \ldots, x_n$, and which has a continuation $\widetilde{f}_n(x)$ that possesses generalized derivatives on E^n in the Sobolev sense of all orders $|\alpha| \leqslant m$. Moreover, $\|f_n - \widetilde{f}_{n-1}\|$ satisfies (22) and (23). Consequently, it follows from (22), (23), etc., that

$$\Big\|f_0 - f_n - \sum_{|\alpha| \leqslant m} c_\alpha x^\alpha\Big\|_{B_\mu^m(Q)} \leqslant C \sum_{i=2}^n \sum_{\theta_i=0}^1 \Big\|\Delta_1 f_0\Big(\cdot + \sum_2^n \theta_i e_i\Big)\Big\|_{B_\mu^m(Q)}.$$

If we now set $g(x) = f_n(x) + \Sigma_{|\alpha| \leqslant m} c_\alpha x^\alpha$ we see that in order to complete the proof of the lemma it is enough to prove that

$$\sum_{i=2}^n \sum_{\theta_i=0}^1 \Big\|\Delta_1 f_0\Big(\cdot + \sum_2^n \theta_i e_i\Big)\Big\|_{B_\mu^m(Q)} \leqslant \varepsilon(h) h^m \|f^{(m)}\|_{\mathfrak{Q}\#h}, \tag{24}$$

where $\epsilon(h) \longrightarrow 0$ as $h \longrightarrow 0$.

In view of the definition of f_0 it is clear that if one expands $\Delta_1 f_0(x + \Sigma_2^n \theta_i e_i)$ as an algebraic sum of the values of f, then a "majority" of the terms will annihilate each other (precisely as in the proof of Lemma 4). Hence with the help of Lemma 1 we obtain (24) with

$$\varepsilon(h) = C'\tau\left(\operatorname{mes}^{\frac{1}{n}}(\partial\Omega_h + hQ^{(2)})\right).$$

REMARK. In the establishment of the results presented above, a conspicuous role was played by the invariance of the norm under a cellular rearrangement of the function. The result can be carried over also to some spaces constructed by means of norms that are not invariant for arbitrary rearrangements, but invariant under cellular rearrangements. We shall here construct a Morrey space of this type (for the concept of Morrey spaces, cf. for instance [8]).

We denote by $L_{p,a}(\Omega, h, q)$, $1 \leqslant p < \infty$, $0 \leqslant a < 1$, $h > 0$, $1 \leqslant q \leqslant \infty$, the space of functions defined on a Jordan measurable domain Ω with the norm

$$\|f\|_{{}_1L_{p,a}(\Omega,h,q)} = h^{\frac{n}{p}a - \frac{n}{p} + \frac{n}{q}} \sup_{0<v\leqslant h} \left\{ \sum_{\gamma h\in\Omega^h} \left(v^{-\frac{n}{p}a} \sup_{Q(v)\subset h\gamma+hQ} \|f\|_{L_p(\Omega\cap Q(v))} \right)^q \right\}^{\frac{1}{q}},$$

which in the case $q = \infty$ has to be interpreted as the supremum

$$h^{\frac{n}{p}a - \frac{n}{p}} \sup v^{-\frac{n}{p}a} \|f\|_{L_p(\Omega\cap Q(v))},$$

taken over all cubes $Q(v) \subset h\gamma + hQ$ (here $Q(v)$ is an arbitrary dilatation vQ), over all $\gamma \in h^{-1}\Omega^h$ and all $v \in (0, h]$. A normalizing factor is chosen in such a way that $\|1\|_{{}_1L_{p,a}(Q,h,q)} = 1$. The norms (25) for different q and $h > 0$ are equivalent, due to the finiteness of the number of h-cells in Ω^h.

Let us mention two more examples of normings that have a "cellular" character:

$$\|f\|_{{}_2L_{p,a}(\Omega,h,q)} = h^{\frac{n}{p}a - \frac{n}{p} + \frac{n}{q}} \left\{ \sum_{\gamma h\in\Omega^h} \sup_{\substack{Q(v)\subset h\gamma+hQ \\ 0<v\leqslant h}} v^{-\frac{n}{p}aq} \|f\|^q_{L_p(\Omega\cap Q(v))} \right\}^{\frac{1}{q}},$$

$$\|f\|_{{}_3L_{p,a}(\Omega,h)} = \sup_{\Gamma\subset h^{-1}\Omega^h} \sup_{0<v\leqslant h} \sup_{y_\gamma\in(h-v)Q} (|\Gamma|v^n)^{-\frac{a}{p}} \|f\|_{L_p\left(\Omega\cap \bigcup_{\gamma\in\Gamma}(h\gamma+y_\gamma+vQ)\right)}.$$

We shall denote by ${}_1L^m_{p,a}(\Omega, h, q)$, ${}_2L^m_{p,a}(\Omega, h, q)$, ${}_3L^m_{p,a}(\Omega, h)$ the spaces of all functions that possess generalized derivatives $D^\alpha f$, $|\alpha| \leqslant m$, with norms, respectively,

$$\|f^{(m)}\|_{{}_1L_{p,a}(\Omega,h,q)}, \quad \|f^{(m)}\|_{{}_2L_{p,a}(\Omega,h,q)}, \quad \|f^{(m)}\|_{{}_3L_{p,a}(\Omega,h)}.$$

The parameter q will be omitted in reference to these spaces in the case $q = p$, and the parameter h in the case $h = 1$.

It is clear that we have the following imbedding and identity relations between the various spaces:

$${}_iL^m_{p,a}(\Omega, h) \subset L^m_p(\Omega), \qquad {}_iL^m_{p,0}(\Omega, h) = L^m_p(\Omega).$$

If we understand $\widetilde{L}^m_{p,a}$ as the space of functions of the form $f(x) = \widetilde{f}(x) + P_m(x)$, where $\widetilde{f}(x)$

is a 1-period function possessing derivatives $D^\alpha\tilde{f}$ of all orders $|\alpha| \leqslant m$ on E^n, and where $P_m(x)$ is a polynomial of degree m, with $\tilde{L}^m_{p,a}$ having the seminorm defined by

$$\|f\|_{\tilde{L}^m_{p,a}} = \|f\|_{{}_iL^m_{p,a}(Q)},$$

then it is possible to demonstrate the following asymptotic behavior of the norms of partition functionals of l with boundary layer in ${}_iL^m_{p,a}(\Omega, h)$:

$$\|l^h\|_{{}_1L^{m*}_{p,a}(\Omega,\, h)} = \|l\|_{\tilde{L}^{m*}_{p,a}} h^m \mu^{1-\frac{1}{p}} + o(h^m), \qquad p > 1,$$

$$\|l^h\|_{{}_3L^{m*}_{p,a}(\Omega,\, h)} = \|l\|_{\tilde{L}^{m*}_{p,a}} h^m \mu^{1-\frac{1-a}{p}} + o(h^m), \quad p + a > 1.$$

Moreover, a set of linear functionals $\{l^h\}$ will be called a set of partitions of the functional with boundary layer in ${}_iL^m_{p,a}(\Omega, h)$ $(i = 1, 3)$ if

$$l^h = \sum_{\gamma h \in \Omega^h} l^h_\gamma = \sum_{\gamma h \in \Omega_h} l\left(\frac{x}{h} - \gamma\right) + \sum_{\gamma h \in \Omega^h \setminus \Omega_h} l^h_\gamma ,$$

and if

$$l^h_\gamma \in {}_iL^m_{p,\,\theta_i a}(\Omega \cap (h\gamma + hQ^{(N)}),$$

$$\|l^h_\gamma\|_{{}_iL^{m*}_{p,\,\theta_i a}}(\Omega \cap (h\gamma + hQ^{(N)})) \leqslant Kh^{m+n-n\frac{1-\theta_i a}{p}}.$$

for some K and $\theta_1 \in (0, 1)$, for $\theta_3 = 1$, and for all $\gamma h \in \Omega^h \setminus \Omega_h$.

BIBLIOGRAPHY

1. S. M. Nikol'skiĭ, *Quadrature formulas,* Fizmatgiz, Moscow, 1958; 2nd rev. ed., "Nauka", Moscow, 1974; English transl., Hindustan, Delhi, 1964.

2. S. L. Sobolev, *Introduction to the theory of cubature formulas,* "Nauka", Moscow, 1974. (Russian)

3. V. I. Polovinkin, *Sequences of functionals with a boundary layer,* Sibirsk. Mat. Ž. **15** (1974), 413–429; English transl., Siberian Math. J. **15** (1974).

4. ———, *Cubature formulas in Sobolev spaces,* Doctoral Dissertation, Inst. Math. Siberian Branch Acad. Sci. USSR, Novosibirsk, 1975. (Russian)

5. David W. Boyd, *Indices of function spaces and their relationship to interpolation,* Canad. J. Math. **21** (1969), 1245–1254.

6. S. G. Kreĭn et al., *Functional analysis,* 2nd rev. ed., "Nauka", Moscow, 1972; English transl. of 1st ed., Noordhoff, Groningen, 1972.

7. M. A. Krasnosel'skiĭ and Ja. B. Rutickiĭ, *Convex functions and Orlicz spaces,* Fizmatgiz, Moscow, 1959; English transl., Noordoff, Groningen, 1961.

8. O. V. Besov, V. P. Il'in and S. M. Nikol'skiĭ, *Integral representations of functions and imbedding theorems,* "Nauka", Moscow, 1974. (Russian)

9. B. S. Mitjagin, *An interpolation theorem for modular spaces,* Mat. Sb. **66 (108)** (1965), 473–482. (Russian)

10. A. P. Calderón, *Spaces between L^1 and L^∞ and the theorem of Markinkiewicz,* Studia Math. **26** (1966), 273–299.

11. Kôsaku Yosida, *Functional analysis,* Academic Press, New York; Springer-Verlag, Berlin, 1965.

Translated by J. FRIBERG

Труды Мат. Инст. Стеклов.
143 (1977)

Proceedings of the Steklov
Institute of Mathematics
1980, Issue 1

THE CONTROLLABILITY PROBLEM FOR LINEAR SYSTEMS

UDC 517.9

V. I. BLAGODATSKIH

ABSTRACT. An algorithm for the numerical solution of the controllability problem for linear systems is presented. An estimate of the error of computations performed with the aid of this algorithm is given.

Bibliography: 4 titles.

§1. Statement of the problem

Let E^n be the n-dimensional Euclidean space of the variable $x = (x_1, \ldots, x_n)$ with the norm $\|x\| = \sqrt{\Sigma_1^n x_i^2}$, and let $\Omega(E^n)$ be the metric space of all nonempty compact subsets of E^n with the Hausdorff metric

$$h(F, G) = \min \{d : F \subset S_d(G), \quad G \subset S_d(F)\},$$

where $S_d(M)$ denotes the closed d-neighborhood of the set M in the space E^n. The number $|F| = h(\{0\}, F)$ will be called the *modulus of the set* $F \in \Omega(E^n)$. An arbitrary mapping F: $E^1 \longrightarrow \Omega(E^n)$ will be called a *multimapping* $F(t)$. We shall say that a multimapping $F(t)$ is *measurable* if the set $\{t: M \cap F(t) \neq \varnothing\}$ is Lebesgue measurable for any closed set $M \subset E^n$.

We consider a *controlled object* (or system) whose behavior is described by the linear differential equation

$$\dot{x} = Ax + u, \tag{1}$$

where A is a given $n \times n$ square matrix, x is the state vector of the system, and u is the control vector. Let $U(t)$ be a measurable multimapping U: $E^1 \longrightarrow \Omega(E^n)$ such that $|U(t)| \leqslant k(t)$, where $k(t)$ is a measurable function on the interval $I = [t_0, t_1]$. We denote by k the number $\int_{t_0}^{t_1} k(s)\,ds$. Henceforth, all integrals are understood in the Lebesgue sense. An arbitrary single valued and Lebesgue measurable branch of the multimapping $U(t)$ will be called an *admissible control* $u(t)$ on the time interval I. This means that $u(t)$ is a measurable function such that $u(t) \in U(t)$ for all $t \in I$.

A set $M \subset E^n$ is called *convex* if, for any numbers $\alpha, \beta \geqslant 0$ with $\alpha + \beta = 1$ and any $x, x' \in M$, the vector $\alpha x + \beta x'$ also belongs to M. Let there be given two

---

1980 *Mathematics Subject Classification*. Primary 49E15.

convex sets $M_0, M_1 \in \Omega(E^n)$. The system is said to be *controllable on the time interval I from the set M_0 into the set M_1* if there exists an admissible control $u(t)$ such that the corresponding solution $x(t)$ of (1) satisfies $x(t_0) \in M_0$ and $x(t_1) \in M_1$.

The purpose of this work is to study the controllability problem for (1), i.e., to find out whether the system is controllable. Subsequently, an algorithm for the numerical solution of this problem will also be given.

§2. Auxiliary results

The results of this section are auxiliary with respect to our controllability problem. However, they are also of significant importance in themselves.

We consider two operations in the space $\Omega(E^n)$, namely, the algebraic sum of two sets and the multiplication of a set by a number. The sum $F = F_1 + F_2$ of two sets F_1 and F_2 is the set

$$F = \{f = f_1 + f_2 : f_1 \in F_1,\ f_2, \in F_2\}. \tag{2}$$

The product $F = \lambda F_1$ of the set F_1 by the number λ is the set

$$F = \{f = \lambda f_1 : f_1 \in F_1\}.$$

With these operations, a d-neighborhood of a set $M \in \Omega(E^n)$ can be represented in the form

$$S_d(M) = M + dS_1(0), \tag{3}$$

where $S_1(0)$ is the unit ball in E^n. We note that the set $\Omega(E^n)$ with the operations of addition and multiplication just introduced is not a linear space.

Consider a function $c(F, \cdot)\colon E^n \to E^1$ for a set $F \in \Omega(E^n)$. We define this function by

$$c(F, \psi) = \max_{f \in F} (f, \psi), \tag{4}$$

and call it the *support function of the set F.* Here, (f, ψ) denotes the inner product of the vectors f and ψ. Support functions are widely used in optimal control problems. Their properties are sufficiently well investigated, and some of them can be found in [1]. For example, the support function of the unit ball has the form

$$c(S_1(0), \psi) = \|\psi\|. \tag{5}$$

Let us consider certain properties of support functions.

PROPERTY 1. *A support function $c(F, \psi)$ is positively homogeneous in ψ, i.e., for any number $\lambda \geqslant 0$ and any vector $\psi \in E^n$*

$$c(F, \lambda\psi) = \lambda c(F, \psi).$$

The proof follows directly from the definition of support function (4).

PROPERTY 2. *For any vectors $\psi_1, \psi_2 \in E^n$, the support function $c(F, \psi)$ satisfies*

$$c(F, \psi_1 + \psi_2) \leqslant c(F, \psi_1) + c(F, \psi_2).$$

The proof follows directly from (4) and from properties of a maximum.

PROPERTY 3. *For any sets* $F_1, F_2 \in \Omega(E^n)$ *and any vector* $\psi \in E^n$,

$$c(F_1 + F_2, \psi) = c(F_1, \psi) + c(F_2, \psi).$$

The proof follows directly from (2) and (4).

By Properties 1 and 3, the support function of the ball with radius r and center at a point a has the form (see (3) and (5))

$$c(S_r(a), \psi) = c(a + rS_1(0), \psi) = (a, \psi) + r\|\psi\|. \tag{6}$$

The *convex hull* conv F of a set F is defined as the minimal convex set containing F.

PROPERTY 4. *If sets* $F_1, F_2 \in \Omega(E^n)$ *satisfy the condition* $F_1 \subset F_2$, *then for any vector* $\psi \in E^n$

$$c(F_1, \psi) \leqslant c(F_2, \psi). \tag{7}$$

Further, if (7) *holds for any* $\psi \in E^n$, *then* conv $F_1 \subset$ conv F_2.

The proof is contained in [1].

COROLLARY. *If* $f \in F$, *then for any vector* $\psi \in E^n$

$$(f, \psi) \leqslant c(F, \psi). \tag{8}$$

Further, if (8) *holds for any* $\psi \in E^n$, *then* $f \in$ conv F.

PROPERTY 5. *Let* A *be an* $n \times m$ *matrix, let* $F \in \Omega(E^n)$, *and let* $c(F, \psi)$ *be the support function of the set* F. *Then the support function of the image of* F *under the linear mapping* A, *i.e., of the set*

$$AF = \{g \in E^m; \quad g = Af, \ f \in F\}, \tag{9}$$

has the form

$$c(AF, \psi) = c(F, A^*\psi).$$

This follows directly from the definition (4).

COROLLARY. *For any real number* λ *and any vector* $\psi \in E^n$,

$$c(\lambda F, \psi) = c(F, \lambda\psi).$$

PROPERTY 6. *If sets* $F_1, F_2 \in \Omega(E^n)$ *have a nonempty intersection, then for any vector* $\psi \in E^n$

$$c(F_1, \psi) + c(F_2, -\psi) \geqslant 0. \tag{10}$$

Further, if (10) *holds for any* $\psi \in E^n$, *then* F_1 *and* F_2 *satisfy*

$$\operatorname{conv} F_1 \cap \operatorname{conv} F_2 \neq \varnothing.$$

PROOF. Let $F_1 \cap F_2 \neq \varnothing$ and $f \in F_1 \cap F_2$. Then $f \in F_1$, $f \in F_2$, and, by (2),

$$0 = f - f = f + (-f) \in F_1 + (-F_2).$$

Hence, by the corollaries to Properties 4 and 5 and by Property 3, we obtain that for any vector ψ

$$0 = (0, \psi) \leqslant c(F_1 + (-F_2), \psi)$$
$$= c(F_1, \psi) + c(-F_2, \psi) = c(F_1, \psi) + c(F_2, -\psi).$$

Thus the first part of Property 6 is proved. The second part is proved in a similar way.

PROPERTY 7. *The support function $c(F, \psi)$ of a set $F \in \Omega(E^n)$ satisfies*

$$|c(F, \psi) - c(F, \psi')| \leqslant |F| \cdot \|\psi - \psi'\| \tag{11}$$

for any vectors $\psi, \psi' \in E^n$.

PROOF. By the definition of $|F|$ we have $F \subset S_{|F|}(0)$. By Property 4 and (6), it follows that

$$c(F, \psi) \leqslant c(S_{|F|}(0), \psi) = |F| \cdot \|\psi\| \tag{12}$$

for any vector $\psi \in E^n$. By Property 2, we write

$$c(F, \psi) = c(F, \psi - \psi' + \psi') \leqslant c(F, \psi - \psi') + c(F, \psi).$$

Hence, by (12), we obtain from this inequality

$$c(F, \psi) - c(F, \psi') \leqslant c(F, \psi - \psi') \leqslant |F| \cdot \|\psi - \psi'\|.$$

Interchanging ψ and ψ', we obtain the similar inequality

$$c(F, \psi') - c(F, \psi) \leqslant |F| \cdot \|\psi - \psi'\|.$$

These two inequalities imply (11).

We now define the integral of a multimapping $F\colon E^1 \longrightarrow \Omega(E^n)$. The set

$$\int_{t_0}^{t_1} F(t)\,dt = \left\{ \int_{t_0}^{t_1} f(t)\,dt : f(t) \in F(t) \right\} \tag{13}$$

is called *the integral on the time interval $I = [t_0, t_1]$ of the multimapping $F(t)$.*

LEMMA 1. *Let a multimapping $F\colon E^1 \longrightarrow \Omega(E^n)$ be measurable, and let there exist a function $k(t)$ summable on $[t_0, t_1]$ and such that $|F(t)| \leqslant k(t)$. Then the integral $\int_{t_0}^{t_1} F(t)\,dt$ is a nonempty compact convex subset of E^n.*

The proof is contained, e.g., in [2].

LEMMA 2. *If a multimapping $F(t)$ satisfies the requirements of Lemma 1, then*

$$c\left(\int_{t_0}^{t_1} F(t)\,dt, \psi\right) = \int_{t_0}^{t_1} c(F(t), \psi)\,dt$$

for any vector $\psi \in E^n$.

The proof follows from the properties of measurable functions and from the definition (13) of the integral.

The set of all points of E^n that can be attained along the solutions of (1) starting at the initial state $x(t_0) \in M_0$ at the time t_1 under all admissible controls $u(t)$ is called the *attainable set* $X(t_1; t_0, M_0)$ *of the controlled system* (1) *on the time interval* $I = [t_0, t_1]$ *from the initial set* M_0. The solution $x(t)$ of (1) which corresponds to an admissible control $u(t)$ can be represented in the form

$$x(t) = e^{(t-t_0)A}x(t_0) + \int_{t_0}^{t_1} e^{(t_1-s)} u(s)\, ds.$$

Hence, taking (2), (9), and (13) into account, one can represent the attainable set in the form

$$X(t_1; t_0, M_0) = e^{(t_1-t_0)A}M_0 + \int_{t_0}^{t_1} e^{(t_1-s)A} U(s)\, ds.$$

It follows from Lemma 1 that the attainable set $X(t_1; t_0, M_0)$ is a nonempty compact convex subset of E^n. Further, by Properties 3 and 5 and Lemma 2, the support function of an attainable set has the form

$$c(X(t_1, t_0, M_0), \psi) = c(M_0, e^{(t_1-t_0)A^*}\psi) + \int_{t_0}^{t_1} c(U(s), e^{(t_1-s)A^*}\psi)\, ds. \tag{14}$$

§3. Controllability

Let $\psi(t, \psi)$ be the solution of the *adjoint system*

$$\psi = -A^*\psi \tag{15}$$

with the initial condition $\psi(t_0, \psi) = \psi$. Then we define a function $\varphi\colon E^n \longrightarrow E^1$ by

$$\varphi(\psi) = c(M_0, \psi(t_0, \psi)) + \int_{t_0}^{t_1} c(U(s), \psi(s, \psi))\, ds + c(M_1, -\psi(t_1, \psi)). \tag{16}$$

THEOREM 1. *The system* (1) *is controllable on the time interval* $I = [t_0, t_1]$ *from the set* M_0 *into the set* M_1 *if and only if*

$$\varphi(\psi) \geqslant 0 \tag{17}$$

for any vector ψ *of the unit sphere* $S = \{\psi\colon \|\psi\| = 1\}$ *in* E^n.

PROOF. Obviously, the system is controllable if and only if

$$X(t_1; t_0, M_0) \cap M_1 \neq \varnothing.$$

By Property 6 and (14), this is equivalent to the fact that the condition

$$c(M_0, e^{(t_1-t_0)A^*}\psi) + \int_{t_0}^{t_1} c(U(s), e^{(t_1-s)A^*}\psi)\, ds + c(M_1, -\psi) \geqslant 0 \tag{18}$$

holds for any vector $\psi \in E^n$.

We set

$$\psi(t)=e^{(t_1-t)A^*}\psi.$$

Clearly, this function $\psi(t)$ is a solution of the adjoint system (15) and satisfies

$$\psi(t_0)=e^{(t_1-t_0)A^*}\psi, \qquad \psi(t_1)=\psi.$$

Further, since the matrix $e^{(t_1-t_0)A^*}$ is nonsingular, (18) is equivalent to the fact that (17) holds for any vector $\psi(t_0)$. Since support functions are positive homogeneous (Property 1), this is equivalent to the fact that (17) holds for any $\psi \in S$. Theorem 1 is proved.

Theorem 1 allows one to find a time interval on which a system is controllable. Let us illustrate this by an example.

EXAMPLE. Let (1) have the form

$$\begin{aligned} \dot{x}_1 &= x_2, \\ \dot{x}_2 &= -x_1+u, \qquad |u|\leqslant 1, \end{aligned} \tag{19}$$

and let the sets M_0 and M_1 be given by

$$M_0=\{x=(x_1,\ x_2): x_1=-5, \quad |x_2|\leqslant 1\}, \qquad M_1=\{0\}.$$

The support functions of M_0, M_1, and U have the forms

$$\begin{aligned} &c(M_0,\ \psi)=-5\psi_1+|\psi_2|, \\ &c(M_1,\ \psi)=0, \qquad c(U,\ \psi)=|\psi_2|, \end{aligned}$$

and an arbitrary solution $\psi(t)$ of the adjoint system (15) with the initial condition $\psi(t_0)\in S$ has the form

$$\psi_1(t)=\sin(t+\alpha), \qquad \psi_2(t)=\cos(t+\alpha).$$

(17) can be rewritten in the form

$$-5\sin(t_0+\alpha)+|\cos(t_0+\alpha)|+\int_{t_0}^{t_1}|\cos(s+\alpha)|\,ds\geqslant 0.$$

Clearly, if $t_1 - t_0 \geqslant 3\pi$, the relation obtained holds also for any α. By Theorem 1, if an interval $I = [t_0,\ t_1]$ satisfies the condition $t_1 - t_0 \geqslant 3\pi$, the system (19) is controllable on this interval.

Let γ_0 and γ_1 be two nonnegative numbers, and let $\gamma = (\gamma_0, \gamma_1)$. The system (1) will be called *γ-controllable on the time interval I from M_0 into M_1* if it is controllable on this interval from $S_{\gamma_0}(M_0)$ into $S_{\gamma_1}(M_1)$.

THEOREM 2. *The system* (1) *is γ-controllable on the interval* $[t_0,\ t_1]$ *from M_0 into M_1 if and only if*

$$\varphi_\gamma(\psi)=\varphi(\psi)+\gamma_0+\gamma_1\|\psi(t_1,\ \psi)\|\geqslant 0 \tag{20}$$

for all vectors $\psi \in S$.

The proof follows directly from Theorem 1, Property 3, and (6).

Theorem 2 allows one to find a time interval on which a system is γ-controllable. Thus, the system (19) in the above example is γ-controllable on the interval $[t_0, t_1]$ if γ_0 and γ_1 satisfy

$$\gamma_0 + \gamma_1 \geqslant 5 - 2\left[\frac{t_1 - t_0}{\pi}\right].$$

§4. Numerical solution of the controllability problem

In Theorem 1, we have obtained necessary and sufficient conditions for controllability. These conditions are convenient to use for the solution of the controllability problem if the solution of the adjoint system (15) and the support functions of the sets M_0, M_1, and $U(t)$ can be written in an explicit form, as was done in the above example. If this cannot be done, one can use a computer in order to obtain the numerical solution of the controllability problem, i.e., in order to verify (17). The function $\varphi(\psi)$ can be evaluated on a computer only for a finite number of vectors $\psi \in S$.

Let $\widetilde{S}$ be a δ-network from the unit sphere S. Further, let $\widetilde{\varphi}(\psi)$ be the approximate computed value of the function $\varphi(\psi)$. We set

$$\tilde{u} = \min_{\tilde{\psi} \in \tilde{S}} \tilde{\varphi}(\tilde{\psi}). \tag{21}$$

Assume that, for any $\psi \in S$, there exists a vector $\widetilde{\psi} \in \widetilde{S}$ such that the computed value $\widetilde{u}$ differs from the value $u = \min_{\psi \in S} \varphi(\psi)$ by no more than ϵ, i.e.,

$$|\tilde{u} - u| \leqslant \varepsilon. \tag{22}$$

In the next section we shall obtain a relation for the error ϵ.

THEOREM 3. 1) *If $\widetilde{u} \geqslant \epsilon$, then the system* (1) *is controllable on the time interval $I = [t_0, t_1]$ from the set M_0 into the set M_1.*

2) *If $-\epsilon \leqslant \widetilde{u} < \epsilon$, the question of controllability remains open.*

3) *If $\widetilde{u} < -\epsilon$, the system* (1) *is not controllable on I from M_0 into M_1.*

The proof follows directly from Theorem 1, (21) and (22).

Theorem 3 allows one to solve the controllability problem numerically, except in the case when $-\epsilon \leqslant \widetilde{u} < \epsilon$. Therefore, it is natural to try to make the error ϵ sufficiently small. However, it is impossible in principle to set $\epsilon = 0$ in view of computational errors. Therefore, whenever it is impossible to solve the controllability problem, we shall be solving the problem of γ-controllability, trying to choose the numbers γ_0 and γ_1 so small that this problem will be in some sense close to the controllability problem. Clearly, the problem of γ-controllability with $\gamma_0 = \gamma_1 = 0$ is equivalent to the controllability problem.

Again, let $\widetilde{S}$ be a δ-network in the unit sphere S. Further, let $\widetilde{\varphi}_\gamma(\psi)$ be the computed approximate value of $\varphi_\gamma(\psi)$. We set

$$\tilde{u}_\gamma = \min_{\tilde{\psi} \in \tilde{S}} \tilde{\varphi}_\gamma(\tilde{\psi}). \tag{23}$$

Assume that, for any $\psi \in S$, there exists a vector $\widetilde{\psi} \in \widetilde{S}$ such that the computed value $\widetilde{u}_\gamma$ differs from $u_\gamma = \min_{\psi \in S} \varphi_\gamma(\psi)$ by no more than ϵ_γ, i.e.,

$$|\tilde{u}_\gamma - u_\gamma| \leqslant \varepsilon_\gamma. \tag{24}$$

In the next section we shall obtain a relation between the errors ϵ_γ and ϵ.

THEOREM 4. 1) *If* $\tilde{u}_\gamma \geqslant \epsilon_\gamma$, *then the system* (1) *is* γ*-controllable on the time interval* $I = [t_0, t_1]$ *from the set* M_0 *into the set* M_1.

2) *If* $-\epsilon_\gamma \leqslant \tilde{u}_\gamma < \epsilon_\gamma$, *the question of* γ*-controllability remains open.*

3) *If* $\tilde{u}_\gamma < -\epsilon_\gamma$, *the system* (1) *is not* γ*-controllable on* I *from* M_0 *into* M_1.

The proof follows directly from Theorem 2, (23) and (24).

§5. An estimate of computational errors

Let ϵ_0 and ϵ_1 be the errors in the computation of the support functions $c(M_0, \psi)$ and $c(M_1, \psi)$, respectively. Further, let $\epsilon_2(t)$ be the error in the computation of the solution $\psi(t, \psi)$ of the adjoint system (15). Assume that $\epsilon_2(t)$ is continuous on the interval I. Let $\epsilon_3(t)$ be the error in the computation of the support function $c(U(t), \psi)$. Assume that the function $\epsilon_3(t)$ is summable on I, and set $\epsilon_3 = \int_{t_0}^{t_1} \epsilon_3(s)\, ds$. Let ϵ_4 be the error of integration, and let ϵ_5 be the error in finding a minimum of a function.

THEOREM 5. *If* $\epsilon_0, \epsilon_1, \epsilon_2(t), \epsilon_3, \epsilon_4, \epsilon_5$, *and* δ *satisfy*

$$\varepsilon_0 + \varepsilon_1 + \varepsilon_3 + \varepsilon_4 + \varepsilon_5 + |M_1|\,\varepsilon_2(t_1) + \int_{t_0}^{t_1} k(s)\,\varepsilon_2(s)\,ds + [|M_0| + (|M_1| + k)\,e^{\|A\|(t_1-t_0)}]\,\delta \leqslant \varepsilon, \tag{25}$$

then the error in the computation of $u = \min_{\psi \in S} \varphi(\psi)$ *is no more than* ϵ.

PROOF. Approximate values of all numbers will be denoted by the same letter with a tilde. We shall evaluate all numbers entering the expression for $\varphi(\psi)$ by (16).

Since the support function $c(M_0, \psi)$ is evaluated with the error ϵ_0, we have

$$|\tilde{c}(M_0, \psi) - c(M_0, \psi)| \leqslant \varepsilon_0. \tag{26}$$

Further, by Property 7

$$|c(M_0, \psi') - c(M_0, \psi)| \leqslant |M_0| \cdot \|\psi' - \psi\| \tag{27}$$

for any vectors $\psi', \psi \in E^n$.

In a similar way, for $C(M_1, \psi)$ we have

$$|\tilde{c}(M_1, \psi) - c(M_1, \psi)| \leqslant \varepsilon_1, \tag{28}$$

$$|c(M_1, \psi') - c(M_1, \psi)| \leqslant |M_1| \cdot \|\psi' - \psi\| \tag{29}$$

for any vectors $\psi', \psi \in E^n$.

It is well known from ordinary differential equations that the following estimates hold for the solution $\psi(t, \psi)$ of the system (15) with the initial condition $\psi(t_0, \psi) = \psi$:

$$\|\psi(t, \psi') - \psi(t, \psi)\| \leqslant e^{\|A\|(t-t_0)}\,\|\psi' - \psi\|; \tag{30}$$

$$e^{-\|A\|(t-t_0)}\|\psi\| \leqslant \|\psi(t, \psi)\| \leqslant e^{\|A\|(t-t_0)}\|\psi\|. \tag{31}$$

Here $\|A\|$ is the norm of the matrix A, defined by

$$\|A\| = \max_{\|x\|=1}\|Ax\|.$$

Since the solution $\psi(t, \psi)$ is computed with the error $\epsilon_2(t)$, we have

$$\|\tilde{\psi}(t, \psi) - \psi(t, \psi)\| \leqslant \varepsilon_2(t) \tag{32}$$

for any $\psi \in E^n$. From (29) and (30) for $\psi', \psi \in E^n$ we obtain

$$\begin{aligned} &|c(M_1, -\psi(t_1, \psi')) - c(M_1, -\psi(t_1, \psi))| \\ &\qquad \leqslant |M_1| \cdot \|\psi(t_1, \psi') - \psi(t_1, \psi)\| \leqslant |M_1| e^{\|A\|(t_1-t_0)}\|\psi' - \psi\|. \end{aligned} \tag{33}$$

Further, by (28), (29) and (32) we have

$$\begin{aligned} &|\tilde{c}(M_1, -\tilde{\psi}(t_1, \psi)) - c(M_1, -\psi(t_1, \psi))| \\ &\qquad \leqslant |\tilde{c}(M_1, -\tilde{\psi}(t_1, \psi)) - c(M_1, -\tilde{\psi}(t_1, \psi))| \\ &\qquad + |c(M_1, -\tilde{\psi}(t_1, \psi)) - c(M_1, -\psi(t_1, \psi))| \\ &\qquad \leqslant \varepsilon_1 + |M_1| \varepsilon_2(t_1). \end{aligned} \tag{34}$$

By Property 7,

$$|c(U(t), \psi') - c(U(t), \psi)| \leqslant k(t)\|\psi' - \psi \tag{35}$$

for any vectors $\psi', \psi \in E^n$. Hence by (30)

$$\begin{aligned} &|c(U(t), \psi(t, \psi')) - c(U(t), \psi(t, \psi))| \\ &\qquad \leqslant k(t)\|\psi(t, \psi') - \psi(t, \psi)\| \leqslant k(t) e^{\|A\|(t-t_0)}\|\psi' - \psi\|. \end{aligned} \tag{36}$$

Since $c(U(t), \psi)$ is computed with the error $\epsilon_3(t)$, we have

$$|\tilde{c}(U(t), \psi) - c(U(t), \psi)| \leqslant \varepsilon_3(t). \tag{37}$$

Further, (37), (35), and (32) give us

$$\begin{aligned} &|\tilde{c}(U(t), \tilde{\psi}(t, \psi)) - c(U(t), \psi(t, \psi))| \leqslant |\tilde{c}(U(t), \tilde{\psi}(t, \psi)) \\ &\quad - c(U(t), \tilde{\psi}(t, \psi))| + |c(U(t), \tilde{\psi}(t, \psi)) - c(U(t), \psi(t, \psi))| \\ &\quad \leqslant \varepsilon_3(t) + k(t)\|\tilde{\psi}(t, \psi) - \psi(t, \psi)\| \leqslant \varepsilon_3(t) + k(t)\varepsilon_2(t). \end{aligned} \tag{38}$$

Since the approximate value of the integral $⨍_{t_0}^{t_1} h(t)\,dt$ is computed with the error ϵ_4, we obtain

$$\left| ⨍_{t_0}^{t_1} h(t)\,dt - \int_{t_0}^{t_1} h(t)\,dt \right| \leqslant \varepsilon_4. \tag{39}$$

Thus by (16), (26), (34), (39), and (38) we have

$$
\begin{aligned}
|\tilde\varphi(\psi)-\varphi(\psi)| &\leqslant |\tilde c(M_0,\psi)-c(M_0,\psi)|+|\tilde c(M_1,-\tilde\psi(t_1,\psi))-c(M_1,-\psi(t_1,\psi))| \\
&+\left|\widetilde{\int_{t_0}^{t_1}}\tilde c(U(s),\tilde\psi(s,\psi))\,ds-\int_{t_0}^{t_1}c(U(s),\psi(s,\psi))\,ds\right|\leqslant \varepsilon_0+\varepsilon_1+|M_1|\varepsilon_2(t_1) \\
&+\left|\widetilde{\int_{t_0}^{t_1}}\tilde c(U(s),\tilde\psi(s,\psi))\,ds-\int_{t_0}^{t_1}\tilde c(U(s),\tilde\psi(s,\psi))\,ds\right| \qquad (40)\\
&+\left|\int_{t_0}^{t_1}\tilde c(U(s),\tilde\psi(s,\psi))\,ds-\int_{t_0}^{t_1}c(U(s),\psi(s,\psi))\,ds\right| \\
&\leqslant \varepsilon_0+\varepsilon_1+|M_1|\varepsilon_2(t_1)+\varepsilon_4+\varepsilon_3+\int_{t_0}^{t_1}k(s)\varepsilon_2(s)\,ds.
\end{aligned}
$$

Now, let the minimum of the function $\varphi(\psi)$ be attained at a vector $\psi^*\in S$, i.e., let

$$\min_{\psi\in S}\varphi(\psi)=\varphi(\psi^*). \tag{41}$$

By the definition of the δ-network $\widetilde S$, there exists a $\widetilde\psi^*\in\widetilde S$ such that

$$\|\tilde\psi^*-\psi^*\|\leqslant\delta. \tag{42}$$

Taking (41) into account, we obtain

$$\varphi(\tilde\psi^*)\geqslant\min_{\tilde\psi\in\tilde S}\varphi(\tilde\psi)\geqslant\min_{\psi\in S}\varphi(\psi)=\varphi(\psi^*).$$

From this inequality, by (16), (27), (33), (36), and (42), we derive the estimate

$$
\begin{aligned}
|\min_{\tilde\psi\in\tilde S}\varphi(\tilde\psi)-\min_{\psi\in S}\varphi(\psi)| &\leqslant\|\varphi(\tilde\psi^*)-\varphi(\psi^*)\|\leqslant|c(M_0,\tilde\psi^*)-c(M_0,\psi^*)| \\
&+|c(M_1,-\psi(t_1,\tilde\psi^*))-c(M_1,-\psi(t_1,\psi^*))| \qquad (43)\\
&+\int_{t_0}^{t_1}|c(U(s),\psi(s,\tilde\psi^*))-c(U(s),\psi(s,\psi^*))|\,ds \\
&\leqslant(|M_0|+|M_1|e^{\|A\|(t_1-t_0)}+ke^{\|A\|(t_1-t_0)})\,\delta.
\end{aligned}
$$

Since the operation of taking a minimum of a function is carried out with the error ϵ_5, we have

$$\left|\widetilde{\min_{\tilde\psi\in\tilde S}}\tilde\varphi(\tilde\psi)-\min_{\tilde\psi\in\tilde S}\tilde\varphi(\tilde\psi)\right|\leqslant\varepsilon_5. \tag{44}$$

It is not difficult to verify that

$$\left|\min_{\tilde\psi\in\tilde S}\tilde\varphi(\tilde\psi)-\min_{\tilde\psi\in\tilde S}\varphi(\tilde\psi)\right|\leqslant\max_{\tilde\psi\in\tilde S}|\tilde\varphi(\tilde\psi)-\varphi(\tilde\psi)|. \tag{45}$$

Thus by (44), (45), (40), and (43) we have

$$|\tilde{u}-u| = \left|\widetilde{\min_{\tilde\psi\in\tilde S}}\tilde\varphi(\tilde\psi) - \min_{\psi\in S}\varphi(\psi)\right| \leqslant \left|\widetilde{\min_{\tilde\psi\in\tilde S}}\tilde\varphi(\tilde\psi) - \min_{\tilde\psi\in\tilde S}\tilde\varphi(\tilde\psi)\right|$$
$$+\left|\min_{\tilde\psi\in\tilde S}\tilde\varphi(\tilde\psi) - \min_{\tilde\psi\in\tilde S}\varphi(\tilde\psi)\right| + \left|\min_{\tilde\psi\in\tilde S}\varphi(\tilde\varphi) - \min_{\psi\in S}\varphi(\psi)\right| \leqslant \varepsilon_5+\varepsilon_0+\varepsilon_1$$
$$+|M_1|\varepsilon_2(t_1)+\varepsilon_4+\int_{t_0}^{t_1}\varepsilon_3(s)\,ds+\int_{t_0}^{t_1}k(s)\,\varepsilon_2(s)\,ds$$
$$+(|M_0|+|M_1|e^{\|A\|(t_1-t_0)}+ke^{\|A\|(t_1-t_0)})\,\delta.$$

By the assumption (25) of the theorem, $|\tilde{u}-u| \leqslant \epsilon$. Thus Theorem 5 is proved.

REMARK. If all the operations on the computer were carried out without an error, it follows from Theorem 5 that, in order to obtain the error ϵ in the expression $u = \min_{\psi\in S}\varphi(\psi)$, one should choose the subdivision parameter δ from the condition

$$[|M_0|+(|M_1|+k)\,e^{\|A\|(t_1-t_0)}]\,\delta \leqslant \varepsilon.$$

We shall now derive a relation for the errors ϵ_γ and ϵ.

THEOREM 6. *In addition to the assumptions of Theorem 5, let ϵ_6 be the error in the computation of the norm of a vector. If the parameters ϵ_0, ϵ_1, $\epsilon_2(t)$, ϵ_3, ϵ_4, ϵ_5, ϵ_6, ϵ and δ satisfy*

$$\varepsilon+\gamma_1(\varepsilon_6+\varepsilon_2(t_1)+e^{\|A\|(t_1-t_0)}\delta)\leqslant\varepsilon_\gamma,$$

then the error in computing $u_\gamma = \min_{\psi\in S}\varphi_\gamma(\psi)$ does not exceed ϵ_γ.

The proof can be carried out in accordance with the scheme given in proving Theorem 5.

§6. Conclusion

Assume that we know the computational error ϵ (in order to obtain this error, we have to choose all the computational parameters in accordance with Theorem 5). Assume that we have numerically obtained the approximate value $\tilde{u}$ of $u = \min_{\psi\in S}\varphi(\psi)$. If $\tilde{u} \geqslant \epsilon$, then the system (1) is controllable by Theorem 3. If $\tilde{u} < \epsilon$, then one can only guarantee that the system is γ-controllable, where the parameters γ_0 and γ_1 must satisfy

$$\tilde{u}+\gamma_0+\gamma_1 e^{-\|A\|(t_1-t_0)} \geqslant \varepsilon+\varepsilon_5+\gamma_1(2\varepsilon_6+2\varepsilon_2(t_1)+e^{\|A\|(t_1-t_0)}\delta).$$

This assertion is a consequence of Theorems 4 and 6 and of the inequality

$$\tilde{u}_\gamma \geqslant \tilde{u}-\varepsilon_5+\gamma_0-\varepsilon_6\gamma_1-\varepsilon_2(t_1)\,\gamma_1-\gamma_1 e^{-\|A\|(t_1-t_0)}. \tag{46}$$

Let us prove (46). Making use of (44), (20), (32) and (31), and of the properties of a minimum, we obtain the chain of inequalities

$$\tilde{u}_\gamma = \widetilde{\min_{\tilde{\psi}\in\tilde{S}}} \tilde{\varphi}_\gamma(\tilde{\psi}) \geqslant -\varepsilon_5 + \min_{\tilde{\psi}\in\tilde{S}} \tilde{\varphi}_\gamma(\tilde{\psi}) \geqslant -\varepsilon_5 + \tilde{u} + \gamma_0 - \varepsilon_6\gamma_1 + \gamma_1 \min_{\tilde{\psi}\in\tilde{S}} \|\tilde{\psi}(t_1, \tilde{\psi})\|$$
$$\geqslant \tilde{u} - \varepsilon_5 + \gamma_0 - \varepsilon_6\gamma_1 - \varepsilon_2(t_1)\gamma_1 + \gamma_1 \min_{\tilde{\psi}\in\tilde{S}} \|\psi(t_1, \tilde{\psi})\| \geqslant$$
$$\geqslant \tilde{u} - \varepsilon_6 + \gamma_0 - \varepsilon_6\gamma_1 - \varepsilon_2(t_1)\gamma_1 + \gamma_2 e^{-\|A\|(t_1-t_0)}.$$

This proves (46).

The proposed algorithm for the numerical solution of the controllability problem differs from the ones which were known before (see, for example, [3]) insofar as exact estimates of the error have been obtained in this algorithm. The basic result of this work was published without proof in [4].

The author would like to express his gratitude to F. P. Vasil'ev for valuable discussions.

BIBLIOGRAPHY

1. R. Tyrrell Rockafellar, *Convex analysis,* Princeton Univ. Press, Princeton, N. J., 1970.

2. Henry Hermes and Joseph P. LaSalle, *Functional analysis and time optimal control,* Academic Press, New York, 1969.

3. F. P. Vasil'ev and R. P. Ivanov, *Some approximate methods for the solution of time-optimal problems in Banach spaces with phase restrictions,* Dokl. Akad. Nauk SSSR **195** (1970), 526–529; English transl. in Soviet Math. Dokl. **11** (1970).

4. V. I. Blagodatskih, *Numerical solution of controllability problems for linear systems,* Second All-Union Sem. Numer. Methods in Math. Programming, Abstracts of Reports, Kharkov, 1976, pp. 77–81. (Russian)

Translated by K. MAKOWSKI

Труды Мат. Инст. Стеклов.
143 (1977)

Proceedings of the Steklov
Institute of Mathematics
1980, Issue 1

ON SOME PROPERTIES OF ORTHOGONAL SYSTEMS OF CONVERGENCE

UDC 517.5

B. S. KAŠIN

ABSTRACT. In this paper the properties of systems of convergence are studied. One of the main results is the following

THEOREM. *There exists a complete orthonormal system* $\{\varphi_n(x)\}_1^\infty$ *in* $L^2(0, 1)$ *for which the satisfaction of the condition* $\Sigma_1^\infty c_n^2 < \infty$ *is necessary and sufficient for the convergence of the series* $\Sigma_1^\infty c_n\varphi_n(x)$.

Bibliography: 17 titles.

An orthonormal system $\{\varphi_n(x)\}_1^\infty$ on the interval $[0, 1]$ is said to be a *system of convergence* if every series

$$\sum_{n=1}^{\infty} c_n\varphi_n(x) \tag{1}$$

whose coefficients satisfy

$$\sum_{n=1}^{\infty} c_n^2 < \infty \tag{2}$$

converges almost everywhere on $[0, 1]$. An orthonormal system $\{\varphi_n(x)\}$ is called a *system of strict convergence* when the series (1) converges almost everywhere if and only if condition (2) is satisfied.

The Rademacher system ([2], Chapter II, §2) and the system $\{\sin 2^k\pi x\}_1^\infty$ are well-known examples of systems of strict convergence.

Given a system of convergence $\Phi = \{\varphi_n(x)\}_1^\infty$, we can define the partial sum majorant operator T_Φ acting from l_2 to L^0 (the space of all measurable almost everywhere finite functions) thus: if $a = \{a_n\}_1^\infty \in l_2$, then

$$T_\Phi(a) = f(x) = \sup_{1 \leqslant k < \infty} \left| \sum_{n=1}^{k} a_n\varphi_n(x) \right|.$$

We study the properties of T_Φ in the first section of this paper. In the second we prove

THEOREM 1. *There exists a complete orthonormal system of strict convergence* $\{\psi_n(x)\}_1^\infty$ *in* $L^2(0, 1)$.

Theorem 1 answers a question posed by Ul'janov [3] and strengthens the author's result in [5]. The system constructed in Theorem 1 has the property that for any function $f(x) \in L^2(0, 1)$ there is a unique series $\{\psi_n(x)\}$ such that $f(x) = \Sigma_1^\infty c_n\psi_n(x)$ a.e.

1980 *Mathematics Subject Classification.* Primary 42C15.

We need the next result, which is also of independent interest, in proving Theorem 1.

THEOREM 2. *Let L be an arbitrary closed linear manifold in* $L^2(0, 1)$. *Then in L one can find an orthonormal basis* $\{\varphi_k(x)\}$ *that is a system of convergence.*

The results of this paper were published without proof in [11].

§1.

THEOREM 3. *There exists an orthonormal system of convergence* $\Phi = \{\varphi_n(x)\}_1^\infty$ *on* $[0, 1]$ *such that for any set* $E \subset (0, 1)$, $\mu E > 0$,

$$T_\Phi(a) \equiv \sup_{1 \leqslant k < \infty} \left| \sum_{n=1}^{k} a_n \varphi_n(x) \right| \notin L^2(E)$$

for some $a = \{a_n\} \in l_2$ *that depends on E.*

Before proving Theorem 3 we mention some results connected with it. We need a few definitions.

Let B be a Banach space. An operator $G\colon B \longrightarrow L^0(0, 1)$ is convex if for any $a, b \in B$ and any $\lambda \in (-\infty, \infty)$ we have

1) $|G(a + b)| \leqslant |G(a)| + |G(b)|$ a.e. on $[0, 1]$;

2) $|G(\lambda a)| = |\lambda| \cdot |G(a)|$ a.e. on $[0, 1]$.

A set of functions $Q \subset L^0(0, 1)$ is bounded in measure if for any $\epsilon > 0$ there is an $R > 0$ such that

$$\mu\{x : |f(x)| \geqslant R\} \leqslant \varepsilon$$

for any $f(x) \in Q$.

An operator $G\colon B \longrightarrow L^0(0, 1)$ is bounded if the image $G(S)$ of the unit ball S of B is bounded in measure.

A typical example of a bounded convex operator is T_Φ, the partial sum majorant ($\Phi = \{\varphi_n(x)\}$ being an orthonormal system of convergence).

Olevskiĭ proved [6] that there are a system of convergence $\Phi_0 = \{\varphi_n(x)\}$ and an element $a \in l_2$ such that

$$T_{\Phi_0}(a) \notin \bigcup_{p>0} L^p(0, 1).$$

Nikišin (see [7] and also [8]) proved

THEOREM A. *Let* $G\colon l_p \longrightarrow L^0(0, 1)$ *be a bounded convex*(1) *operator. Then for each* $\epsilon > 0$ *there is a set* $E_\epsilon \subset [0, 1]$ *with* $\mu E_\epsilon > 1 - \epsilon$ *such that the operator G is of weak type* (p, q) *on* E_ϵ *for* $q = \min(2, p)$.

It follows directly from Theorem A that for any orthonormal system of convergence $\Phi = \{\varphi_n(x)\}$ the operator T_Φ is a bounded operator from l_2 to $L^p(E_\epsilon)$ ($E_\epsilon \subset [0, 1]$, $\mu E_\epsilon > 1 - \epsilon$) for any $p < 2$.

Nikišin has constructed a bounded linear operator $G\colon l_p \longrightarrow L^0(0, 1)$ for each p, $1 \leqslant p < 2$, such that for any set $E \subset [0, 1]$ with $\mu E > 0$ we have

(1) Nikišin calls *sublinear* the operators that we call convex.

$$\sup_{\|a\|_{l_p}=1} \left(\int_E (G(a))^p \, dx \right)^{1/p} = \infty.$$

By the same token Theorem A cannot be strengthened for $1 \leqslant p < 2$, even for linear operators.

If $p \geqslant 2$ and $G: l_p \longrightarrow L^0(0, 1)$ is a bounded linear operator, then it is known (see [13] or [14]) that Theorem A can be strengthened.

THEOREM B. *Let $p \geqslant 2$, and let $G: l_p \longrightarrow L^0(0, 1)$ be a bounded linear operator. Then for any $\epsilon > 0$ there are a set $E_\epsilon \subset [0, 1]$ with $\mu E_\epsilon > 1 - \epsilon$ and a constant C_ϵ such that*

$$\| G(a) \|_{L^2(E_\epsilon)} \leqslant C_\epsilon \| a \|_{l_p}.$$

Theorem B can be obtained by applying Theorem A and the theorem of Grothendieck [4] that for any bounded linear operator $G: l_p \longrightarrow L^1(0, 1)$ $(p \geqslant 2)$ we have $G(a) = H(a) f_0(x)$, where $H: l_p \longrightarrow L^2(0, 1)$ is a bounded operator and the function $f_0(x)$ does not depend on a.

It follows from Theorem B that the examples constructed by Nikišin for $1 \leqslant p < 2$ do not extend to the case $p \geqslant 2$. The case $p = 2$ is of particular interest. The question of settling Theorem A even for $p = 2$ has remained open. It follows directly from Theorem 3 of this paper that Theorem A cannot be strengthened and that the requirement that G be linear in Theorem B (and in Grothendieck's theorem) is essential.

We pass on to the proof of Theorem 3. We need some auxiliary systems of functions to construct the required system $\{\varphi_n(x)\}$.

Let $N = 2^{2q} > 100$ (q integral). We define a system of functions $\{f_j^N(x)\}_{j=1}^N$ on $[0, 1 - N^{-1/2}]$ by putting, for $x \in [(i-1)/n, i/n)$, $i = 1, 2, \ldots, N - \sqrt{N}$,

$$f_j^N(x) = \begin{cases} 0 & \text{for } 1 \leqslant j \leqslant 2\sqrt{N}, \quad N - 2\sqrt{N} \leqslant j \leqslant N, \\ \sqrt{N}/(i-j) & \text{for } 2\sqrt{N} < j < N - 2\sqrt{N}, \quad i \neq j \text{ and } |i-j| < \sqrt{N}, \\ 0 & \text{for } 2\sqrt{N} < j < N - 2\sqrt{N}, \quad i = j \text{ or } |i-j| \geqslant \sqrt{N}. \end{cases} \tag{3}$$

We note some properties of the system $\{f_j^N(x)\}$. It follows directly from (3) that

$$\int_0^{1-N^{-1/2}} f_j^N(x)\, dx = 0, \quad j = 1, 2, \ldots, N, \tag{4}$$

$$\mu \left\{ x \in [0, 1]: \sum_{j=1}^N f_j^N(x) \neq 0 \right\} < \frac{10}{\sqrt{N}}. \tag{5}$$

It is easy to check that

$$\mu \left\{ x \in [0, 1]: \sup_{1 \leqslant r \leqslant N} \left| \sum_{j=1}^r \frac{1}{\sqrt{N}} f_j^N(x) \right| \geqslant \frac{1}{4} \ln N \right\} \geqslant 1 - \frac{10}{\sqrt{N}}. \tag{6}$$

It is well known (see [5] for example) that the system $\{f_j^N(x)\}$ satisfies

$$\left\|\sum_{j=1}^{N} c_j f_j^N(x)\right\|_{L^2(0,1)} \leqslant C\left(\sum_{j=1}^{N} c_j^2\right)^{1/2},$$

where C is independent of N.

It follows from this and Schur's theorem ([2], Chapter III, §2) that one can extend $f_j^N(x)$ to $[1 - N^{-1/2}, 1]$ so that

$$\int_0^1 f_i^N(x) f_j^N(x)\,dx = \begin{cases} B & \text{for } i = j, \\ 0 & \text{for } i \neq j \end{cases} \tag{7}$$

(the constant B being independent of N).

Further, it is easy to see that the extensions can be chosen so that

a)

$$\int_0^1 f_j^N(x)\,dx = 0 \quad \text{for } j = 1, 2, \ldots, N; \tag{8}$$

b) the functions $f_j^N(x)$ are piecewise constant on $[0, 1]$ with interval of constancy of length $(2N)^{-2}$.

We define another system $\{g_j^N(x)\}_{j=1}^N$ on the line $-\infty < x < +\infty$ by first putting

$$g_j^N(x) = \begin{cases} \dfrac{1}{\sqrt{B}\,\log_2 N}\, 2^{k/2} f_j^N\left(2^k\left(x - \dfrac{1}{2^k}\right)\right) & \text{for} \\ 2^{-k} < x \leqslant 2^{-(k-1)}, & 1 \leqslant k \leqslant (\log_2 N)^2; \\ 0 \quad \text{for } 0 < x \leqslant 2^{-(\log_2 N)^2}, \end{cases} \tag{9}$$

for $x \in [0, 1]$ and $j = 1, \ldots, N$, and then putting $g_j^N(x + r) = g_j^N(x)$ for integral r.

For $j = 1, \ldots, N$ we have (see (7) and (9))

$$\text{a)} \int_0^1 g_j^N(x)\,dx = 0;$$

$$\text{b)} \int_0^1 g_i^N(x)\, g_j^N(x)\,dx = 0 \qquad \text{for } i \neq j; \tag{10}$$

$$\text{c)} \int_0^1 (g_j^N(x))^2\,dx = \sum_{k=1}^{\log_2^2 N} (\log_2 N)^{-2} = 1.$$

It also follows from (7) and (9) that for $k = 1, 2, \ldots, (\log_2 N)^2$

$$\int_{(2^{-k},\, 2^{-k+1}]} \left(\sum_{j=1}^{N} c_j g_j^N(x)\right)^2 dx \leqslant \left(\sum_{j=1}^{N} c_j^2\right)(\log_2 N)^{-2}. \tag{11}$$

Further, if one puts

$$g_0^N(x) = \sup_{1 \leqslant r \leqslant N} \left|\sum_{j=1}^{r} \frac{1}{N^{1/2}}\, g_j^N(x)\right|,$$

it is easy to verify, using (6), that for any set $A \subset (2^{-k}, 2^{-k+1}]$ with measure $|A| \leqslant 2^{-(k+1)}$ the integral(2)

$$\int_{(2^{-k},\, 2^{-k+1}]-A} (g_0^N(x))^2\,dx \geqslant C > 0 \quad \text{for } 1 \leqslant k \leqslant \log_2^2 N \tag{12}$$

---

(2) In the sequel C, C' and c will denote various, in general different, absolute constants.

and consequently

$$\left(\int_0^1 (g_0^N(x))^2\,dx\right)^{1/2} \geqslant \left(\sum_{k=1}^{\log_2^2 N} \int_{2^{-k}}^{2^{-k+1}} (g_0^N(x))^2\,dx\right)^{1/2} \geqslant C\log_2 N. \tag{12'}$$

We prove another property of the system $\{g_j^N(x)\}$.

LEMMA 1. *For any numbers $N = 2^{2q}$ and $y > 0$ and any sequence $a = \{a_j\}_1^N$,*

$$\mu\left\{x\in[0,\ 1]: g_a(x) \equiv \sup_{1\leqslant r\leqslant N}\left|\sum_{j=1}^r a_j g_j^N(x)\right| \geqslant y\right\} \leqslant \frac{C}{y^2}\left(\sum_{j=1}^N a_j^2\right) + \frac{C}{\sqrt{N}}.$$

PROOF. Let a set of intervals Δ_{ik} with $1 \leqslant i \leqslant N-\sqrt{N}$ and $1 \leqslant k \leqslant (\log_2 N)^2$ be defined so that

$$\Delta_{ik} = \left(\frac{1}{2^k} + \frac{i-1}{2^k N},\ \frac{1}{2^k} + \frac{i}{2^k N}\right].$$

Then

$$[0,\ 1] = \bigcup_{k,\,i} \Delta_{ik} + E, \qquad \mu E \leqslant \frac{2}{\sqrt{N}}. \tag{13}$$

For $x \in [0, 1]$ we put

$$F_a(x) \equiv \frac{1}{\sqrt{B}\,\log_2 N} \sup_{1\leqslant r\leqslant N}\left|\sum_{j=1}^r a_j f_j^N(x)\right|.$$

By (7) and the Menšov-Rademacher lemma ([2], Chapter V, §3) we have

$$\int_0^1 (F_a(x))^2\,dx \leqslant C\sum_{j=1}^N a_j^2. \tag{14}$$

It follows from the definition of the functions $g_j^N(x)$ (see (9)) that

$$g_j^N(x) = \frac{2^{k/2}}{\sqrt{B}\,\log_2 N} f_j\left(\frac{i-1/2}{N}\right)$$

for $x \in \Delta_{ik}$, whence for $x \in \Delta_{ik}$ we have

$$g_a(x) = 2^{k/2} F_a\left(\frac{i-1/2}{N}\right).$$

Consequently for fixed i

$$\mu\left\{x : x \in \bigcup_{1\leqslant k\leqslant \log_2^2 N} \Delta_{ik},\quad g_a(x) \geqslant y\right\} \tag{15}$$

$$\leqslant \sum_{k:\,2^{k/2}F_a\left(\frac{i-1/2}{N}\right)\geqslant y} \frac{1}{N\cdot 2^k} \leqslant \frac{2}{N}\,\frac{F_a^2\left(\frac{i-1/2}{N}\right)}{y^2}.$$

It follows from (14) that

$$\frac{1}{N}\sum_{i=1}^N F_a^2\left(\frac{i-1/2}{N}\right) \leqslant C\sum_{j=1}^N a_j^2,$$

so, summing (15) over all i, $1 \leqslant i \leqslant N-\sqrt{N}$, we obtain

$$\mu\left\{x : x \in \bigcup_{k,\,i} \Delta_{ik},\ g_a(x) > y\right\} \leqslant \frac{C}{y^2}\sum_{j=1}^N a_j^2,$$

which, together with (13), proves Lemma 1.

Now we construct an orthonormal system $\{\varphi_n(x)\}_1^\infty$ to satisfy Theorem 3. To do this we take a sequence of integers $\{s_k\}_1^\infty$ such that $s_1 = 0$ with the numbers $N_k = s_{k+1} - s_k$ $(k = 1, 2, \ldots)$ of the form $N_k = 2^{2^q}$ (q an integer > 10) such that for all k

$$N_k^3 \cdot 2^{(\log_2 N_k)^2} < N_{k+1}. \tag{16}$$

Then for $x \in [0, 1]$ and $s_k < n \leqslant s_{k+1}$ we put

$$\varphi_n(x) = g_{n-s_k}^{N_k}(N_k x). \tag{17}$$

In view of (10a) we have

$$\int_0^1 \varphi_n(x)\,dx = 0.$$

Further (see (16), (17), (8) and (9)), for $s_k < n \leqslant s_{k+1}$ the functions $\varphi_n(x)$ are piecewise constant with interval of constancy of length

$$\frac{1}{N_k}\frac{1}{N_k^2}2^{-(\log_2 N_k)^2},$$

while at the same time the period of the function $\varphi_n(x)$ with $s_{n+1} < n \leqslant s_{n+2}$ is equal to N_{k+1}^{-1}. It follows (considering (16) and also (10) b)) that the functions $\varphi_n(x)$ form an orthonormal system on $[0, 1]$.

We claim that $\{\varphi_n(x)\}$ is a system of convergence. In fact, suppose

$$\sum_{n=1}^{\infty} a_n\varphi_n(x), \qquad \sum_{n=1}^{\infty} a_n^2 < \infty. \tag{18}$$

It is easy to see that the sequence $S_{s_k}(x)$, where

$$S_{s_k}(x) = \sum_{n=1}^{s_k} a_n\varphi_n(x), \qquad k = 2, 3, \ldots,$$

converges almost everywhere. It is thus enough to show that for any $y > 0$

$$\sum_{k=1}^{\infty} \mu\left\{x:\ \sup_{s_k<r\leqslant s_{k+1}}\left|\sum_{n=s_k+1}^{r} a_n\varphi_n(x)\right| > y\right\} < \infty. \tag{19}$$

But by (17) and Lemma 1 we have

$$\mu\left\{x:\ \sup_{s_k<r\leqslant s_{k+1}}\left|\sum_{n=s_k+1}^{r} a_n\varphi_n(x)\right| > y\right\} \leqslant \frac{C}{y^2}\sum_{n=s_k+1}^{s_{k+1}} a_n^2 + \frac{C}{N_k^{1/2}}. \tag{20}$$

Since $\Sigma_1^\infty N_k^{-1/2} < \infty$ (see (16)), and, by hypothesis,

$$\sum_{k=1}^{\infty}\left(\sum_{n=s_k+1}^{s_{k+1}} a_n^2\right) < \infty,$$

the inequality (20) implies (19).

To complete the proof of Theorem 3 we must construct a series of the form (18) for each set $E \subset [0, 1]$ (which we can take to be closed) with $\mu E > 0$ such that

$$\int_E \left(\sup_{1 \leqslant r < \infty} \left| \sum_{n=1}^{r} a_n \varphi_n(x) \right| \right)^2 dx = \infty. \tag{21}$$

For a given integer $m > 10$ we find a finite set of disjoint intervals $\{I_l\}_1^p$ with dyadic rational endpoints, such that

$$E \subset \bigcup_{l=1}^{p} I_l \quad \text{and} \quad \mu E \left(1 + \frac{1}{m}\right) > \sum_{l=1}^{p} \mu I_l.$$

For $k = 1, 2, \ldots$ and $\nu = 1, 2, \ldots, (\log_2 N_k)^2$ we define the set

$$J_{k,\nu} = \bigcup_{m=1}^{N_k} \left(\frac{m-1}{N_k} + \frac{1}{2^\nu N_k}, \ \frac{m-1}{N_k} + \frac{1}{2^{\nu-1} N_k} \right]. \tag{22}$$

It is clear that the measure

$$\mu J_{k\nu} = 2^{-\nu}. \tag{23}$$

It is easy to see that for $\nu = 1, 2, \ldots, [\frac{1}{2} \log_2 m]$

$$\overline{\lim_{k \to \infty}} \, \mu (J_{k\nu} \cap E) \geqslant {}^2/{}_3 \mu J_{k\nu} \cdot \mu E.$$

Using this estimate, (12), and the definition of $\varphi_n(x)$ (see (17)), we easily obtain that

$$\overline{\lim_{k \to \infty}} \int_{J_{k,\nu} \cap E} \sup_{s_k < r \leqslant s_{k+1}} \left(\sum_{n=s_k+1}^{r} \frac{1}{N_k^{1/2}} \varphi_n(x) \right)^2 dx \geqslant C(E) > 0,$$

$$\nu = 1, 2, \ldots, [{}^1/{}_2 \log_2 m].$$

On summing these inequalities over ν, $1 \leqslant \nu \leqslant [\frac{1}{2} \log_2 m]$, we obtain

$$\begin{aligned} &\overline{\lim_{k \to \infty}} \int_E \sup_{s_k < r \leqslant s_{k+1}} \left(\sum_{n=s_k+1}^{r} \frac{1}{N_k^{1/2}} \varphi_n(x) \right)^2 dx \\ &\geqslant \overline{\lim_{k \to \infty}} \sum_{\nu=1}^{[{}^1/{}_2 \log_2 m]} \int_{E \cap J_{k,\nu}} \sup_{s_k < r \leqslant s_{k+1}} \left(\sum_{n=s_k+1}^{r} \frac{1}{N_k^{1/2}} \varphi_n(x) \right)^2 dx \geqslant C(E) \log_2 m. \end{aligned} \tag{24}$$

Since the integer m can be arbitrarily large, it follows from (24) that

$$\overline{\lim_{k \to \infty}} \int_E \sup_{s_k < r \leqslant s_{k+1}} \left(\sum_{n=s_k+1}^{r} \frac{1}{N_k^{1/2}} \varphi_n(x) \right)^2 dx = \infty.$$

Using this, it is not hard to define numbers $\{a_n\}_1^\infty$ such that (21) is satisfied. This proves Theorem 3.

In conclusion we note an inequality to be used in §2.

It follows from (11), (17) and the definition of $J_{k,\nu}$ that for $\nu = 1, 2, \ldots, (\log_2 N_k)^2$ we have

$$\int_{J_{k,\nu}} \left(\sum_{n=s_k+1}^{s_{k+1}} c_n \varphi_n(x) \right)^2 dx \leqslant \log_2^{-2} N_k \sum_{n=s_k+1}^{s_{k+1}} c_n^2. \tag{25}$$

§2.

First we prove Theorem 1, stated at the beginning of this paper. Our proof has much in common with the proof of Theorem 1 of [5]. We need a number of lemmas.

LEMMA 2 (see [15], Chapter 2, §7, for example). *Let the sets $E_k \subset [0, 1]$, $k = 1, 2, \ldots$, be independent, let $\Sigma_1^\infty \mu E_k = \infty$, and define*

$$\overline{\lim_{k\to\infty}} E_k = \bigcap_{n=1}^{\infty} \bigcup_{k=n}^{\infty} E_k.$$

Then $\mu \overline{\lim}_{k\to\infty} E_k = 1$.

LEMMA 3 (see [1], Chapter V, Theorem 8.27). *There exist absolute constants $C_1, C_2 \in (0, 1)$ such that for any polynomial $P(x) = \Sigma_1^N c_k r_k(x)$ in the Rademacher system the measure*

$$\mu\left\{x \in [0, 1] : |P(x)| \geqslant C_1 \left(\int_0^1 P^2(x)\, dx\right)^{1/2}\right\} \geqslant C_2.$$

Further, let $P'(x) = u(x) + Q(x)$, where $Q(x) = \Sigma_p^{p'} c_k r_k(x)$ and $u(x)$ is some function in the Walsh system with $u(x) \not\equiv 1$ and $u(x) = \Pi_{i=1}^{p-1} (r_i)^{\epsilon_i}$ ($\epsilon_i = 0, 1; i = 1, \ldots, p-1$). Then

$$\mu\left\{x \in [0, 1] : |P'(x)| \geqslant C_1 \left(\int_0^1 (P'(x))^2\, dx\right)^{1/2}\right\} \geqslant C_2. \tag{26}$$

Indeed, (26) follows easily from Lemma 3 and the fact that there is a measure-preserving transformation $\tau(x)$ of $[0, 1]$ such that $u(\tau(x)) = r_1(x)$ and $r_k(\tau(x)) = r_k(x)$ for $k \geqslant p$.

LEMMA 4 (see [1], Chapter V, Lemma 8.26). *Let $C \geqslant 1$. For any function $f(x)$ such that $\int_0^1 |f(x)|\, dx \geqslant 1$ and $(\int_0^1 f^2(x)\, dx)^{1/2} \leqslant C$, we have*

$$\mu\left\{x : |f(x)| \geqslant \frac{1}{4}\right\} \geqslant \frac{1}{8C^2}.$$

For each $N \geqslant 10$ we define the orthonormal matrix $A_N = \{a_{ji}\}_{i,j=1}^N$ thus:

$$a_{ji} = \begin{cases} 1 - (N-1)^{-1} & \text{for } i = j < N, \\ 0 & \text{for } i = j = N, \\ -(N-1)^{-1} & \text{for } i \neq j, \quad i, j < N, \\ (N-1)^{-1/2} & \text{for } i = N \quad \text{or} \quad j = N, \quad i \neq j. \end{cases} \tag{27}$$

The matrices A_N were first applied by Olevskiĭ in [9] to transform unbounded orthonormal systems to bounded systems while preserving their fundamental properties.

We need some new properties of the system $\{f_j^N(x)\}_{j=1}^N$ defined in §1.

First, a direct calculation shows easily that for $2\sqrt{N} < j < k < N - 2\sqrt{N}$ we have

$$\int_{j/N}^{1-N^{-1/2}} f_j^N(x)\, dx = \int_{k/N}^{1-N^{-1/2}} f_k^N(x)\, dx \geqslant C \frac{\log_2 N}{N^{1/2}}. \tag{28}$$

LEMMA 5. *Let a series of numbers* $a_1, \ldots, a_N$ *be given, and let*

$$\frac{1}{N^{1/2}}\left|\sum_{j=1}^{N} a_j\right| \geqslant \gamma \left(\sum_{j=1}^{N} a_j^2\right)^{1/2}. \tag{29}$$

Then for $N > N(\gamma)$

$$\mu\left\{x \in [0, 1 - N^{-1/2}] : \left|\sum_{j=1}^{M(x)} a_j f_j^N(x)\right| \geqslant C_\gamma \log_2 N \cdot \left(\sum_{j=1}^{N} a_j^2\right)^{1/2}\right\} \geqslant C'_\gamma,$$

where the constants C_γ, C'_γ *and* N_γ *depend only on* γ, *and* $M(x) = i$ *for* $x \in ((i-1)/N, i/N)$, $1 \leqslant i \leqslant N - \sqrt{N}$.

PROOF. Using (29) and the definitions of $M(x)$ and $f_j^N(x)$ (see (3)), we have

$$\begin{aligned}\int_0^{1-N^{-1/2}} \left|\sum_{j=1}^{M(x)} a_j f_j^N(x)\right| dx &\geqslant \left|\int_0^{1-N^{-1/2}} \sum_{j=1}^{M(x)} a_j f_j^N(x)\, dx\right| \\ &= \left|\sum_{j=2\sqrt{N}+1}^{N-2\sqrt{N}-1} a_j \int_{j/N}^{1-N^{-1/2}} f^N(x)\, dx\right| \geqslant C\left|\sum_{j=2\sqrt{N}+1}^{N-2\sqrt{N}-1} a_j\right| \frac{\log_2 N}{N}.\end{aligned} \tag{30}$$

Since by the Cauchy inequality

$$\left|\sum_{j=1}^{2\sqrt{N}} a_j\right| \leqslant \left(\sum_{j=1}^{2\sqrt{N}} a_j^2\right)^{1/2} (2\sqrt{N})^{1/2} \leqslant \left(\sum_{j=1}^{N} a_j^2\right)^{1/2} (2\sqrt{N})^{1/2}$$

and, similarly,

$$\left|\sum_{j=N-2\sqrt{N}}^{N} a_j\right| \leqslant \left(\sum_{j=1}^{N} a_j^2\right)^{1/2} (2\sqrt{N})^{1/2},$$

it follows that

$$\begin{aligned}\frac{1}{\sqrt{N}}\left|\sum_{j=2\sqrt{N}+1}^{N-2\sqrt{N}-1} a_j\right| &\geqslant \left|\sum_{j=1}^{N} a_j\right| \frac{1}{\sqrt{N}} - 4N^{-1/4}\left(\sum_{j=1}^{N} a_j^2\right)^{1/2} \\ &\geqslant (\gamma - 4N^{-1/4})\left(\sum_{j=1}^{N} a_j^2\right)^{1/2} \geqslant \frac{1}{2}\left(\sum_{j=1}^{N} a_j^2\right)^{1/2} \quad \text{for } N > N_\gamma.\end{aligned}$$

Thus it follows from (30) that for $N > N_\gamma$

$$\int_0^{1-N^{-1/2}} \left|\sum_{j=1}^{M(x)} a_j f_j^N(x)\right| dx \geqslant C\left(\sum_{j=1}^{N} a_j^2\right)^{1/2} \log_2 N. \tag{31}$$

Since the functions $\{B^{-1/2} f_j^N(x)\}_{j=1}^N$ (where B is an absolute constant) are orthonormal on $[0, 1]$ (see (7)), we can use the Men′šov-Rademacher lemma (see [2], Chapter V, §3) to obtain

$$\left(\int_0^{1-N^{-1/2}} \left(\sum_{j=1}^{M(x)} a_j f_j^N(x)\right)^2 dx\right)^{1/2} \leqslant C'\left(\sum_{j=1}^{N} a_j^2\right)^{1/2} \log_2 N. \tag{32}$$

Now on applying Lemma 4 to the function

$$f(x)=\left|\sum_{j=1}^{M(x)} a_j f_j^N(x)\right|\left[C\left(\sum_{j=1}^{N} a_j^2\right)^{1/2}\log_2 N\right]^{-1},$$

we obtain Lemma 5 by virtue of (31) and (32).

Let C_γ and C'_γ be the constants of Lemma 5, and let $k = 1, 2, \ldots$ and $1 \leqslant \nu \leqslant (\log_2 N_k)^2$. We define the set

$$E(k, \nu, \{a_n\}) \equiv \left\{x : x \in J_{k,\nu}, \sup_{s_k < r \leqslant s_{k+1}} \left|\sum_{n=s_k+1}^{r} a_n\varphi_n(x)\right| \geqslant C_\gamma 2^{\nu/2}\left(\sum_{n=s_k+1}^{s_{k+1}} a_n^2\right)^{1/2}\right\}.$$

From Lemma 5 and the definition of $\{g_j^N(x)\}_1^N$ and $\{\varphi_n(x)\}_1^\infty$ (see (9), (17) and (22)) we obtain for $k = 1, 2, \ldots$ and

$$\frac{1}{N_k^{1/2}}\left|\sum_{n=s_k+1}^{s_{k+1}} a_n\right| \geqslant \gamma\left(\sum_{n=s_k+1}^{s_{k+1}} a_n^2\right)^{1/2} \tag{33}$$

that

$$\mu E(k, \nu, \{a_n\}) \geqslant C'_\gamma \mu J_{k,\nu}. \tag{34}$$

We note that the functions $\varphi_n(x)$, $s_k < n < s_{k+1}$, have period N_k^{-1} and are constant on intervals of the form

$$\left(\frac{p-1}{N_{k+1}}, \frac{p}{N_{k+1}}\right), \quad 1 \leqslant p \leqslant N_{k+1}$$

(see (16), (17)), so that

a) the set $E(k, \nu, \{a_n\})$ is periodic on $[0, 1]$ with period N_k^{-1}, and

b) $$E(k, \nu, \{a_n\}) = \bigcup_{i=1}^{i_0}\left(\frac{p_i-1}{N_{k+1}}, \frac{p_i}{N_{k+1}}\right), \quad 1 < p_1 < \ldots < p_{i_0} \leqslant N_{k+1}. \tag{35}$$

We pass on to construct the desired system $\{\psi_n(x)\}$. First we extend the system $\{\varphi_n(x)\}$ (defined on $[0, 1]$ in §1) to $[0, 2]$ by setting

$$\varphi_n(x) = r_n(x), \quad x \in [1, 2], \quad n = 1, 2, \ldots, \tag{36}$$

where $\{r_n(x)\}$, as before, is the Rademacher system extended from $[0, 1]$ to the whole real axis by periodicity with period 1. Let $\{u_k(x)\}_1^\infty$ be the Walsh-Paley system ($u_k(x) = u_k(x + 1)$). It is known (see [16]) that $\{u_k(x)\}$ is a system of convergence. We extract the Rademacher functions $r_n(x)$, $1 \leqslant n < \infty$, from the system $\{u_k(x)\}$, enumerating the remaining functions and denoting the system obtained by $\{u'_k(x)\}_1^\infty$.

It is clear that, like the Walsh-Paley system, $\{u'_k(x)\}$ is a system of convergence.

Further, from Theorem 2 stated in the Introduction (see the end of the paper for the proof of Theorem 2) it follows at once that there exists an orthonormal system of convergence $\{y_k(x)\}_1^\infty$ ($x \in [0, 1]$) such that $\{\varphi_n(x)\} \cup \{y_k(x)\}$ forms a complete orthonormal system on $[0, 1]$.

Now we use $\{y_k(x)\}$ and $\{u'_k(x)\}$ to enlarge the system $\{\varphi_n(x)\}_1^\infty$ ($x \in [0, 2]$) in a specific way. To do this we append to the functions $\varphi_n(x)$ the functions $v_k(x)$, $1 \leqslant k < \infty$, defined as follows for $j = 1, 2, \ldots$:

$$v_{3j}(x)=\begin{cases}\frac{1}{\sqrt{2}}\varphi_j(x), & x\in[0,1],\\ -\frac{1}{\sqrt{2}}r_j(x), & x\in(1,2],\end{cases}\qquad v_{3j-1}(x)=\begin{cases}\frac{1}{\sqrt{2}}y_j(x), & x\in[0,1],\\ \frac{1}{\sqrt{2}}u'_j(x), & x\in(1,2],\end{cases}$$

$$v_{3j-2}(x)=\begin{cases}\frac{1}{\sqrt{2}}y_j(x), & x\in[0,1],\\ -\frac{1}{\sqrt{2}}u'_j(x), & x\in(1,2].\end{cases}$$

It is easy to see that the functions in $\{\varphi_n(x)\}_1^\infty \cup \{v_k(x)\}_1^\infty$ form a complete orthonormal system in $L^2(0, 1)$. Finally we define the desired system $\{\psi_n(x)\}_1^\infty$.(3) Let the sequences $\{N_k\}$ and $\{s_k\}$ $(N_k = s_{k+1} - s_k)$ be as in §1 (see (16)), and put

$$Q_1=0,\qquad Q_{k+1}=Q_k+N_k+1,\qquad k=1,2,\ldots.$$

We define the functions $\psi_n(x)$ with $Q_k < n \leqslant Q_{k+1}$ as follows: we consider the matrix $A_{N_k+1} = \{a_{ji}\}$ (see (27)), and for $x \in [0, 2]$ and $n = Q_k + q$ $(1 \leqslant q \leqslant N_k + 1)$ we put

$$\psi_n(x)=\sum_{r=1}^{N_k} a_{qr}\varphi_{s_k+r}(x)+a_{q,N_k+1}v_k(x). \tag{37}$$

The resulting system $\{\psi_n(x)\}_1^\infty$ is a complete orthonormal system on $[0, 2]$, since the matrices A_{N_k+1} are orthonormal and all the functions $\varphi_n(x)$ and $v_k(x)$ can be expressed in terms of the functions $\psi_n(x)$.

Using the form of the matrices A_{N_k+1} (see (27)) and the fact that $\{\varphi_n(x)\}$ and $\{v_k(x)\}$ are systems of convergence, it is easy to prove that $\{\psi_n(x)\}_1^\infty$ is also a system of convergence. So to prove Theorem 1 it is enough to show that any series

$$\sum_{n=1}^{\infty} c_n\psi_n(x),\qquad \sum_{n=1}^{\infty} c_n^2=\infty, \tag{38}$$

diverges on a set of positive measure.

We fix a series of the form (38), and for $k = 1, 2, \ldots$ we put

$$\beta_k=\left(\sum_{n=Q_k+1}^{Q_{k+1}} c_n^2\right)^{1/2},\qquad d_k=\frac{1}{(N_k+1)^{1/2}}\left|\sum_{n=Q_k+1}^{Q_{k+1}} c_n\right|.$$

By Cauchy's inequality, $\beta_k \geqslant d_k$. Let the absolute constant $C_0 = C_1C_2^{1/2}(1/20)$, where C_1 and C_2 are as in Lemma 3.

We divide the integers $k \geqslant 1$ into two groups. The first group, S_1, contains all k such that

$$C_0\beta_k\leqslant d_k. \tag{39}$$

The remaining k belong to S_2.

First we show that if

$$\overline{\lim_{k\to\infty,\,k\in S_1}}\beta_k>0, \tag{40}$$

(3)We define $\{\psi_n(x)\}$ on $[0, 2]$ and transform to $[0, 1]$ by using a similarity.

then (38) does not converge in measure on $[1, 2]$.

In fact, it follows from properties of A_{N_k+1} and from (37) that

$$\left| \sum_{n=Q_k+1}^{Q_{k+1}} c_n\psi_n(x) - \frac{1}{(N_k+1)^{1/2}} \left(\sum_{n=Q_k+1}^{Q_{k+1}} c_n \right) v_k(x) - P(x) \right| \leqslant \frac{C\beta_k}{(N_k+1)^{1/2}}, \tag{41}$$

where $P(x) = \sum_{n=s_k+1}^{s_{k+1}} \alpha_n r_n(x)$ is a polynomial in the Rademacher system for $x \in [1, 2]$, and

$$v(x) = r_{n_1}(x) \cdot r_{n_2}(x) \cdot \ldots \cdot r_{n_j}(x)$$

is a function in the Walsh system; and it is clear from the construction of the system $\{\psi_n(x)\}$ that $n_j < s_k$ for sufficiently large k.

Consequently, by (26),

$$\mu \left\{ x \in [1, 2] : \left| \frac{1}{(N_k+1)^{1/2}} \left(\sum_{n=Q_k+1}^{Q_{k+1}} c_n \right) \cdot v_k(x) - P(x) \right| > C_1 d_k \right\} \geqslant C_2. \tag{42}$$

Since $C_0\beta_k \leqslant d_k$ for $k \in S_1$, it follows from (41) and (42) that for sufficiently large $k \in S_1$

$$\mu \left\{ x \in [1, 2] : \left| \sum_{n=Q_k+1}^{Q_{k+1}} c_n\psi_n(x) \right| \geqslant C\beta_k \right\} \geqslant C'.$$

This last inequality shows immediately that (38) diverges if (40) holds. It is thus enough to show that (38) diverges under the assumption that

$$\lim_{k\to\infty,\, k\in S_1} \beta_k = 0. \tag{43}$$

There are two possible cases for a series of the form (38):

$$1) \sum_{k \in S_1} \beta_k^2 = \infty, \qquad 2) \sum_{k \in S_1} \beta_k^2 < \infty.$$

Suppose that case 1) occurs. If we put

$$S_1' = \{k : k \in S_1,\ \beta_k \geqslant 1/k\},$$

then

$$\sum_{k \in S_1'} \beta_k^2 = \infty. \tag{44}$$

For $k \in S_1'$ let

$$H_k(x) = \frac{v_k(x)}{N_k^{1/2}} \left| \sum_{n=Q_k+1}^{Q_{k+1}-1} c_n \right|, \qquad \tilde{H}_k(x) = \frac{v_k(x)}{N_k^{1/2}} \sup_{Q_k < r \leqslant Q_{k+1}} \left| \sum_{n=Q_k+1}^{r} c_n \right|.$$

Since

$$\frac{1}{N_k^{1/2}} \sup_{Q_k < r \leqslant Q_{k+1}} \left| \sum_{n=Q_k+1}^{r} c_n \right| \leqslant \frac{1}{N_k^{1/2}} \sum_{n=Q_k+1}^{Q_{k+1}} |c_n| \leqslant 2\beta_k \leqslant \frac{2d_k}{C_0} \leqslant C \frac{1}{N_k^{1/2}} \left| \sum_{n=Q_k+1}^{Q_{k+1}} c_n \right|,$$

for $k \in S_1'$, it follows that

$$|H_k(x)| \leqslant |\tilde{H}_k(x)| \leqslant C |H_k(x)|. \tag{45}$$

We take a number ν_k, $k \in S_1'$, such that

$$\sqrt{2^{\nu_k - 1}} \leqslant \frac{1}{\beta_k} \leqslant \sqrt{2^{\nu_k}}. \tag{46}$$

By (44),

$$\lim_{k\to\infty,\, k\in S_1'} \nu_k = \infty.$$

Further, since $\beta_k \geqslant 1/k$ for $k \in S_1'$, we get

$$\nu_k \leqslant C \log_2 k \leqslant \log_2 N_k \qquad \text{for } k \geqslant k_0.$$

Consequently we can consider the set J_{k,ν_k} (the sets $J_{k,\nu}$ are defined by (22) for $k = 1, 2, \ldots$ and $1 \leqslant \nu \leqslant (\log_2 N_k)^2$).

Let there exist a number $y > 0$ such that

$$\mu\left[\overline{\lim_{k\to\infty,\, k\in S_1'}} \{x \in J_{k,\nu_k} : |H_k(x)| > y\}\right] > 0. \tag{47}$$

We claim that in this case the sequence of partial sums $\Sigma_1^{Q_k - 1} c_n \psi_n(x)$ diverges on a set of positive measure.

Further, for $k = 1, 2, \ldots$ and $1 \leqslant m \leqslant N_k + 1 = Q_{k+1} - Q_k$ let

$$\alpha_m^k = c_{Q_k + m}. \tag{48}$$

Using the properties of $A_{N_k + 1}$ (see (27)), we have

$$\sum_{m=1}^{N_k} \alpha_m^k \psi_{m+Q_k}(x) = \sum_{m=1}^{N_k} \alpha_m^k \varphi_{m+s_k}(x) + H_k(x) + \Delta_k(x), \tag{49}$$

where

$$|\Delta_k(x)| \leqslant \left| \frac{C}{N_k} \left(\sum_{m=1}^{N_k} \alpha_m^k \right) \sum_{m=1}^{N_k} \varphi_{m+s_k}(x) \right| \leqslant C \frac{1}{\sqrt{N_k}} \left(\sum_{m=1}^{N_k} |\alpha_m^k| \right) \left| \sum_{m=1}^{N_k} \frac{1}{\sqrt{N_k}} \varphi_{m+s_k}(x) \right|.$$

By (25), and since $\lim_{k\to\infty} \beta_k = 0$, $k \in S_1$ (see (43)), we have

$$\sum_{k\in S_1'} \int_{J_{k\nu}} |\Delta_k(x)|\, dx + \int_{J_{k\nu}} \left| \sum_{m=1}^{N_k} \alpha_m^k \varphi_{m+s_k}(x) \right| dx \leqslant C \sum_{k=1}^{\infty} (\log_2 N_k)^{-1} < \infty.$$

So, for any $y > 0$,

$$\mu\left[\overline{\lim_{k\in S_1'}} \left\{x\colon \left| \sum_{m=1}^{N_k} \alpha_m^k \varphi_{m+s_k}(x) + \Delta_k(x) \right| > \frac{y}{2},\ x \in J_{k,\nu_k} \right\}\right] = 0. \tag{50}$$

From (49) and (50) (see (47), too) divergence of (38) follows in case 1) under the assumption (47). But if case 1) occurs, then for any $y > 0$

$$\mu\left[\overline{\lim_{k\to\infty,\, k\in S_1'}} \{x \in J_{k,\nu_k}\colon\ |H_k(x)| > y\}\right] = 0, \tag{51}$$

so it again follows from the definition of $A_{N_k + 1}$ that

$$\sup_{1\leqslant r\leqslant N_k} \left| \sum_{m=1}^{r} \alpha_m^k \psi_{Q_k + m}(x) \right| = \sup_{1\leqslant r\leqslant N_k} \left| \sum_{m=1}^{r} \alpha_m^k \varphi_{s_k + m}(x) \right| + \Delta_k'(x),$$

where

$$|\Delta_k'(x)|\leqslant|\tilde{H}_k(x)|+C\left(\sum_{m=1}^{N_k}|\alpha_m^k|\right)\frac{1}{N_k^{1/2}}\left|\sum_{n=s_k}^{s_{k+1}-1}\frac{1}{N_k^{1/2}}\varphi_n(x)\right|\equiv|\tilde{H}_k(x)|+F_k(x). \tag{52}$$

Using (25) and the fact that, by (43) and (48),

$$\frac{1}{N_k^{1/2}}\sum_{m=1}^{N_k}|\alpha_m^k|\leqslant\beta_k\leqslant 1 \quad \text{for} \quad k\geqslant k_0,$$

we obtain

$$\sum_{k\in S_1'}\int_{J_{k,\nu_k}}|F_k(x)|\,dx\leqslant C\sum_{k=1}^{\infty}(\log_2 N_k)^{-1}<\infty. \tag{53}$$

Further, by (45), $|\tilde{H}_k(x)|\leqslant C|H_k(x)|$, and it follows (see (52)) from (53) and (47) that for any $y>0$

$$\mu\,[\overline{\lim_{k\to\infty,\ k\in S_1'}}\{x:|\Delta_k'(x)|>y\}]=0. \tag{54}$$

Further, for $k\in S_1$ it follows from (34) (see (33), too) that

$$\mu\left\{x\in J_{k,\nu_k}:\sup_{1\leqslant r\leqslant N_k}\left|\sum_{m=1}^{r}\alpha_m^k\varphi_{s_k+m}(x)\right|\geqslant C\cdot 2^{\nu/2}\beta_k\right\}\geqslant C\mu J_{k,\nu_k}.$$

So, considering (46), (23) and (48), we obtain

$$\mu G_k\equiv\mu\left\{x\in J_{k,\nu_k}:\sup_{1\leqslant r\leqslant N_k}\left|\sum_{m=1}^{r}\alpha_m^k\varphi_{s_k+m}(x)\right|>C\right\}\geqslant C\beta_k^2. \tag{55}$$

But since the sets G_k are independent (see (35)), and (by (55) and (44))

$$\sum_{k\in S_1'}\mu G_k\geqslant C\sum_{k\in S_1'}\beta_k^2=\infty,$$

we obtain from Lemma 2 and (55) that for some $y>0$

$$\mu\left[\overline{\lim_{k\to\infty,\ k\in S_1'}}\left\{x:\sup_{1\leqslant r\leqslant N_k}\left|\sum_{m=1}^{r}\alpha_m^k\varphi_{s_k+m}(x)\right|\geqslant y\right\}\right]>0. \tag{56}$$

From (56), (54), (51) and (49) we have

$$\mu\left[\overline{\lim_{k\to\infty,\ k\in S_1'}}\left\{x:\sup_{Q_k<r<Q_{k+1}}\left|\sum_{n=Q_k+1}^{r}c_n\psi_n(x)\right|>\frac{y}{2}\right\}\right]>0.$$

Thus divergence of (38) is proved completely for case 1).

Suppose case 2) occurs; that is, $\sum_{k\in S_1}\beta_k^2<\infty$. In this case (see (38))

$$\sum_{k\in S_2}\beta_k^2=\infty$$

and we claim that (38) does not converge in measure on [1, 2]. We note that for case 2) the argument is the same as the corresponding part of our paper [5].

We take a number $k \in S_2$. By (37) and the fact that A_{N_k+1} is orthonormal we have

$$\sum_{n=Q_k+1}^{Q_{k+1}} c_n\psi_n(x) = \sum_{n=s_k+1}^{s_{k+1}} \gamma_n\varphi_n(x) + \omega_k v_k(x),$$
$$\sum_{n=s_k+1}^{s_{k+1}} \gamma_n^2 + \omega_k^2 = \sum_{n=Q_k+1}^{Q_{k+1}} c_n^2. \tag{57}$$

From the definition of A_{N_k+1} it again follows that

$$|\omega_k| \leqslant \frac{2}{(N_k+1)^{1/2}} \left| \sum_{n=Q_k+1}^{Q_{k+1}} c_n \right| + \frac{2}{(N_k+1)^{1/2}} \max_{Q_k<n\leqslant Q_{k+1}} |c_n| \leqslant 2d_k + \frac{2}{(N_k+1)^{1/2}} \beta_k.$$

From these estimates and the fact that

$$2d_k \leqslant \frac{C_1\sqrt{C_2}}{10}\beta_k \leqslant \frac{1}{10}\beta_k,$$

for $k \in S_2$, we have, for sufficiently large $k \in S_2$,

$$\sum_{n=s_k+1}^{s_{k+1}} \gamma_n^2 \geqslant {}^1/_2\beta_k^2;$$
$$\left(\sum_{n=s_k+1}^{s_{k+1}} \gamma_n^2\right) \frac{C_1^2C_2}{20} \geqslant \frac{C_1^2C_2}{40}\beta_k^2 \geqslant \left(2d_k + \frac{2\beta_k}{(N_k+1)^{1/2}}\right)^2 \geqslant \omega_k^2. \tag{58}$$

We take a sequence of integers $\{k_j\}_1^\infty$ such that (58) holds for $k \geqslant k_1$ $(k \in S_2)$ and

$$\sum_{k=k_j}^{k_{j+1}-1} \beta_k^2 = \sum_{n=Q_{k_j}+1}^{Q_{k_{j+1}}} c_n^2 \geqslant j. \tag{59}$$

We consider the sum

$$\sum_{n=Q_{k_j}+1}^{Q_{k_{j+1}}} c_n\psi_n(x) = I'_j(x) + I''_j(x), \tag{60}$$

where

$$I'_j(x) = \sum_{k\in S_1\cap[k_j,\,k_{j+1})} \sum_{n=Q_k+1}^{Q_{k+1}} c_n\psi_n(x);$$
$$I''_j(x) = \sum_{k\in S_2\cap[k_j,\,k_{j+1})} \sum_{n=Q_k+1}^{Q_{k+1}} c_n\psi_n(x).$$

Since $\Sigma_{k\in S_1}\beta_k^2 < \infty$, we have

$$\sum_{j=1}^{\infty} \int_0^2 (I'_j(x))^2\,dx \leqslant \sum_{k\in S_1} \beta_k^2 \leqslant C < \infty. \tag{61}$$

Further (see (57) and (60)),

$$I''_j(x) = \sum_{k\in S_2\cap[k_j,\,k_{j+1})} \left(\sum_{n=s_k+1}^{s_{k+1}} \gamma_n\varphi_n(x) + \omega_k v_k(x)\right).$$

In addition, by (58),

$$D_j \equiv \sum_{k\in S_2\cap[k_j,\,k_{j+1})} \left(\sum_{n=s_k+1}^{s_{k+1}} \gamma_n^2\right) \geqslant \frac{j}{2}$$

and

$$D_j \geqslant \left(\frac{C_1^2 C_2}{20}\right)^{-1} \left(\sum_{k \in S_2 \cap [k_j,\, k_{j+1})} \omega_k^2\right).$$

By (36),

$$\varphi_n(x) = \frac{1}{\sqrt{2}} r_n(x)$$

for $x \in [1, 2]$ and $n = 1, 2, \ldots$, whence the sum $I_j''(x)$ can be represented in the form

$$I_j''(x) = P_j(x) + \sum_{k \in S_2 \cap [k_j,\, k_{j+1})} \omega_k v_k(x), \tag{62}$$

where $P_j(x)$ is a polynomial in the Rademacher system with

$$\int_1^2 P_j^2(x)\, dx = {}^1/_2 D_j \geqslant {}^1/_4 j,$$

and

$$\sum_{k \in S_2 \cap [k_j,\, k_{j+1})} \omega_k^2 \leqslant D_j \frac{C_1^2 C_2}{20}.$$

Since the functions $v_k(x)$ are orthogonal and $\int_0^2 v_k^2(x)\, dx = 1$, we have

$$\int_0^2 \left(\sum_{k \in S_2 \cap [k_j,\, k_{j+1})} \omega_k v_k(x)\right)^2 dx = \sum_{k \in S_2 \cap [k_j,\, k_{j+1})} \omega_k^2 \leqslant D_j \frac{C_1^2 C_2}{20}.$$

Thus, by Čebyšev's inequality,

$$\mu\left\{x \in [1,\ 2] : \left|\sum_{k \in S_2 \cap [k_j,\, k_{j+1})} \omega_k v_k(x)\right| \geqslant D_j \frac{C_1}{3}\right\} \leqslant \frac{C_2}{2}. \tag{63}$$

By Lemma 3 we have (see (62))

$$\mu\left\{x \in [1,\ 2] : |P_j(x)| \geqslant \frac{\sqrt{D_j}}{2} C_1\right\} \geqslant C_2. \tag{64}$$

From (63) and (64) it follows (see (62)) that

$$\mu\left\{x \in [1,2] : |I_j''(x)| \geqslant \frac{\sqrt{D_j}\, C_1}{6} \geqslant \frac{\sqrt{j}\, C_1}{12}\right\} \geqslant \frac{C_2}{2}.$$

It follows from this last inequality (on considering (61) and (60)) that the series (38) does not converge in measure on $[1, 2]$. Theorem 1 is fully proved.

We need the following lemma in proving Theorem 2.

LEMMA 6. *Let $\{f_k(x)\}_1^N$ be an orthonormal set of functions on* $[0, 1]$. *Define a set of functions $\{\psi_k(x)\}_1^{N^2}$ for $k = Ni + j$ $(0 \leqslant i < N;\ 1 \leqslant j \leqslant N)$ by putting*

$$\psi_k(x) = \frac{1}{\sqrt{N}} f_j(x).$$

Then for any numbers $a_1, \ldots, a_{N^2}$

$$F(\{a_k\}) \equiv \int_0^1 \sup_{1 \leqslant n \leqslant N^2} \left(\sum_{k=1}^n a_k \psi_k(x)\right)^2 dx \leqslant 4\left(\sum_{k=1}^{N^2} a_k^2\right).$$

PROOF. We have

$$\sup_{1\leqslant n\leqslant N^2}\left(\sum_{k=1}^{n} a_k\psi_k(x)\right)^2 \leqslant 2\sup_{0<i<N}\left(\sum_{k=1}^{Ni} a_k\psi_k(x)\right)^2 + 2\sup_{i,\nu:\,Ni<\nu\leqslant N(i+1)}\left(\sum_{k=Ni}^{\nu} a_k\psi_k(x)\right)^2.$$

Thus

$$F(\{a_k\})\leqslant 2\int_0^1 \sup_{0<i<N}\left(\sum_{k=1}^{Ni} a_k\psi_k(x)\right)^2 dx + 2\int_0^1 \sup_{i,\nu:\,Ni<\nu\leqslant N(i+1)}\left(\sum_{k=Ni}^{\nu} a_k\psi_k(x)\right)^2 dx$$

$$\leqslant 2\left(\sum_{i=0}^{N-1}\left\|\sum_{k=Ni+1}^{N(i+1)} a_k\psi_k(x)\right\|_{L^2}\right)^2 + 2\sum_{i=1}^{N-1}\int_0^1 \sup_{\nu:\,Ni<\nu\leqslant N(i+1)}\left(\sum_{k=Ni+1}^{\nu} a_k\psi_k(x)\right)^2 dx.$$

For any i, $0\leqslant i\leqslant N-1$, we obtain

$$\left\|\sum_{k=Ni+1}^{N(i+1)} a_k\psi_k(x)\right\|_{L^2} = \left(\frac{1}{N}\sum_{k=Ni+1}^{N(i+1)} a_k^2\right)^{1/2}$$

from the orthonormality of the $f_j(x)$, and, further,

$$\int_0^1 \sup_{\nu:\,Ni<\nu\leqslant N(i+1)}\left(\sum_{k=Ni+1}^{\nu} a_k\psi_k(x)\right)^2 dx \leqslant \left(\sum_{k=Ni+1}^{N(i+1)} |a_k|\cdot\|\psi_k(x)\|_{L^2}\right)^2 \leqslant \left(\sum_{k=Ni+1}^{N(i+1)} \frac{|a_k|}{\sqrt{N}}\right)^2.$$

Thus

$$F(\{a_k\})\leqslant 2\left(\sum_{i=0}^{N-1} N^{-1/2}\left(\sum_{k=Ni+1}^{N(i+1)} a_k^2\right)^{1/2}\right)^2 + 2\sum_{i=0}^{N-1}\left(\sum_{k=Ni+1}^{N(i+1)} N^{-1/2}|a_k|\right)^2.$$

Using Cauchy's inequality, we obtain

$$F(\{a_k\})\leqslant 2\sum_{k=1}^{N^2} a_k^2 + 2\sum_{k=1}^{N^2} a_k^2 = 4\left(\sum_{k=1}^{N^2} a_k^2\right)$$

from the last estimate. This proves Lemma 6.

Now we prove Theorem 2. The result is obvious when the manifold L is finite-dimensional, so we can assume that L has an infinite orthonormal basis $\{f_k(x)\}$, which we can replace by a basis $\{\varphi_k(x)\}$ which is a system of convergence. It is known (see [12]) that we can extract a subsequence $\{u_k(x)\} = \{f_{n_k}(x)\}$, $k = 1, 2, \ldots$, from $\{f_k(x)\}$ so that for any choice of numbers $\{a_k\}\in l_2$ and any permutation $\{p_k\}_1^\infty$ of the integers we have

$$\sup_{1\leqslant r<\infty}\left|\sum_{k=1}^{r} a_k u_{p_k}(x)\right| < \infty \text{ a.e.} \tag{65}$$

and, in addition, this choice can be made(4) so that

(4) This follows, for example, from the theorem of Marcinkiewicz cited above.

$$\int_0^1 \sup_{1\leqslant r<\infty} \left(\sum_{k=1}^{r} a_k u_k(x)\right)^2 dx \leqslant C \sum_{k=1}^{\infty} a_k^2. \tag{66}$$

We enumerate arbitrarily the functions remaining after extracting the system $\{u_k(x)\}$ from $\{f_k(x)\}$, and denote the set of functions so obtained by $\{f'_k(x)\}_1^\infty$.

By a well-known theorem of Marcinkiewicz (see [10]) there is a sequence $\{m_s\}_1^\infty$, $m_1 = 0$, of integers such that for any sequence $\{a_k\} \in l_2$

$$\int_0^1 \left(\sup_{1<s<\infty} \left|\sum_{k=1}^{m_s} a_k f'_k(x)\right|\right)^2 dx \leqslant C \sum_{k=1}^{\infty} a_k^2. \tag{67}$$

Now we construct the required system $\{\varphi_k(x)\}$. Let $\{\nu_s\}_1^\infty$ be the sequence of integers such that

$$\nu_1 = 0, \qquad (\nu_{s+1} - \nu_s) = (m_{s+1} - m_s)^2 + (m_{s+1} - m_s),$$

and $\{\mu_s\}_1^\infty$ the sequence of integers such that

$$\mu_1 = 0, \quad \mu_{s+1} - \mu_s = (m_{s+1} - m_s)^2.$$

For each $s = 1, 2, \ldots$ we define the functions $\varphi_k(x)$ with $\nu_s < k \leqslant \nu_{s+1}$ thus: we consider the matrix $A_{m_{s+1}-m_s+1}$ (see (27)), and for $\nu_s < k \leqslant \nu_{s+1}$ and $k = \nu_s + j(m_{s+1} - m_s) + i$, $0 \leqslant j \leqslant m_{s+1} - m_s$, $1 \leqslant i \leqslant m_{s+1} - m_s$, we put

$$\varphi_k(x) = \sum_{r=1}^{m_{s+1}-m_s} a_{jr} u_{\mu_s+i(m_{s+1}-m_s)+r}(x) + a_{j,\, m_{s+1}-m_s+1} f'_{m_s+i}(x). \tag{68}$$

We remark that it follows from (68) that for $\nu_s < k \leqslant \nu_{s+1}$ the functions $\varphi_k(x)$ are linear combinations of the functions $u_k(x)$ with $\mu_s < k \leqslant \mu_{s+1}$ and $f'_k(x)$ with $m_s < k \leqslant m_{s+1}$.

It follows from this remark and orthonormality of A_N (see (27)) that the system $\{\varphi_k(x)\}$ forms an orthonormal basis in L. In addition it follows directly from this remark, (65) and (67) that for any series

$$\sum_{k=1}^{\infty} a_k \varphi_k(x), \qquad \sum_{k=1}^{\infty} a_k^2 < \infty,$$

the sequence of partial sums $S_{\nu_s}(x)$ converges almost everywhere.

Thus to prove the theorem it is enough to show that

$$\lim_{s\to\infty} G_s(x, \{a_k\}) \equiv \lim_{s\to\infty} \sup_{\nu_s<n\leqslant\nu_{s+1}} \left|\sum_{k=\nu_s+1}^{n} a_k \varphi_k(x)\right| = 0 \text{ a.e.} \tag{69}$$

By (68) and the properties of A_N (see (27)) the function $\varphi_k(x)$ for $k = \nu_s + j(m_{s+1} - m_s) + i$ can be represented in the form

a) for $0 \leqslant j < m_{s+1} - m_s$

$$\varphi_k(x) = (m_{s+1} - m_s)^{-1/2} f'_{m_s+i}(x) + u_{\mu_s+i(m_{s+1}-m_s)+j}(x)$$
$$- (m_{s+1} - m_s)^{-1/2} \sum_{r=1}^{m_{s+1}-m_s} (m_{s+1} - m_s)^{-1/2} u_{\mu_s+i(m_{s+1}-m_s)+r}(x);$$

b) for $j = m_{s+1} - m_s$

$$\varphi_k(x) = (m_{s+1} - m_s)^{-1/2} \sum_{r=1}^{m_{s+1}-m_s} u_{\mu_s+i(m_{s+1}-m_s)+r}(x).$$

It follows from (65) and b) that

$$\int_0^1 \sup_{\nu_s+(m_{s+1}-m_s)^2<n\leqslant\nu_{s+1}} \left(\sum_{k=\nu_s+(m_{s+1}-m_s)^2+1}^{n} a_k\varphi_k(x)\right)^2 dx \leqslant C \sum_{k=\nu_s+1}^{\nu_{s+1}} a_k^2. \tag{70}$$

Further, because the functions

$$g_i(x) \equiv \sum_{r=1}^{m_{s+1}-m_s} (m_{s+1}-m_s)^{-1/2} u_{\mu_s+i(m_{s+1}-m_s)+r}(x)$$

$$(1\leqslant i< m_{s+1}-m_s)$$

form an orthonormal system we can use Lemma 6 twice to obtain

$$G_s(x, \{a_k\}) \leqslant \tilde{G}_s(x, \{a_k\}) + 2 \sup_{\nu_s<n\leqslant\nu_s+(m_{s+1}-m_s)^2} \left|\sum_{k=\nu_s}^{n} a_k u_{p_k}(x)\right|, \tag{71}$$

where $\widetilde{G}_s(x, \{a_k\})$ is some function such that

$$\int_0^1 \tilde{G}_s(x, \{a_k\})\, dx \leqslant C \sum_{k=\nu_s+1}^{\nu_{s+1}} a_k^2, \tag{72}$$

and the numbers p_k ($\nu_s < k \leqslant \nu_s + (m_{s+1} - m_s^2)$) form a permutation of the set of integers r such that $\mu_s < r \leqslant \mu_{s+1}$.

The required inequality (69) follows from (71), (72) and (65). This proves Theorem 2.

REMARK. If the manifold L contains an orthonormal system $\{f_{n_k}(x)\}_{k=1}^{\infty}$ such that

$$\mu\{x\in[0, 1]: |f_{n_k}(x)| > \varepsilon_k\} < \varepsilon_k \text{ with } \lim_{k\to\infty}\varepsilon_k = 0,$$

then Theorem 2 follows easily from Lemma 6 and the above-cited theorem of Marcinkiewicz.

In conclusion we state two results without proof; the second may be considered the 'finite-dimensional' analog of Theorem 1.

I. Let O^n be the group of orthogonal matrices of order n and μ_n normalized left Haar measure on it.

We define a function $f(G)$ on this group by putting, for $G = \{g_{ij}\}_{i,j=1}^n$,

$$f(G) = \sup_{\{a_i\}:\ \sum_{i=1}^n a_i^2=1,\ \{n_j\}:\ 1\leqslant n_j\leqslant n} \left(\sum_{j=1}^{n}\left(\sum_{i=1}^{n_j} a_i g_{ij}\right)^2\right)^{1/2}.$$

Then there exist absolute constants $C > 0$ *and* $\gamma > 0$ *such that*

$$\mu_n\{G\in O^n: f(G) > y\} < (C\exp(-\gamma y^2))^n \quad y\geqslant 0.$$

II. It follows easily from I and Theorem 3 of our paper [17] that *for any* $n > 1$ *there is a matrix* $G = \{g_{ij}\} \in O^n$ *such that*

a) $f(G) \leqslant C$, *and*

b) *for any set of numbers* $\{a_i\}_1^n$ *with* $\Sigma_1^n a_i^2 = 1$ *there is a number* m $(1 \leqslant m \leqslant n)$ *such that*

$$\sum_{j:\ \left|\sum_{i=1}^m a_i g_{ij}\right| > cn^{-1/2}} 1 \geqslant cn.$$

BIBLIOGRAPHY

1. A. Zygmund, *Trigonometric series*, 2nd rev. ed., Vol. I, Cambridge Univ. Press, New York, 1959.

2. Stefan Kaczmarz and Hugo Steinhaus, *Theorie der Orthogonalreihen*, PWN, Warsaw, 1935; reprint, Chelsea, New York, 1951.

3. P. L. Ul'janov, *Solved and unsolved problems in the theory of trigonometric and orthogonal series*, Uspehi Mat. Nauk **19** (1964), no. 1 (115), 3–69; English transl. in Russian Math. Surveys **19** (1964).

4. A. Grothendieck, *Résumé de la théorie métrique des produits tensoriels topologiques*, Bol. Soc. Mat. São Paulo **8** (1953), 1–79 (1956).

5. B. S. Kašin, *On a complete orthonormal system*, Mat. Sb. **99** (**141**) (1976), 356–365; English transl. in Math. USSR Sb. **28** (1976).

6. A. M. Olevskiĭ, *On an orthonormal system and its applications*, Mat. Sb. **71** (**113**) (1966), 297–336; English transl. in Amer. Math. Soc. Transl. (2) **76** (1968).

7. E. M. Nikišin, *Resonance theorems and series of functions*, Doctoral Dissertation, Moscow State Univ., Moscow, 1971. (Russian) (Abstract published in [8])

8. ———, *Resonance theorems and series of functions*, Mat. Zametki **10** (1971), 583–595; English transl. in Math. Notes **10** (1971). (Abstract of [7])

9. A. M. Olevskiĭ, *Fourier series of continuous functions relative to bounded orthonormal systems*, Izv. Akad. Nauk SSSR Ser. Mat. **30** (1966), 387–432; English transl. in Amer. Math. Soc. Transl. (2) **67** (1968).

10. Józef Marcinkiewicz, *Sur la convergence des séries orthogonales*, Studia Math. **6** (1936), 39–45; reprinted in his *Collected papers*, PWN, Warsaw, 1964.

11. B. S. Kašin, *On orthogonal systems of convergence*, Dokl. Akad. Nauk SSSR **228** (1976), 285–286; English transl. in Soviet Math. Dokl. **17** (1976).

12. J. Komlós, *Every sequence converging to 0 weakly in L^2 contains an unconditional convergence sequence*, Ark. Mat. **12** (1974), 41–49.

13. E. M. Nikišin, *On systems of convergence in measure for l_2*, Mat. Zametki **13** (1973), 337–340; English transl. in Math. Notes **13** (1973).

14. Bernard Maurey, *Sur une application de la théorie des opérateurs p-sommants*, C. R. Acad. Sci. Paris Sér. A-B **274** (1972), A1304–A1307.

15. John Lamperti, *Probability*, Benjamin, New York, 1966.

16. Pierre Billard, *Sur la convergence presque partout des séries de Fourier-Walsh des fonctions de l'espace $L^2(0, 1)$*, Studia Math. **28** (1966/67), 363–388.

17. B. S. Kašin, *The orders of the widths of some classes of smooth functions*, Uspehi Mat. Nauk **32** (1977), no. 1 (193), 191–192. (Russian)

Translated by P. G. SPAIN

Труды Мат. Инст. Стеклов.
143 (1977)

Proceedings of the Steklov
Institute of Mathematics
1980, Issue 1

ON BASES AND MULTIPLIERS IN THE SPACES $B^r_{p,\theta}(\Pi)$

UDC 517.5

P. I. LIZORKIN

ABSTRACT. In the paper we construct an unconditional basis for the Besov space $B^r_{p,\theta}(\Pi)$ (periodic case). This basis has several additional properties. In particular, the norm in it admits a convenient equivalent description through the absolute values of the expansion coefficients relative to the basis. In the paper this circumstance is used for the characterization of Fourier multipliers in Besov spaces. We also obtain simpler sufficient conditions for the multipliers.

Bibliography: 8 titles.

The system of exponential functions $\{\exp(ikx)\}^{\infty}_{-\infty}$, with the natural ordering, forms a (conditional) Schauder basis in the Besov space $B^r_{p,\theta}(\Pi)$ (see the definition in §1). In the paper we shall construct an unconditional basis $\{E_k(x)\}^{\infty}_{-\infty}$ of $B^r_{p,\theta}(\Pi)$ by extending a basic block-system of exponential functions.

The basis $\{E_k(x)\}$ has several additional properties. In particular, the norm in $B^r_{p,\theta}(\Pi)$ admits a convenient equivalent description through the absolute values of the expansion coefficients relative to this basis (see §4, Theorem 1). In the paper the latter circumstance is used for the characterization of the Fourier multipliers from $B^r_{p,\theta}(\Pi)$ to $B^r_{p',\theta}(\Pi)$ (Theorem 2). We also obtain simpler and more convenient sufficient conditions for the multipliers.

It should be noted that in the nonperiodic multidimensional case similar results have been obtained by Triebel [1]. Our results can be carried over without much difficulty to the multidimensional periodic case. We restrict ourselves to the one-dimensional case in order to have simple formulas. The technique used in the periodic case is significantly different from that of [1].

§1. The spaces $B^r_{p,\theta}$

The theory of the spaces $B^r_{p,\theta}$ is set forth in full generality in the book of S. M. Nikol'skiĭ [2]. Here we reproduce the definitions for the periodic case in the form to be used later on (see also [3] and [4]). Let the complex-valued periodic function $f(x)$ be represented by the Fourier series

$$f(x)=\sum_{k=-\infty}^{\infty} C_k e^{ikx}. \tag{1}$$

Set

$$f(x)=f^-(x)+C_0+f^+(x), \tag{2}$$

1980 *Mathematics Subject Classifications.* Primary 46E35, 46B15, 42A45.

where

$$f^{\pm}(x)=\sum_{k=1}^{\infty} C_{\pm k}e^{\pm ikx}.$$

Consider the decomposition of $f^+(x)$ into packets $\delta_m^+(f, x)$ (or $f_m^+(x)$ for short):

$$f^+(x)=\sum_{m=1}^{\infty}\delta_m^+(f, x)=\sum_{m=1}^{\infty} f_m^+(x), \tag{3_1}$$

where

$$f_m^+(x)=\sum_{k=2^{m-1}}^{2^m-1} C_k e^{ikx}, \quad m\geqslant 1.$$

Similarly,

$$f^-(x)=\sum_{k=1}^{\infty}\delta_m^-(f, x)=\sum_{k=1}^{\infty} f_m^-(x), \tag{3_2}$$

where

$$f_m^-(x)=\sum_{k=-2^m+1}^{-2^{m-1}} C_k e^{ikx}=\sum_{k=2^{m-1}}^{2^m-1} C_{-k}e^{-ikx}, \quad m\geqslant 1.$$

Functions $f(x)$ of period 2π may also be regarded as functions defined on the unit circle Π, and in the sequel we shall often do so. We shall also consider 2π-periodic generalized functions. In this case the transition to considering functions defined on Π has some advantages: since Π is compact, the space $D = D(\Pi)$ of test functions simply consists of all functions that are infinitely differentiable on Π (with the appropriate topology). We denote by $D'(\Pi) = D'$ the space of all 2π-periodic generalized functions (generalized functions on Π). The family of all 2π-periodic functions that are pth power integrable over a period will be denoted, as usual, by $L_p(\Pi) = L_p$; for the norm in L_p we write

$$\|f\|_p=\left\{\int_0^{2\pi}|f(x)|^p\,dx\right\}^{1/p}, \quad 1\leqslant p<\infty.$$

It is well known that for functions $f\in D'$ the series (1) is weakly convergent to f; that is,(1)

$$\lim_{A,\, B\to+\infty}\left\langle\sum_{-A}^{B} C_k e^{ikx},\quad \varphi(x)\right\rangle=\langle f, \varphi\rangle\,(\forall\varphi\in D).,$$

and if $f\in L_p$ then the series (1) is convergent to f in the norm of L_p, $1<p<\infty$.

DEFINITION 1. The space $B_{p,\theta}^r(\Pi) = B_{p,\theta}^r$, $-\infty<r<\infty$, $1<p<\infty$, $1\leqslant\theta\leqslant\infty$, consists of all 2π-periodic generalized functions $f(x)$ that may be represented by a weakly convergent series (1) and have finite norm

$$|f, B_{p,\theta}^r|=|C_0|+\left\{\sum_{m=1}^{\infty}2^{mr}\|f_m^+\|_p^\theta\right\}^{1/\theta}+\left\{\sum_{m=1}^{\infty}2^{mr}\|f_m^-\|_p^\theta\right\}^{1/\theta}, \quad 1\leqslant\theta<\infty; \tag{4_1}$$

(1) $\langle f, \varphi\rangle$ stands for the value of the functional $f\in D'$ on the element $\varphi\in D$.

$$|f,\ B^r_{p,\infty}| = |C_0| + \sup_{m\geqslant 1} 2^{mr}\|f^+_m\|_p + \sup_{m\geqslant 1} 2^{mr}\|f^-_m\|. \qquad (4_2)$$

REMARKS. 1.1. The above definition makes sense also for $p = 1$ and $p = \infty$, but for these values of p we obtain spaces different from the Besov spaces $B^r_{1,\theta}$ and $B^r_{\infty,\theta}$. Later on, we shall have to do with the space obtained from Definition 1 by setting $p = 1$; we denote it by $B^{*r}_{1,\theta}$. It may be shown that the following continuous embedding is valid:

$$B^{*r}_{1,\theta} \subset B^r_{1,\theta}.$$

In order to unify notation, sometimes we shall write $B^{*r}_{p,\theta}$ instead of $B^r_{p,\theta}$ also for $p \neq 1$.

1.2. For $r > 0$ the elements of $B^*_{p,\theta}$ are ordinary functions belonging to L_p. In this case the convergence of series (1) in Definition 1 may be taken in the L_p norm.

1.3. We shall write the norm in $B^r_{p,\theta}$ in a more condensed form. To this end, we set

$$f_m(x) = f^+_m(x) + f^-_m(x) = \sum_{2^{m-1}\leqslant |k| \leqslant 2^m-1} C_k e^{ikx}.$$

Then, writing $f_0(x) = C_0$, we have

$$|f,\ B^r_{p,\theta}| \sim \left\{|C_0|^\theta + \sum_{m=1}^{\infty} 2^{mr}\|f^+_m + f^-_m\|^\theta_p\right\}^{1/\theta} = \left\{\sum_{m=0}^{\infty} 2^{mr}\|f_m\|^\theta_p\right\}^{1/\theta}. \qquad (4_1')$$

Similar expressions for the norm are to be found in the book [2] (see also [3] and [4]). For real-valued $f(x)$ the functions $f_m(x)$ are real-valued and may be expressed by trigonometrical functions.

Let the set I_m of integers k be defined as follows:

$$I_m = \begin{cases} \{k;\quad 2^{m-1} \leqslant k \leqslant 2^m - 1\}, & m > 0, \\ \{k;\quad k = 0\}, & m = 0, \\ \{k;\quad -2^m + 1 \leqslant k \leqslant -2^{|m|-1}\} & m < 0. \end{cases}$$

If f is represented by the series (1), set

$$f^*_m(x) = \sum_{k\in I_m} C_k e^{ikx}.$$

For positive m the function f^*_m coincides with f^+_m, and for negative m it coincides with f^-_m. Clearly, using this notation, the norm in $B^r_{p,\theta}$ is given by

$$\left\{\sum_{m=-\infty}^{\infty} 2^{|m|r\theta}\|f^*_m\|^\theta_p\right\}^{1/\theta}. \qquad (4_1'')$$

Similar norms equivalent to (4_2) may be introduced in $B^r_{p,\infty}$.

1.4. Let Ξ_m be the linear hull of the functions e^{ikx}, $k \in I_m$. We shall consider Ξ_m as a subspace of $B^r_{p,\theta}$ (with the induced norm). Then from the definition of $B^r_{p,\theta}$ it immediately follows that

$$B^r_{p,\theta} = \bigcup_{m=-\infty}^{\infty} \oplus\, \Xi_m,$$

a direct sum of finite-dimensional subspaces. Moreover, the subspaces

$$\Xi_m = \Xi_m \cup \Xi_{-m}, \qquad m = 0, 1, \ldots,$$

form a Schauder basis of subspaces for $B^r_{p,\theta}$. This means that each $f \in B^r_{p,\theta}$ is uniquely representable by the series

$$f(x) = \sum_{k=0}^{\infty} (P_k f)(x),$$

convergent in the norm of $B^r_{p,\theta}$, where $P_k f$ is the projection of f to Ξ_k: $P_k f = f_k$. In many problems a representation of this kind plays the same role as the expansion relative to an ordinary Schauder basis. Essentially, the construction of a Schauder basis for $B^r_{p,\theta}$ carried out below relies on the fact that the Schauder basis of subspaces $\{\Xi_m\}_0^\infty$ is unconditional. Note that in L_p this is no more the case: the system $\{\Xi_m\}$ forms a conditional basis of subspaces in it.

§2. A basis of $B^r_{p,\theta}$. Construction of a block basis

We first note that the system $\{e^{ikx}\}_{-\infty}^{\infty}$ indexed in the natural order

$$\varphi_0(x) = 1, \ \varphi_1(x) = e^{ix}, \qquad \varphi_2 = e^{-ix}, \ldots, \varphi_{2k-1} = e^{ikx}, \qquad \varphi_{2k} = e^{-ikx}, \ldots, \tag{5}$$

constitutes a basis of $B^r_{p,\theta}$ if $\theta \neq \infty$.

For, let $f \in B^r_{p,\theta}$. Then the expansion

$$f = \sum_{-\infty}^{\infty} C_k e^{ikx}$$

is convergent in the weak sense and the norm (4_1) is finite. Let us prove that the series

$$\sum_{\sigma=0}^{\infty} d_\sigma \varphi_\sigma, \qquad d_{2k-1} = C_k, \qquad d_{2k} = C_{-k}, \tag{6}$$

strongly converges to f. It is easy to see that this series satisfies the Cauchy convergence principle in the $B^r_{p,\theta}$ norm. In fact, the expression

$$\sum_{\sigma=n}^{n+m} d_\sigma \varphi_\sigma \tag{7}$$

can be represented as a sum of packets, the extreme ones being possibly incomplete. According to the theorem of Riesz the incomplete packets can be estimated in the L_p norm by the corresponding complete packets. Thus the $B^r_{p,\theta}$ norm of the sum (7) can be estimated by the segments of the series occurring in (4_1), and, consequently, the norm

$$\left\| \sum_{\sigma=n}^{n+m} d_\sigma \varphi_\sigma \right\|_{B^r_{p,\theta}}$$

is arbitrarily small for sufficiently large n. Therefore the series (6) is convergent in the $B^r_{p,\theta}$ norm. By the uniqueness of the limit of a weakly convergent sequence(2), the

(2) For $r > 0$ it is sufficient to note that (6) converges to f in the L_p metric.

sum of its series is equal to f:

$$f(x)=\sum_{\sigma=0}^{\infty} d_\sigma \varphi_\sigma.$$

Obviously, the coefficients of this expansion are unique. Thus we have proved

PROPOSITION 1. *The system* $\{e^{ikx}\}_{-\infty}^{\infty}$ *with the natural ordering is a basis of the space* $B^r_{p,\theta}$, $-\infty<r<\infty$, $1<p<\infty$, $1\leqslant\theta<\infty$.

We shall now use the following result:

PROPOSITION 2. *The one-sided exponential polynomial*

$$U_n(x)\equiv\sum_{k=0}^{n} C_k e^{ikx}$$

of odd order n may be represented by the interpolation formula

$$U_n(x)\equiv\sum_{k=0}^{n} C_k e^{ikx}=\frac{2}{n+1}\sum_{k=0}^{n} U_n(t_n^{(s)})\,Д_n^{(s)}(x), \tag{8}$$

where $Д_n^{(s)}(x)=Д_n^{+}(x-t_n^{(s)})$ *is the "shifted" one-sided Dirichlet kernel,*

$$Д_n^{+}(x)=\frac{1}{2}\sum_{k=0}^{n} e^{ikx}=\frac{1}{2}\frac{1-e^{i(n+1)x}}{1-e^{ix}}=e^{inx/2}\frac{\sin[(n+1)x/2]}{2\sin(x/2)}, \tag{9}$$

and the interpolation points $t_n^{(s)}$ *are defined by*

$$t_n^{(s)}=\frac{2s+1}{n+1}\pi,\qquad s=0,1,\dots,n. \tag{10}$$

PROOF. Indeed, the functions $[2/(n+1)]\,Д_n^{(s)}(x)$ are the fundamental interpolation polynomials corresponding to the points $t_n^{(s)}$ (see [8]):

$$\frac{2}{n+1}D_n^{(k)}(t_n^{(l)})=\delta_{kl}.$$

They form an (orthogonal) basis in the space of all polynomials of degree not exceeding n. These considerations yield relation (8) immediately.

The packet $f_m^+(x)$ multiplied by $e^{-i2^{m-1}x}$ becomes a one-sided exponential polynomial of degree $2^{m-1}-1$. Making use of Proposition 2, we obtain

$$f_m^+(x)=\sum_{s=0}^{2^{m-1}-1} f_m^+(t_{2^{m-1}-1}^{(s)})\,E_{2^{m-1}-1}^{(s)+}(x), \tag{11}$$

where

$$E_{2^{m-1}-1}^{(s)+}(x)=E_{2^{m-1}-1}^{+}(x-t_{2^{m-1}-1}^{(s)}),\qquad s=0,1,\dots,2^{m-1}-1, \tag{12}$$

$$E_{2^{m-1}-1}^{+}(x)=\frac{1}{2^{m-2}}e^{i2^{m-1}}Д_{2^{m-1}-1}(x)=\frac{1}{2^{m-1}}\sum_{k=2^{m-1}}^{2^m-1} e^{ikx}. \tag{13}$$

Similar expressions can be obtained for the packet

$$f^{-}(x)=\sum_{s=0}^{2^{m-1}-1} f_m^{-}(t_{2^{m-1}-1}^{(s)})\, E_{2^{m-1}-1}^{(s)-}(x), \tag{14}$$

where

$$E_{2^{m-1}-1}^{(s)}(x)=E_{2^{m-1}-1}^{-}(x-t_{2^{m-1}-1}^{(s)}), \qquad s=0,\ 1,\ \ldots,\ 2^{m-1}-1, \tag{15}$$

$$E_{2^{m-1}-1}^{-}(x)=\frac{1}{2^{m-2}}\, e^{-i(2^m-1)x} Д_{2^{m-1}-1}(x)=\frac{1}{2^{m-1}}\sum_{k=-2^m+1}^{-2^{m-1}} e^{ikx}. \tag{16}$$

From (12), (13), (15) and (16) one sees that the functions E with index + and − are conjugate to each other. The system of E-functions just introduced can be provided with a unified numbering. We represent the positive number $n\geqslant 1$ in the form

$$n=2^{m-1}+s, \qquad s=0,\ 1,\ \ldots,\ 2^{m-1}-1, \qquad m\geqslant 1,$$

and set

$$E_0(x)=1, \quad E_1(x)=e^{ix}, \quad E_n(x)=E_{2^{m-1}-1}^{(s)+}(x). \tag{17}$$

For negative n we set

$$-n=2^{m-1}+s, \qquad s=0,\ 1,\ \ldots,\ 2^{m-1}-1, \qquad m\geqslant 1,$$
$$E_{-1}(x)=e^{-ix}, \qquad E_n(x)=E_{2^{m-1}-1}^{(s)-}(x).$$

So we have defined the "E-system" of functions $E_n(x)$. With its help, (11) can be written in the form

$$f_m^{+}(x)=\sum_{n=2^{m-1}}^{2^m-1}\Gamma_n E_n(x)=\sum_{s=0}^{2^{m-1}-1}\Gamma_{2^{m-1}+s}E_{2^{m-1}+s}, \tag{18}$$

where the coefficients Γ_n are values of $f_m^{+}(x)$ at the points of interpolation:

$$\Gamma_{2^{m-1}+s}=f_m^{+}(t_{2^{m-1}-1}^{(s)})=\sum_{k=2^{m-1}}^{2^m-1} C_k e^{ik\frac{2s+1}{2^{m-1}}\pi}. \tag{19}$$

(14) can be rewritten in a similar way. Later on we shall see that the Γ_n may be represented as the Fourier coefficients of f relative to the system $E_n(x)$ (see §3). In particular, from (18) we obtain

$$e^{in_0x}=\sum_{s=0}^{2^{m-1}-1} e^{in_0\frac{2s+1}{2^{m-1}}}E_{2^{m-1}+s}(x), \qquad 2^{m-1}\leqslant n\leqslant 2^m-1.$$

In view of (2) and (3), the expansion (1) may be formally rewritten as

$$f(x)=\sum_{-\infty}^{\infty}\Gamma_n E_n. \tag{20}$$

For functions $f\in B_{p,\theta}^{r}$ relation (20) is the expansion in the unconditional basis

$\{E_n(x)\}$. We shall prove this in §4 after considering in §3 the simplest properties of the system $\{E_n\}$.

First, however, we point out the essence of the constructions performed so far. The system $\{E_n\}$ arose as a result of extending the block-system that consisted of the blocks (see (12) and (15))

$$E^{(s)\pm}_{2^{m-1}-1}(x)=\frac{1}{2^{m-1}}\sum_{k=2^{m-1}}^{2^m-1} e^{\pm iku}, \qquad u=\left(x-\frac{2s+1}{2^{m-1}}\pi\right),$$

which were, in turn, linear combinations of elements of the initial basis $\{e^{ikx}\}$. It is well known [5] that the set of all such blocks constitutes a basic sequence in our space and can be extended to a basis. We have just carried out such an extension. These considerations enable us to outline the way of constructing other systems with similar properties. Let us be given a lacunary sequence $\{N_m\}$ of integers $(N_m/N_{m-1} \geqslant d > 1)$. Consider the blocks

$$\frac{1}{2^{N_m-N_{m-1}}}\sum_{k=N_{m-1}}^{N_m-1} e^{\pm ikx}.$$

The translations of these blocs by $t^{(s)}_{N_m-N_{m-1}-1}$, $s = 0, 1, \dots, N_m - N_{m-1} - 1$, form in their totality a system $\{\mathscr{E}_n\}$ analogous to $\{E_n\}$.

§3. The orthonormal system $\{e_n\}_{-\infty}^{\infty}$

We present an explicit expression for the function E_n. Since $E_{-n} = E_n$, we may restrict ourselves to the case $n > 0$. By the definitions given above (see (17), (12), (13), and also (10)), for $n = 2^{m-1} + s \geqslant 2$, $s = 0, 1, \dots, 2^{m-1} - 1$, we have (setting $x - (2s + 1)\pi/2^{m-1} = u$)

$$E_n(x)=E^{+}_{2^{m-1}-1}\left(x-t^{(s)}_{2^{m-1}-1}\right)=\frac{1}{2^{m-1}}\sum_{2^{m-1}}^{2^m-1} e^{iku}=e^{i\frac{3\cdot 2^{m-1}-1}{2}u}\,\frac{\sin(2^{m-2}u)}{2^{m-1}\sin(u/2)}. \tag{21}$$

Separating the real and imaginary parts, we obtain

$$C_n(x)=\frac{1}{2^{m-1}}\sum_{2^{m-1}}^{2^m-1}\cos ku=\cos\left(\frac{3\cdot 2^{m-1}-1}{2}u\right)\cdot\frac{\sin(2^{m-2}u)}{2^{m-1}\sin(u/2)},$$

$$S_n(x)=\frac{1}{2^{m-1}}\sum_{k=2^{m-1}}^{2^m-1}\sin kx=\sin\left(\frac{3\cdot 2^{m-1}-1}{2}u\right)\cdot\frac{\sin(2^{m-2}u)}{2^{m-1}\sin(u/2)},$$

$$E_n(x)=C_n(x)+iS_n(x).$$

The functions E_n are orthogonal to each other; that is,

$$(E_n, E_m)=\int_0^{2\pi} E_n(x)\overline{E_m(x)}\,dx=0 \qquad \text{for } n\neq m. \tag{22}$$

To verify this, divide the functions E_n into series:

0th series $E_0(x)=1$;

1st series $E_{-1}(x)=e^{-ix},\quad E_1(x)=e^{ix}$;

2nd series $E_{-3}(x),\ E_{-2}(x),\ E_2(x),\ E_3(x)$;

. . . .

mth series $E_{-2^m+1},\ \ldots,\ E_{-2^{m-1}},\ E_{2^{m-1}},\ \ldots,\ E_{2^m-1}.$

So the mth series contains the 2^m functions

$$E_n,\ E_{-n};\qquad n=2^{m-1}+s,\qquad s=0,\ 1,\ \ldots,\ 2^{m-1}-1.$$

The orthogonality of E-functions belonging to different series is obvious (use the representation of $E_n(x)$ by exponential functions). If $E_{n_1}(x)$ and $E_{n_2}(x)$ belong to one and the same series and the indices n_1 and n_2 have opposite signs, then again there is no doubt about the orthogonality of E_{n_1} and E_{n_2} (by the same reasoning). Finally, let

$$n_1=2^{m-1}+s_1,\qquad n_2=2^{m-1}+s_2,\qquad s_1\neq s_2.$$

We have (see (21))

$$\int_0^{2\pi} E_{n_1}(x)\overline{E(x)}\,dx=\left(\frac{1}{2^{m-1}}\right)^2\int_0^{2\pi}\sum_{2^{m-1}}^{2^m-1} e^{ik\left(x-\frac{2s_1+1}{2^{m-1}}\pi\right)}\overline{\sum_{2^{m-1}}^{2^m-1} e^{il\left(x-\frac{2s_2+1}{2^{m-1}}\pi\right)}}\,dx$$

$$=\frac{\pi}{2^{2m-3}}\sum_{2^{m-1}}^{2^m-1} e^{ik\frac{s_2-s_1}{2^{m-2}}\pi}=Ce^{i\frac{3\cdot 2^{m-1}-1}{2}u}\,\frac{\sin(2^{m-2}u)}{\sin(u/2)}=0\qquad\left(u=\frac{s_2-s_1}{2^{m-2}}\pi\right).$$

This completes the proof of the orthogonality relation (22).

A similar calculation yields the expression for the norm of E_n:

$$\|E_n\|=\left(\int_0^{2\pi} E_n(x)^2\,dx\right)^{1/2}=\sqrt{2\pi}\,2^{-\frac{m-1}{2}}.$$

In the sequel, besides $\{E_n(x)\}_{-\infty}^{\infty}$, we shall also use the orthonormal system $\{\mathbf{e}_n(x)\}_{-\infty}^{\infty}=\{\mathbf{e}_n\}_{-\infty}^{\infty}$, where

$$\mathbf{e}_n(x)=\frac{E_n(x)}{\|E_n\|}.$$

Note that the function system

$$1,\ \{C_n(x),\ S_n(x)\}_1^{\infty},$$

composed of the real and imaginary parts of the functions $E_n(x)$, is also an orthogonal system.

By the foregoing, (20) may be regarded as the expansion of $f(x)$ in a Fourier series with respect to $\{E_n\}$. Consequently,

$$\Gamma_n=\frac{1}{\|E_n\|^2}(f,\ E_n)=\frac{2^{m-2}}{\pi}(f,\ E_n).$$

If we write the same expansion in terms of the orthonormal system $\{\mathbf{e}_n\}_{-\infty}^{\infty}$,

$$f(x)=\sum_{n=-\infty}^{\infty}\gamma_n\mathbf{e}_n(x),$$

then

$$\gamma_n = (f, \mathbf{e}_n) = \frac{\Gamma_n}{\|E_n\|}, \qquad \Gamma_n = \frac{\sqrt{2\pi}}{2^{(m-1)/2}} \gamma_n.$$

We have already given the expression of Γ_n through the Fourier coefficients of f with respect to the system $\{e^{ikx}\}$ (see (19)). It is easy to find the reversion of (19):

$$C_k = \frac{1}{2^{m-1}} \sum_{s=0}^{2^{m-1}-1} \Gamma_{2^{m-1}+s} e^{-ik\frac{2s+1}{2^{m-1}}\pi}, \qquad k = 2^{m-1}, 2^{m-1}+1, \ldots, 2^m - 1 \tag{23}$$

(the expression of E_n using exponential functions must be substituted in (18) and the coefficients of e^{ikx} collected).

As the functions $\mathbf{e}_n(x)$ are finite linear combinations of exponential functions, the system $\{\mathbf{e}_n\}$ shares many properties of the system $\{e^{ikx}/\sqrt{2\pi}\}$. We shall not dwell on this. We shall rather deal with the property that sharply distinguishes the first system from the second: the system $\{\mathbf{e}_n\}$ is unbounded. This easily follows from (21) and the expression of $\|E_n\|$. Moreover, the system $\{\mathbf{e}_n\}$ is of Haar type. In fact, it can be split into series of the same type as the Haar system, and each function $\mathbf{e}_k(x)$ has one "peak" within a period. It seems that these very properties are responsible for $\{E_k(x)\}$ being an unconditional basis of $B^r_{p,\theta}$.

§4. An unconditional basis of $B^r_{p,\theta}$

Theorem 7.10 of [8], Chapter X, asserts that the L_p norm of the exponential polynomial

$$U_n(x) = \sum_{k=0}^{n} C_k e^{ikx}$$

is equivalent, for $1 < p < \infty$, to the quantity

$$(n+1)^{-1/p} \left\{ \sum_{s=0}^{n} |U_n(t_n^{(s)})|^p \right\}^{1/p},$$

where $t_n^{(s)}$ are equidistant points of interpolation, $t_n^{(s)} = x_0 + 2\pi s/(n+1)$, $s = 0, 1, \ldots, n$ (for $1 \leqslant p \leqslant \infty$ a one-sided estimate holds).

As a corollary of this theorem we obtain

PROPOSITION 3. *For the packet $f_m^+(x)$ of the Fourier series* (1) *the estimates*

$$A_p \|f_m^+(\cdot)\|_p \leqslant \left(\frac{1}{2^m} \sum_{2^{m-1}}^{2^m-1} |\Gamma_n|^p \right)^{1/p} \leqslant B_p \|f_m^+(\cdot)\|_p \tag{24}$$

hold, where the constants $0 < A_p \leqslant B_p < \infty$ do not depend on m.

PROOF. By the interpolation formula (11) and the subsequent definitions we have

$$f_m^+(x) = \sum_{s=0}^{2^{m-1}-1} f_m^+(t_{2^{m-1}-1}^{(s)}) E_{2^{m-1}-1}^{(s)+}(x) = \sum_{2^{m-1}}^{2^m-1} \Gamma_n E_n(x).$$

Thus

$$\Gamma_n = f_m^+(t_{2^{m-1}-1}^{(s)}), \qquad n = 2^{m-1}+s, \quad s = 0, 1, \ldots, 2^{m-1}-1. \tag{25}$$

As

$$f_m^+(x) = \sum_{2^{m-1}}^{2^m-1} C_k e^{ikx} = e^{i2^{m-1}x} \sum_{s=0}^{2^{m-1}-1} C'_s e^{isx} = e^{i2^{m-1}x} U_{2^{m-1}-1}(x),$$

where $C'_s = C_{2^{m-1}+s}$, it follows that

$$f_m^+(t^{(s)}) = e^{i2^{m-1}t^{(s)}} U_{2^{m-1}-1}(t^{(s)}) \; (t^{(s)} = t_{2^{m-1}-1}^{(s)}).$$

Applying Theorem 7.10 of [8] mentioned above, we obtain

$$\|f_m^+(\cdot)\|_p = \|U_{2^{m-1}-1}(\cdot)\|_p \sim \left(\frac{1}{2^{m-1}}\right)^{1/p} \left\{\sum_{s=0}^{2^{m-1}-1} |U_{2^{m-1}}(t^{(s)})|^p\right\}^{1/p}$$

$$= \left(\frac{1}{2^{m-1}}\right)^{1/p} \left\{\sum_{s=0}^{2^{m-1}-1} |f_m^+(t^{(s)})|^p\right\}^{1/p} = \left\{\sum_{2^{m-1}}^{2^m-1} |\Gamma_n|^p\right\}^{1/p} (2^{m-1})^{-1/p}.$$

Here, as usual, the symbol $\sim$ denotes the equivalence relation of the type (24), and in the last step we used (25). A similar procedure leads to estimates for $f_m^-(x)$. The proof of Proposition 3 is complete.

We are now able to prove one of the main theorems of the paper.

THEOREM 1. *The periodic generalized function $f(x)$ belongs to the space $B_{p,\theta}^r$, $-\infty < r < \infty$, $1 < p < \infty$, $1 \leqslant \theta \leqslant \infty$, if and only if it may be represented by a weakly convergent series*

$$f(x) = \sum_{-\infty}^{\infty} \Gamma_n E_n(x) \tag{26}$$

with coefficients Γ_n satisfying the condition

$$\beta_{p,\theta}^r = \left[\sum_{m=1}^{\infty} 2^{m(r-1/p)\theta} \left\{\sum_{2^{m-1} \leqslant |k| \leqslant 2^m-1} |\Gamma_k|^p\right\}^{\theta/p}\right]^{1/\theta} < \infty, \qquad 1 \leqslant \theta < \infty; \tag{27_1}$$

$$\beta_{p,\theta}^r = \sup_{m \geqslant 1} 2^{m(r-1/p)} \left\{\sum_{2^{m-1} \leqslant |k| \leqslant 2^m-1} |\Gamma_k|^p\right\}^{1/p} < \infty, \qquad \theta = \infty. \tag{27_2}$$

In this case (26) *converges in the $B_{p,\theta}^r$ norm* (*for $\theta \neq \infty$*), *and the quantity*

$$\{|\Gamma_0|^\theta + (\beta_{p,\theta}^r)^\theta\}^{1/\theta} \tag{28}$$

is equivalent to the $B_{p,\theta}^r$ norm of f ($1 \leqslant \theta \leqslant \infty$).

PROOF. Let $f \in B_{p,\theta}^r$. Then f may be represented by the series

$$f(x) = \sum_{-\infty}^{\infty} C_k e^{ikx},$$

convergent to f weakly, and also in the $B_{p,\theta}^r$ norm (for $\theta \neq \infty$) if the natural summation is used (see §2). Then, however, (26) is also convergent in the $B_{p,\theta}^r$ norm under the natural summation, which consists in setting

$$\Phi_0(x) = E_0(x), \qquad \Phi_{2k-1}(x) = E_k(x), \qquad \Phi_{2k} = E_{-k}, \qquad k = 1, 2, \ldots,$$

and considering the convergence of the partial sums

$$\sum_{k=0}^{n} \Gamma_k^* \Phi_k.$$

This follows by the reasoning applied in §2 (convergence by packets and the possibility of neglecting the "extreme" packets). As certain subsequences of the partial sums of (6) and (26) coincide, the two series have the same sum. Therefore the system $\{\Phi_n\}$ is a (conditional) basis of $B^r_{p,\theta}$ for $1 \leqslant \theta < \infty$. Obviously, (26) will converge weakly to $f(x)$ for all θ, $1 \leqslant \theta \leqslant \infty$. In view of the expression (4) for the norm in $B^r_{p,\theta}$, and of inequalities (24), we obtain for $1 \leqslant \theta \leqslant \infty$ the equivalence relation

$$|f, B^r_{p,\theta}| \sim \{|\Gamma_0|^\theta + (\beta^r_{p,\theta})^\theta\}^{1/\theta}. \tag{29}$$

Conversely, let $f(x)$ be represented by a weakly convergent series (26), and suppose that (27) is fulfilled. Then (26) is convergent in the norm (28) for $1 \leqslant \theta < \infty$, and, consequently, also in the $B^r_{p,\theta}$ norm. Furthermore (since the trigonometric polynomials belong to $B^r_{p,\theta}$ and $B^r_{p,\theta}$ is complete, $f \in B^r_{p,\theta}$. If $\theta = \infty$, then $f \in B^r_{p,\theta}$ follows from (24) and the definition of $B^r_{p,\infty}$.

The proof of Theorem 1 is complete. An immediate consequence of the theorem is the following important assertion.

COROLLARY 1. *The system $\{E_n(x)\}_{-\infty}^{\infty}$ constitutes an unconditional basis of the space $B^r_{p,\theta}$, $-\infty < r < \infty$, $1 < p < \infty$, $1 \leqslant \theta < \infty$.*

To prove this it is sufficient to recall that one of the criteria for the unconditional convergence of Σx_n in the Banach space X requires the convergence of the series $\Sigma \epsilon_n x_n$ for arbitrary $\epsilon_n = \pm 1$.

Let us stress that the system $\{E_n(x)\}$ is a basis simultaneously in all spaces $B^r_{p,\theta}$ indicated. Recall that it consists of orthogonal functions (in the L_2 sense). An important property, far from being trivial, of the basis $\{E_n\}$ is that the norm in $B^r_{p,\theta}$ may be expressed (up to equivalence) by the absolute values of the coefficients Γ_n in the expansion

$$f(x) = \sum \Gamma_n E_n(x).$$

We shall make use of this circumstance in §5.

Note that (29) combined with (19) yields the following equivalent expression for the $B^r_{p,\theta}$ norm in terms of the Fourier coefficients:

$$|f, B^r_{p,\theta}|^\theta \sim |C_0|^\theta + \sum_{m=1}^{\infty} 2^{m(r-1/p)} \left\{ \sum_{s=0}^{2^{m-1}-1} \left| \sum_{2^{m-1} \leqslant |k| \leqslant 2^m - 1} C_k e^{ik\frac{2s+1}{2^{m-1}}\pi} \right|^p \right\}^{\theta/p}.$$

Nevertheless, we see that this expression depends not only on the moduli of the C_k, but also takes into account the mutual correlation of the coefficients within the packet f_m. This correlation, as an easy calculation shows, leads to the appearance of the moduli only for $p = 2$:

$$|f,\ B^r_{2,\theta}| \sim \left[|C_0|^\theta + \sum_{m=1}^{\infty} 2^{m\theta r}\left\{\sum_{2^{m-1}\leqslant |k| \leqslant 2^m-1} |C_k|^2\right\}^{\theta/2}\right]^{1/\theta}. \tag{30}$$

For $\theta = 2$ and $r = 0$ we obtain from (30), as expected,

$$|f,\ B^0_{2,2}| \sim \|f\|_{L_2}$$

(that is, $B^0_{2,2}$ and L_2 coincide up to equivalence of the norms).

§5. Multipliers of Fourier transforms in $B^r_{p,\theta}$ and L_p

The equivalence relation (29) enables us to give a complete description of Fourier multipliers in Besov spaces (in the periodic case considered). Let $f \in B^r_{p,\theta}$ and

$$f = \sum_{-\infty}^{\infty} C_k e^{ikx}.$$

Consider the transformation T defined by

$$(Tf)(x) = \sum_{-\infty}^{\infty} \tau_k C_k e^{ikx}, \tag{31}$$

where $\tau = \{\tau_k\}$ is a sequence of complex numbers. We shall say that T is a multiplier transform from $B^r_{p,\theta} = B$ to $B^{r'}_{p',\theta'} = B'$ if for any $f \in B^r_{p,\theta}$ the function Tf belongs to $B^{r'}_{p',\theta'}$ and

$$|Tf,\ B^{r'}_{p',\theta'}| \leqslant C|f,\ B^r_{p,\theta}|, \tag{32}$$

where the constant C does not depend on f. In this case the sequence $\{\tau_k\}$ will be called a $(B^r_{p,\theta}, B^{r'}_{p',\theta'})$ *multiplier* ((B, B') *multiplier* for short). By this definition, a multiplier transform T is a bounded transformation with norm $\|T\|$ not greater than the constant C in (32). We denote the set of all such transformations T by $\mathcal{M}(B^r_{p,\theta}, B^{r'}_{p',\theta'})$ (or $\mathcal{M}(B, B')$ for short). Clearly, $\mathcal{M}(B, B')$ is a subspace of $\mathcal{L}(B, B')$, the space of bounded linear operators from B to B'. The set of all (B, B') multipliers will be denoted by $\mathbf{m}(B, B')$.

The definitions we have stated work for any pair B, B' of Banach spaces of periodic functions. In particular, we shall be interested in $(L_p, L_{p'})$ multipliers.

First of all, let us note the following fact.

PROPOSITION 4. *If the sequence* $\tau = \{\tau_k\}^{\infty}_{-\infty}$ *is an* $(L_p, L_{p'})$ *multiplier* (*that is,* $\tau \in \mathbf{m}(L_p, L_{p'})$), *then* $\tau \in \mathbf{m}(B^r_{p,\theta}, B^r_{p',\theta})$, $1 < p,\ p' < \infty$.

To prove this it is sufficient to note that if $f \in B^r_{p,\theta}$, then

$$f(x) = \sum_{m=0}^{\infty} f_m(x),$$

$$Tf = \sum_{0}^{\infty} Tf_m,$$

$$|Tf,\ B^r_{p',\theta}| = \left(\sum_0^{\infty} 2^{mr\theta}\|Tf_m\|^\theta_{p'}\right)^{1/\theta} \leqslant \|T\|_{L_p\to L_{p'}}\left\{\sum_0^{\infty} 2^{mr\theta}\|f_m\|^\theta_p\right\}^{1/\theta}.$$

(in the last step we used the fact that $\tau \in \mathbf{m}(L_p, L_{p'})$). Therefore

$$|Tf,\ B^r_{p',\theta}| \leqslant \|T\|_{L_p\to L_{p'}} |f,\ B^r_{p,\theta}|.$$

Consequently, $\tau \in \mathbf{m}(B^r_{p,\theta}, B^r_{p',\theta})$ and

$$\|T\|_{B^r_{p,\theta}\to B^r_{p',\theta}} \leqslant \|T\|_{L_p\to L_{p'}}.$$

The proof of Proposition 4 is complete. Thus we have

$$\mathbf{m}(L_p,\ L_{p'}) \subset \mathbf{m}(B^r_{p,\theta},\ B^r_{p',\theta}).$$

It is well known [6] that $\mathbf{m}(L_p, L_{p'})$ contains nonzero elements only if $p \leqslant p'$. A similar statement holds for $\mathbf{m}(B^r_{p,\theta}, B^r_{p',\theta})$ [7]. We are going to give a characterization of the space $\mathbf{m}(B^r_{p,\theta}, B^r_{p',\theta})$, $p \leqslant p'$, from which it will follow that these spaces do not depend on r and θ.

Let us be given a sequence $\{\tau_k\}_{-\infty}^{\infty}$. Consider the formal series

$$\sum_k \tau_k e^{ikx} \sim \mathscr{T}(x)$$

and the corresponding packets

$$\mathscr{T}^{\pm}_m(x) = \sum_{2^{m-1}}^{2^m-1} \tau_{\pm k} e^{\pm ikx}, \qquad m \geqslant 1. \tag{33}$$

THEOREM 2. *The relation*

$$\{\tau_k\} \in \mathbf{m}(B^r_{p,\theta},\ B^r_{p',\theta}), \qquad 1 < p \leqslant p' < \infty, \qquad 1 \leqslant \theta \leqslant \infty, \qquad -\infty < r < \infty,$$

is valid if and only if there exists a constant $A > 0$ such that $|\tau_0| < A$ and such that for $m = 1, 2, \ldots$ and any complex numbers $z_l = z_l^{(m)}$, $l = 0, 1, \ldots, 2^{m-1} - 1$, the inequality

$$\left\{\sum_{l=0}^{2^{m-1}-1} \left|\sum_{s=0}^{2^{m-1}-1} z_l \mathscr{T}^{\pm}_m\left(\frac{l-s}{2^{m-2}}\pi\right)\right|^{p'}\right\}^{1/p'} \leqslant A 2^{m\left(1-\frac{1}{p}+\frac{1}{p'}\right)} \left\{\sum_{l=0}^{2^{m-1}-1} |z_l|^p\right\}^{1/p} \tag{34}$$

holds, where the constant A does not depend on $\{\tau_k\}$ and m. If (34) *holds, then the norm of the corresponding multiplier operator T is majorized by the number A; that is,*

$$\|T\| \leqslant CA,$$

where C does not depend on $\{\tau_k\}$.

PROOF. *Sufficiency.* Let $f(x) = \sum_{-\infty}^{\infty} C_k e^{ikx} \in B^r_{p,\theta}$, and suppose that the conditions of the theorem are fulfilled. Represent f in the form

$$f(x) = \sum_{-\infty}^{\infty} \Gamma_n E_n(x).$$

The segment of this sum that corresponds to the packet $f_m^+(x) = \Sigma C_k e^{ikx}$ may be written as (see (23))

$$f_m^+(x) = \sum_{s=0}^{2^{m-1}-1} \Gamma_{2^{m-1}+s} E_{2^{m-1}+s}(x) = \sum_{k=2^{m-1}}^{2^m-1} \frac{1}{2^{m-1}} \left(\sum_{s=0}^{2^{m-1}-1} \Gamma_{2^{m-1}+s} e^{-ikt^{(s)}_{2^{m-1}-1}} \right) e^{ikx}.$$

The multiplier operator T induced by the sequence $\{\tau_k\}$ has the property that it carries the subspace Ξ_m into itself. To be specific, let $m > 0$. Starting from the definition of T, we obtain

$$(Tf_m^+)(x) = \sum_{k=2^{m-1}}^{2^m-1} \tau_k \left(\frac{1}{2^{m-1}} \sum_{s=0}^{2^{m-1}-1} \Gamma_{2^{m-1}+s} e^{-ikt^{(s)}_{2^{m-1}-1}} \right) e^{ikx}$$

$$= \sum_{l=0}^{2^{m-1}-1} E_{2^{m-1}+l} \left\{ \frac{1}{2^{m-1}} \sum_{s=0}^{2^{m-1}-1} \Gamma_{2^{m-1}+s} \mathcal{T}_m^+ \left(\frac{l-s}{2^{m-2}} \pi \right) \right\} = \sum_{l=0}^{2^{m-1}-1} \Gamma'_{2^{m-1}+l} E_{2^{m-1}+l},$$

where

$$\Gamma'_{2^{m-1}+l} = \frac{1}{2^{m-1}} \sum_{s=0}^{2^{m-1}-1} \Gamma_{2^{m-1}+s} \mathcal{T}_m^+ \left(\frac{l-s}{2^{m-2}} \pi \right).$$

Using condition (34) of Theorem 2, we find that

$$\left(\sum_{l=0}^{2^{m-1}-1} |\Gamma'_{2^{m-1}+l}|^{p'} \right)^{1/p'} = \left\{ \frac{1}{2^{m-1}} \sum_{l=0}^{2^{m-1}-1} \left| \sum_{s=0}^{2^{m-1}-1} \Gamma_{2^{m-1}+s} \mathcal{T}_m^+ \left(\frac{l-s}{2^{m-2}} \pi \right) \right|^{p'} \right\}^{1/p'}$$

$$\leqslant \frac{A}{2^{m\left(\frac{1}{p}-\frac{1}{p'}\right)}} \left\{ \sum_{l=0}^{2^{m-1}-1} |\Gamma^+_{2^{m-1}+l}|^p \right\}^{1/p}.$$

A similar estimate can be given for the corresponding sum with negative indices of the coefficients.

As a consequence of these estimates we have

$$\left\{ \sum_{m=1}^{\infty} 2^{m\theta\left(r-\frac{1}{p'}\right)} \left[\sum_{2^{m-1} \leqslant |k| \leqslant 2^m-1} |\Gamma'_k|^{p'} \right]^{\theta/p'} \right\}^{1/\theta}$$

$$\leqslant A \left\{ \sum_{m=1}^{\infty} 2^{m\theta\left(r-\frac{1}{p}\right)} \left[\sum_{2^{m-1} \leqslant |k| \leqslant 2^m-1} |\Gamma_k|^p \right]^{\theta/p} \right\}^{1/\theta} < \infty.$$

Hence, relying on Theorem 1, we conclude that the series

$$\sum_{-\infty}^{\infty} (Tf_m)(x) = \sum_{-\infty}^{\infty} \Gamma'_n E_n(x)$$

is convergent in the $B^r_{p',\theta}$ norm to a function $F \in B^r_{p',\theta}$. It is clear that

$$F(x) = \sum_{-\infty}^{\infty} \tau_k C_k e^{ikx} = Tf$$

and that (in view of (34) and (29))

$$|Tf,\ B^r_{p,\theta}| \leqslant CA\,|f,\ B^r_{p,\theta}|.$$

Therefore $\|T\| \leqslant CA$.

REMARK. If we denote the restriction of the operator T to the subspace Ξ_m by T_m, then for $m > 0$

$$T_m f_m = \frac{1}{2\pi}\int_0^{2\pi} f_m(y)\,\mathcal{T}^+_m(x-y)\,dy = \sum_{k=2^{m-1}}^{2^m-1} \tau_k C_k e^{ikx}. \tag{35}$$

The operator K_m defined by

$$K_m : f_m \to \{\Gamma_{2^{m-1}+s}\}_{s=0}^{2^{m-1}-1}$$

maps the subspace Ξ_m isomorphically onto $l_p(2^{m-1})$, and the norm of K_m is equivalent to a quantity which does not depend on m (see (24)). Since $T_m f_m \in \Xi_m$, we have

$$K_m(T_m f_m) = \{\Gamma'_{2^{m-1}+s}\}_{s=0}^{2^{m-1}-1}.$$

So there occurs the operator $K_m T_m (K_m)^{-1}$ which takes the vector $\mathbf{\Gamma}_m = \{\Gamma_{2^{m-1}+s}\}$ into the vector $\Gamma'_m = \{\Gamma'_{2^{m-1}+s}\}$. This operator, as previous calculations show, is given by the Toeplitz matrix

$$Q^{(m)} = (q^{(m)}_{ij}),$$

where

$$q^{(m)}_{ij} = \frac{1}{2^{m-1}}\,\mathcal{T}_m\left(\frac{i-j}{2^{m-2}}\pi\right).$$

In these terms, condition (34) expresses the boundedness of the norm of the matrix $Q^{(m)}$ as an operator from $l_p(2^{m-1})$ to $l_{p'}(2^{m-1})$; that is, it can be written in the form

$$\|Q^{(m)}\|_{l_p\to l_{p'}} \leqslant B.$$

These considerations yield

$$\|T\| \leqslant \sup_m \|Q^{(m)}\|_{l_p\to l_{p'}}. \tag{36}$$

Necessity. By what has been said, the sufficiency of condition (34) is due to the fact that it assures the uniform boundedness of the restrictions of the operator T to the subspace Ξ_m.

Let $\tau \in \mathbf{m}(B^r_{p,\theta}, B^r_{p',\theta})$. Then, obviously, the corresponding operator T is (uniformly) bounded on the subspaces Ξ_m. This means that (34) is satisfied (with a constant proportional to the norm of T). Hence the conditions (34) are necessary.

The proof of the theorem is complete.

The criterion for $\{\tau_k\}$ to belong to $\mathbf{m}(B^r_{p,\theta}, B^r_{p',\theta})$ given by Theorem 2 is hard to verify in practice. As an application, let us deduce from it a sufficient condition.

THEOREM 3. *If the sequence* $\tau = \{\tau_k\}$ *is such that the function* $\mathcal{T}(x) = \Sigma_k \tau_k e^{ikx}$ *satisfies the conditions*(3)

$$|\tau_0| \leqslant M, \quad \|\mathcal{T}_m^+(\cdot)\|_1 = \left\| \sum_{k=2^{m-1}}^{2^m-1} \tau_k e^{ikx} \right\|_1 \leqslant M, \quad \|\mathcal{T}_m^-\|_1 \leqslant M, \tag{37}$$

then $\tau \in \mathbf{m}(B^r_{p,\theta}, B^r_{p,\theta})$ *(simultaneously for all* p, θ *and* r, $1 < p < \infty$, $1 \leqslant \theta \leqslant \infty$, $-\infty < \tau < \infty$). *Moreover,* $\|T\| \leqslant \text{const} \cdot M$.

PROOF. Let us show that, under the conditions of Theorem 3, inequality (34) is satisfied (with $p = p'$). Write the left-hand side of (34) in the form

$$J = \left\{ \sum_{l=0}^{2^{m-1}-1} \left| \sum_{\sigma=1}^{l-2^{m-1}+1} z_{l-\sigma} \mathcal{T}_m\left(\frac{\sigma}{2^{m-2}}\pi\right) \right|^p \right\}^{1/p}. \tag{38}$$

The numbers z_s have been defined for $0 \leqslant s \leqslant 2^{m-1} - 1$; define them to be zero for $s \notin [0, 2^{m-1} - 1]$. The numbers $\mathcal{T}_m(\sigma\pi/2^{m-2})$ in (38) are considered for integers $\sigma \in [-(2^{m-1} - 1), 2^{m-1} - 1]$; put them equal to zero for other integers σ. Then in (38) the summation over l and σ may be extended from $-\infty$ to ∞, and Minkowski's inequality may be applied. We obtain

$$J \leqslant \sum_{\sigma=-\infty}^{\infty} \left| \mathcal{T}_m\left(\frac{\sigma}{2^{m-2}}\pi\right) \right| \left(\sum_{-\infty}^{\infty} |z_{l-\sigma}|^p \right)^{1/p} = \sum_{\sigma=-2^{m-1}+1}^{2^{m-1}-1} \left| \mathcal{T}_m\left(\frac{\sigma}{2^{m-2}}\pi\right) \right| \left(\sum_{s=0}^{2^{m-1}-1} |z_s|^p \right)^{1/p}$$

$$= \left\{ \sum_{\sigma=0}^{2^{m-1}-1} \left| \mathcal{T}_m\left(\frac{\sigma}{2^{m-2}}\pi\right) \right| + \sum_{\sigma=-2^{m-1}+1}^{-1} \left| \mathcal{T}_m\left(\frac{\sigma}{2^{m-2}}\pi\right) \right| \right\} \left(\sum_{s=0}^{2^{m-1}-1} |z_s|^p \right)^{1/p}.$$

Replacing in the second term the summation over σ by summation over $\sigma' = \sigma + 2^{m-1}$ and making use of the 2π-periodicity of the function $\mathcal{T}$, we find the estimate

$$J \leqslant 2 \sum_{\sigma=0}^{2^{m-1}-1} \left| \mathcal{T}_m\left(\frac{\sigma}{2^{m-1}}\pi\right) \right| \left(\sum_{s=0}^{2^{m-1}-1} |z_s|^p \right)^{1/p}.$$

But according to Theorem 7.10 of [8], Chapter X,

$$\sum_{\sigma=0}^{2^{m-1}-1} \left| \mathcal{T}_m\left(\frac{\sigma}{2^{m-1}}\pi\right) \right| \leqslant C 2^{m-1} \|\mathcal{T}_m(\cdot)\|_1.$$

We finally obtain

$$J \leqslant 2CM 2^{m-1} \left(\sum_0^{2^{m-1}-1} |z_s|^p \right)^{1/p}.$$

Consequently, (34) is valid with the constant $A = 2CM$.

The proof of Theorem 3 is complete.

REMARK. By Hölder's inequality we have

$$\|\mathcal{T}_m(\cdot)\|_1 \leqslant (2\pi)^{1-1/a} \|\mathcal{T}_m(\cdot)\|_a, \quad 1 < a < \infty.$$

(3) The conditions say that $\mathcal{T} \in \overset{*}{B}{}^0_{1,\infty}$.

Thus the conditions of Theorem 3 will be fulfilled if $\mathscr{T}(x) \in B^0_{a,\infty}$ for some a, $1 < a < \infty$.

In the case $p' > p$ a similar sufficient condition has a simple formulation.

THEOREM 4. *Let $\{\tau_k\}$ be a numerical sequence such that $\mathscr{T}(x) \in B^0_{a,\infty} \equiv H^0_a$. Then $\{\tau_k\} \in \mathbf{m}(B^r_{p,\theta}, B^r_{p',\theta})$ simultaneously for all p, p', θ and r satisfying the relations $1 < p < p' < \infty$, $1/a = 1 - 1/p + 1/p'$, $1 \leqslant \theta \leqslant \infty$ and $-\infty < r < \infty$. Moreover,*

$$\|T\| \leqslant \|\mathscr{T}\|_{H^0_a}.$$

PROOF. Consider the restriction T_m of T to the subspace Ξ_m (see (35)). By Young's inequality we have

$$\|T_m f_m\|_{p'} = \frac{1}{2\pi}\left\| \int_0^{2\pi} \mathscr{T}_m(y) f_m(x-y)\,dy \right\|_{p'}$$
$$\leqslant \|\mathscr{T}_m\|_a \cdot \|f_m\|_p, \quad 1/a = 1-(1/p - 1/p').$$

From $\mathscr{T} \in B^0_{a,\infty} = H^0_a$ it follows that

$$\|\mathscr{T}_m\|_a \leqslant \sup_m \|\mathscr{T}_m\|_a \leqslant \|\mathscr{T}\|_{H^0_a}$$

and therefore

$$|Tf, B^r_{p',\theta}| \leqslant \left\{\sum_m 2^{m\theta r} \|T_m f_m\|^\theta_{p'}\right\}^{1/\theta}$$
$$\leqslant \{\sum 2^{m\theta r} \|\mathscr{T}_m\|^\theta_a \cdot \|f_m\|^\theta_p\}^{1/\theta} \leqslant \|\mathscr{T}\|_{H^0_a} \cdot |f, B^r_{p,\theta}|.$$

The proof is complete.

REMARK. Theorem 4 is contained as a special case in a result of Taibleson [7], who obtained it by a different method.

In conclusion we give a criterion expressed directly in terms of $\{\tau_k\}$.

THEOREM 5. *If the sequence $\{\tau_k\}$ satisfies the conditions*

$$|\tau_k| \leqslant M k^{-(1/p-1/p')},$$
$$\sum_{k=2^{m-1}}^{2^m-1} |\tau_k - \tau_{k+1}| \leqslant M 2^{-m(1/p-1/p')},$$

then $\{\tau_k\} \in \mathbf{m}(B^r_{p,\theta}, B^r_{p',\theta})$ simultaneously for all p, p', θ and r such that $1 < p \leqslant p' < \infty$, $1 \leqslant \theta < \infty$ and $-\infty < r < \infty$. Moreover, $\|T\| \leqslant \mathrm{const} \cdot M$.

PROOF. Application of Abel's transformation to the sum

$$T_m f_m = \sum_{k=2^{m-1}}^{2^m-1} \tau_k C_k e^{ikx}$$

yields

$$T_m f_m = \sum_{s=2^{m-1}}^{2^m-1} \left(\sum_{2^{m-1}}^{s} C_k e^{ikx}\right)(\tau_s - \tau_{s+1}) + \tau_{2^m} \sum_{k=2^{m-1}}^{2^m-1} C_k e^{ikx}. \tag{39}$$

We make use of S. M. Nikol'skiĭ's inequality (see [2], Theorem 3.5.5)

$$\| T_m f_m \|_{p'} \leqslant 2\,(2^{m-1})^{1/p-1/p'} \| T_m f_m \|_p$$

and Minkowski's inequality for the sum (39)

$$\| T_m f_m \|_p \leqslant \sum_{s=2^{m-1}}^{2^m-1} \left\| \sum_{2^{m-1}}^{s} C_k e^{ikx} \right\|_p \cdot |\tau_s - \tau_{s+1}| + \tau_{2^m} \left\| \sum_{k=2^{m-1}}^{2^m-1} C_k e^{ikx} \right\|_p .$$

Relying on the theorem of Riesz according to which

$$\left\| \sum_{2^{m-1}}^{s} C_k e^{ikx} \right\|_p \leqslant C \| f_m \|_p ,$$

we obtain

$$\| T_m f_m \|_{p'} \leqslant 2\,(2^{m-1})^{\frac{1}{p}-\frac{1}{p'}} \left(\sum_{s=2^{m-1}}^{2^m-1} |\tau_s - \tau_{s+1}| + |\tau_{2^m}| \right) C \| f_m \|_p \leqslant \mathrm{const} \cdot M \cdot \| f_m \|_p .$$

This proves the theorem.

BIBLIOGRAPHY

1. Hans Triebel, *Multipliers and unconditional Schauder bases in Besov spaces*, Studia Math. **60** (1976), 145–156.

2. S. M. Nikol'skiĭ, *Approximation of functions of several variables and embedding theorems*, "Nauka", Moscow, 1969; English transl., Springer-Verlag, Berlin and New York, 1974.

3. P. I. Lizorkin, *On the Fourier transform in Besov spaces*, Dokl. Akad. Nauk SSSR **163** (1965), 1318–1321; English transl. in Soviet Math. Dokl. **6** (1965).

4. ———, *Generalized Hölder spaces* $B^{(r)}_{p,\theta}$ *and their relation with the Sobolev spaces* $L^{(r)}_p$, Sibirsk. Mat. Ž. **9** (1968), 1127–1152; English transl. in Siberian Math. J. **9** (1968).

5. Ivan Singer, *Bases in Banach spaces*. I, Springer-Verlag, Berlin and New York, 1970.

6. Lars Hörmander, *Estimates for translation invariant operators in* L^p *space*, Acta Math. **104** (1960), 93–140.

7. Mitchell H. Taibleson, *On the theory of Lipshcitz spaces of distributions on Euclidean n-space*. I, II, III, J. Math. Mech. **13** (1964), 407–479; **14** (1965), 821–839; **15** (1966), 973–981.

8. A. Zygmund, *Trigonometric series*, 2nd rev. ed., Vol. II, Cambridge Univ. Press, New York, 1959.

Translated by J. BOGNÁR

Труды Мат. Инст. Стеклов.
143 (1977)

Proceedings of the Steklov
Institute of Mathematics
1980, Issue 1

THE PROBLEM OF AVOIDING ENCOUNTER IN N-PERSON DIFFERENTIAL GAMES

UDC 517.9

E. F. MIŠČENKO, M. S. NIKOL'SKIĬ AND N. SATIMOV

ABSTRACT. In the paper methods are developed for solving the problem of one evader avoiding encounter with several pursuers. The applicability of the results is demonstrated by a number of examples of general type. In addition to the possibility of evasion, in the paper lower bounds are obtained for the distance from the evader to each of the pursuers in the process of avoiding encounter. A class of problems of the type of the damped oscillation of a mathematical pendulum is studied at the end of the paper.

Bibliography: 17 titles.

Introduction

An example of an N-person differential game is provided by the following process of one controlled object evading a group of other controlled objects. Points $x_1, \ldots, x_r$ belonging to the phase space R^s pursue a point y belonging to the same space. The movements of the points are described by the corresponding differential equations

$$\dot{x}_i = f_i(x_1, x_2, \ldots, x_r, u_i), \quad i = 1, 2, \ldots, r; \quad \dot{y} = g(y, v). \tag{1}$$

In these equations u_i is the control vector of the ith pursuing object, and v is the control vector of the evading object. The point y is considered to be captured if for at least one value of the index i, $i = 1, \ldots, r$, a particular part of the coordinates of x_i coincides with the same part of the coordinates of y. The goal of the evader is to avoid capture.

From the evasion process described we obtain a differential game if we introduce a new phase vector $z = (x_1, \ldots, x_r, y)$ of the direct product $R^{(r+1)s}$ of the phase spaces of the vectors $x_1, \ldots, x_r, y$ and write (1) in the form of a single differential equation

$$\dot{z} = f(z, u, v). \tag{2}$$

The conditions for the capture of the point y by the points $x_1, \ldots, x_r$ distinguish in the space $R^{(r+1)s}$ a terminal set M which is the union of r linear subspaces.

We recall the formulation of the general problem of avoiding encounter (the evasion problem) put forth in [1]. Suppose that a differential game is described by the equation

$$\dot{z} = F(z, u, v), \tag{3}$$

1980 *Mathematics Subject Classification.* Primary 90D26, 90D10.

where $z \in R^n$, R^n is n-dimensional Euclidean space (the phase space), u is the control vector of pursuit, v is the control vector of evasion, $u \in P$, $v \in Q$, P and Q are nonempty compact subsets of R^p and R^q respectively, and $F(z, u, v)$ is a function defined on the set $R^n \times P \times Q$ and is continuous in (z, u, v) jointly. Let, further, $M \subset R^n$ be the terminal set of the game (if the phase point z enters the set M the game is considered finished).

We shall say that from the point $z_0 \in R^n \backslash M$ it is possible to avoid encounter with the set M if for any measurable function $u(t) \in P$ it is possible to construct a measurable variation $v(t)$ of the vector $v \in Q$ such that the solution $z(t)$ of the equation

$$\dot{z} = F(z, u(t), v(t)), \quad z(0) = z_0,$$

is defined for all $t \geqslant 0$ and for no value of the time t, $0 < t < \infty$, does it enter the set M: $z(t) \notin M$ for all $t \in (0, \infty)$. Here in finding the values $v(t)$ at each moment of time $t \geqslant 0$ it is permitted to use only values $u(t)$ at the same moment of time $t \geqslant 0$ and values $z(s)$ of the vector z for $s \in [0, t]$.

In the present paper the value of $v(t)$ is chosen as $V_t(u(t); z(s), 0 \leqslant s \leqslant t)$ where (cf. [1]–[3]): 1) $V_t(\cdot, \cdot)$ is a one-parameter family of mappings of the set $P \times J_t$ into the set Q, where J_t is the set of all possible absolutely continuous functions $h(s)$, $0 \leqslant s \leqslant t$, $h(s) \in R^n$; 2) for an arbitrary measurable function $u(s)$, $s \geqslant 0$, $u(s) \in P$, and an arbitrary absolutely continuous function $h(s)$, $s \geqslant 0$, $h(s) \in R^n$, $V_t(u(t); h(s), 0 \leqslant s \leqslant t)$ is measurable as a function of the variable t; 3) for any $z_0 \in R^n \backslash M$ all solutions of the equation $\dot{z}(t) = F(z(t), u(t), V_t(u(t); z(s), 0 \leqslant s \leqslant t))$, $z(0) = z_0$, extend to the semiaxis $[0, \infty)$, and none of them lies in M for any value of t, $0 < t < \infty$. A description of the family $V_t(\cdot, \cdot)$ is given below. We note that for extensive differential games a rigorous formalization of such basic concepts of game theory as strategy, saddle point, game value, etc. was first given by N. N. Krasovskiĭ and his students (cf. [4]).

In the case where it is possible to avoid encounter with the set M for an arbitrary point $z_0 \in R^n \backslash M$ we say that *evasion is possible in the game* (3).

Many papers have been devoted to the problem of evasion in linear, quasilinear, and nonlinear differential games (cf., for example, [1]–[16]). In the present paper we present certain sufficient conditions for the possibility of evasion in games of evading several pursuers. Theorems 1–7 refer to quasilinear games, while in Theorem 6 a sufficient condition is obtained for the possibility of l-evasion (cf. [14]–[16]), and in Theorem 7 a condition for evasion with phase restrictions is given. In Theorem 8, which pertains to the nonlinear game (3), the terminal set M is a sufficiently sparse set which is no more than countable; the model example of such games is the problem of the damped oscillation of a mathematical pendulum considered by N. N. Krasovskiĭ and A. I. Subbotin (cf. [4]).

Part of the results of the present paper were previously published without proof in [15].

NOTATION. Here R^r is r-dimensional, arithmetic Euclidean space; M_i is a subspace of R^n; L_i is the orthogonal complement of M_i in R^n; $\xi_i(t)$ and $\eta_i(t)$ represent the distance from the point $z(t)$ to M_i and L_i respectively; W_i is a subspace of L_i; π_i is the operator of orthogonal projection in R^n onto W_i; 0 is the zero of the space R^n; S_i is the unit sphere of W_i with center at 0; the index i takes the values $1, \ldots, m$; (z_1, z_2) is the inner product of the vectors $z_1, z_2 \in R^n$; the constants are positive numbers depending only on the game (3) but not depending either on the initial point z_0 or on the controls $u(t)$ and $v(t)$; ξ_i is the distance from the point z to M_i; η_i is the distance from z to L_i; and T is the symbol denoting the transpose of a matrix.

§1. Maneuvering superiority

Here and in §§2 and 3 the problem of evasion is considered for the differential game described by the equation (cf. (3))

$$\dot{z} = Cz + f(u, v), \tag{4}$$

where $z \in R^n$, C is a linear mapping of R^n into itself, u is the control vector of the pursuer and v that of the evader, $u \in P$, $v \in Q$, P and Q are nonempty compact subsets of R^p and R^q respectively, and f is a continuous mapping of $P \times Q$ into R^n. The terminal set $M \subset R^n$ is a sum $\bigcup_1^m M_i$ of subspaces $M_i \subset R^n$.

A. ASSUMPTION 1. There exist natural numbers k_i, two-dimensional subspaces $W_i \subset L_i$, linear mappings $F_i\colon W_i \to R^2$, and continuous mappings $\Phi\colon P \times Q \to R^2$ and $\Phi_i\colon P \times Q \to R^2$ such that:

a) the mapping F_i is injective;

b) each set $F_i\pi_i C^j f(P, Q)$, $j = 0, 1, \ldots, k_i - 2$, consists of a single point;

c) $F_i\pi_i C^{k_i-1} f(u, v) = \Phi(u, v) + \Phi_i(u, v)$ for all $(u, v) \in P \times Q$, and the set $R = \bigcap_{u\in P} \Phi(u, Q)$ contains an interior point.

The set $\{y\colon |y_1|, |y_2| \leqslant \lambda\}$, $\lambda \geqslant 0$, of points $y = (y_1, y_2)^T$ of the plane R^2 we call a square with center at the point 0 and side 2λ. It is obvious that if Assumption 1 is satisfied the set R is compact, and there exist a largest square $\Gamma \subset R^2$ with center at 0 and a vector $l \in R^2$ such that $l + \Gamma \subset R$; let $2c_1$ be the side of Γ. We denote by $\varphi_i^1(t, z_0)$ and $\varphi_i^2(t, z_0)$ the components of the two-dimensional vector-valued function $F_i\pi_i e^{tC} z_0$, $0 \leqslant t \leqslant 1$. As z_0 varies in R^n the functions $\varphi_i^1(t, z_0)$ and $\varphi_i^2(t, z_0)$ obviously form a finite-dimensional family of functions Σ_i'. Let $\varphi_1(t), \ldots, \varphi_{N_i}(t)$ be a basis of the family Σ_i', and let Σ_i be the linear space of functions spanned by the functions $\varphi_1(t), \ldots, \varphi_{N_i}(t)$ and $t, t^2, \ldots, t^{k_i}$. It is known (cf. [2]) that there exists a constant n_i such that an arbitrary nonzero function of the family Σ_i has no more than n_i zeros in the interval $[0, 1]$ counting multiplicity.

ASSUMPTION 2. Suppose that for the game (4): a) Assumption 1 is satisfied; and b) in R^2 there exist vectors l_i such that for the numbers c_1 and $c_2 = \max_{i, u\in P, v\in Q} |\Phi_i(u, v) - l_i|$ and n_i we have the inequality $c_1 > 2^m(1 + n_1) \ldots (1 + n_m)c_2$.

THEOREM 1. *If Assumption 2 is satisfied, then evasion is possible in the game (4). Here the process of avoiding encounter with the set M from the point $z_0 \in R^n \setminus M$ can be controlled in such a way that for all $t \geqslant 0$ the estimate*

$$\xi_i(t) > \begin{cases} \dfrac{c\xi_i^k(0)}{[1+\eta_i(t)]^k}, & if \quad \xi_i(0) < \varepsilon, \\ \dfrac{c\varepsilon^k}{[1+\eta_i(t)]^k}, & if \quad \xi_i(0) \geqslant \varepsilon, \end{cases} \tag{5}$$

holds, where c, ϵ are constants and $k = \max\{k_1, \ldots, k_m\}$.

REMARK 1. Theorem 1 has been proved in [15] for the case in which $\Phi_i(P, Q)$ is for each i a set consisting of a single point.

PROOF. For arbitrary functions $u(t)$ and $v(t)$, defined and measurable on $[0, 1]$, $u(t) \in P$, $v(t) \in Q$, and the corresponding solution $z(t)$, $0 \leqslant t \leqslant 1$, of the equation $\dot{z} = Cz + f(u(t), v(t))$, $z(0) = z_0 \in R^n \backslash M$, we have (cf. [1] and [5])

$$F_i\pi_i z(t) = \varphi_i(t, z_0) + \int_0^t \frac{(t-s)^{k_i-1}}{(k_i-1)!}[\Phi(u(s), v(s)) - l + \Phi_i(u(s), v(s)) - l_i + g_i(t, s)]\, ds, \tag{6}$$

where

$$\varphi_i(t, z_0) = F_i\pi_i e^{tC} z_0 + a_{i0}t + \ldots + a_{i;\, k_i-2}\frac{t^{k_i-1}}{(k_i-1)!} + (l + l_i)\frac{t^{k_i}}{k_i!},$$

$$a_{ij} = F_i\pi_i C^j f(P, Q), \qquad |g_i(t, s)| \leqslant d(t-s),$$

and d is a constant.

Let $u(t)$, $0 \leqslant t \leqslant 1$, be an arbitrary measurable function with values in P, and let ω be an arbitrary fixed point of the square Γ. Each value $\widetilde{v}_\omega$, $0 \leqslant t \leqslant 1$, is chosen as the lexicographic minimum among solutions of the equation $\Phi(u(t), v) = l = \omega$, $v \in Q$. As shown in [5], the function $\widetilde{v}_\omega(t)$, $0 \leqslant t \leqslant 1$, is measurable on $[0, 1]$. In (6) we set $v(t) = \widetilde{v}_\omega(t)$. Using the definition of $\widetilde{v}_\omega(t)$, we obtain for all $t \in [0, 1]$

$$F_i\pi_i z(t) = \varphi_i(t, z_0) - \frac{t^{k_i}}{k_i!}[\omega + h_i^1(t) + h_i^2(t)], \tag{7}$$

where

$$t^{k_i}h_i^1(t) = -k_i\int_0^t (t-s)^{k_i-1}[\Phi_i(u(s), \widetilde{v}_\omega(s)) - l_i]\, ds,$$

$$t^{k_i}h_i^2(t) = -k_i\int_0^t (t-s)^{k_i-1} g_i(t, s)\, ds, \qquad h_i^1(0) = h_i^2(0) = 0.$$

According to Assumption 2, we have $|h_i^1(t)| \leqslant c_2$. Further, it is obvious that $|h_i^2(t)| \leqslant dt$.

By a well-known fact proved by L. S. Pontrjagin (cf. [2]), in partitioning the square Γ into $(2n_1 + 2)^2$ equal small squares, for fixed z_0 there exists a square $\Gamma_1 \subset \Gamma$ with side $2c_1/(2n_1 + 2) = c_1/(1 + n_1)$ the interior of which has no common point with the curve $\varphi_1(t, z_0)k_1!/t^{k_1}$, $0 < t \leqslant 1$ (in this case we say that the curve does not intersect the square). Similarly, in partitioning Γ_1 into $(2n_2 + 2)^2$ equal small squares there exists a square $\Gamma_2 \subset \Gamma_1$ with side $c_1/(1 + n_1)(2 + 2n_2)$ having no intersection with the curve $\varphi_2(t, z_0)k_2!/t^{k_2}$, $0 < t \leqslant 1$, etc. Finally, in partitioning

Γ_{m-1} into $(2n_m + 2)^2$ equal squares there is a square $\Gamma_m \subset \Gamma_{m-1}$ with side $c_1/(1 + n_1)(2 + 2n_2) \ldots (2 + 2n_m)$ having no intersection with the curve $\varphi_m(t, z_0)k_m!/t^{k_m}$, $0 < t \leqslant 1$. Clearly, for $0 \leqslant t \leqslant 1$

$$|h_i^1(t) + h_i^2(t)| \leqslant c_2 + dt < \frac{c_1}{2^m(1+n_1)\ldots(1+n_m)} + dt. \tag{8}$$

Hence there exist constants $\theta \in (0, 1]$ and $\gamma \in [0, 1)$ such that for all $t \in [0, \theta]$

$$|h_i^1(t) + h_i^2(t)| \leqslant \frac{\gamma c_1}{2^m(1+n_1)\cdots(1+n_m)}. \tag{9}$$

Let $\omega = \omega_0 = \omega_0(z_0)$ be the center of Γ_m. According to the above construction, $\Gamma \supset \Gamma_1 \supset \ldots \supset \Gamma_m$. Using (9), from this we find that for all $t \in [0, \theta]$

$$|F_i\pi_i z(t)| = \left|\varphi_i(t, z_0) - \frac{t^{k_i}}{k_i!}[\omega_0 + h_i^1(t) + h_i^2(t)]\right| > \delta t^{k_i}, \tag{10}$$

where δ is some constant. Let $k = \max\{k_1, \ldots, k_m\}$, and let $N = \max\{\|F_1\|, \ldots, \|F_m\|\}$, where $\|F_i\|$ denotes the operator norm of the mapping F_i (according to condition a) of Assumption 1 the numbers $\|F_i\|$ are nonzero, and therefore $N > 0$). Then from the obvious inequalities $|F_i\pi_i z(t)| \leqslant \|F_i\| \cdot |\pi_i z(t)| \leqslant N|\pi_i z(t)|$, $\xi_i(t) \geqslant |\pi_i z(t)|$ and (10) we obtain

$$\xi_i(t) > \delta_1 t^k, \qquad 0 \leqslant t \leqslant \theta, \tag{11}$$

where δ_1 is a constant, and $\xi_i(0) \neq 0$.

In the space R^n we consider the hypersurfaces $\Sigma^1, \ldots, \Sigma^m$ defined by the equations (cf. [2])

$$\xi_1 = \varepsilon, \qquad \xi_2 = \varepsilon, \ldots, \qquad \xi_m = \varepsilon, \qquad \varepsilon = \min\{1, \delta_1\theta^k\},$$

respectively. The surface Σ^i divides R^n into two regions: an interior region Σ_1^i containing M_i and an exterior region Σ_2^i. There are two possible cases: 1) $z_0 \in (\Sigma^i \cup \Sigma_1^i)\backslash M_i$ for some value i_0 of the index i; 2) $z_0 \in \bigcap_1^m \Sigma_2^i$.

In case 1), choosing the function $v_0(t) = \widetilde{v}_{\omega_0}(t)$, $0 \leqslant t \leqslant \theta$ (for the definition of $\widetilde{v}_\omega(t)$, $0 \leqslant t \leqslant 1$, see above), we have (cf. (11)) $\xi_i(t) > \delta_1 t^k$ for all i and $t \in [0, \theta]$, and, moreover, $\xi_i(\theta) > \epsilon$, i.e., $z(\theta) \in \bigcap_1^m \Sigma_2^i$. For $t > \theta$ we set $v_0(t) \equiv v_0$, where v_0 is some fixed vector in Q, up to the first moment t_0 for which $z(t_0) \in \bigcup_1^m \Sigma^i$. For $t_0 < t \leqslant t_0 + \theta$ active control of evasion $v_0(t) = \widetilde{v}_{\omega_0}(t - t_0)$ which is constructed from $u(t - t_0)$ and $z(t_0)$ is used, etc. In case 2) we set $v_0(t) \equiv v_0$ up to the first moment t_1 at which $z(t_1) \in \bigcup_1^m \Sigma^i$. For $t_1 < t \leqslant t_1 + \theta$ active control of evasion $v_0(t) = \widetilde{v}_{\omega_0}(t + t_1)$ constructed from $u(t - t_1)$ and $z(t_1)$ is used. For $t > t_1 + \theta$ passive control $v_0(t) \equiv v_0$ is used up to the first moment t_2 for which $z(t_2) \in \bigcup_1^m \Sigma^i$, after which active control is used for $t_2 < t \leqslant t_2 + \theta$, etc.

Thus, for both cases with the method of control of the vector v described above the process of avoiding encounter with the set M from the point $z_0 \in R^n\backslash M$ continues indefinitely, which, according to the definition, implies the possibility of evasion in the game (4).

We now derive the estimate (5). It is obvious that for arbitrary measurable functions $u(t)$ and $v(t)$ defined on $[0, 1]$, $u(t) \in P$, $v(t) \in Q$, the corresponding solution $z(t)$, $0 \leqslant t \leqslant 1$, of the equation $\dot{z} = Cz + f(u(t), v(t))$, $z(0) = z_0 \in R^n \backslash M$, satisfies the inequality $|z(t) - z_0| \leqslant d_1 t(1 + |z_0|)$, where d_1 is a constant. Suppose that $\xi_i(0) \leqslant 1$ for a given i. Then for $0 \leqslant t \leqslant 1$

$$|\Pi_i z(t) - \Pi_i z_0| \leqslant 2d_1 t\,[1 + \eta_i(0)], \quad |\Pi^i z(t) - \Pi^i z_0| \leqslant 2d_1 t\,[1 + \eta_i(0)],$$

where π_i and π^i denote the operators of orthogonal projection in R^n onto M_i and L_i respectively. From this it follows that

$$\xi_i(t) \geqslant \frac{\xi_i(0)}{2}, \qquad 0 \leqslant t \leqslant \frac{\xi_i(0)}{4d_1[1 + \eta_i(0)]}. \tag{12}$$

On applying the control $\widetilde{v}_{\omega_0}(t)$, $0 \leqslant t \leqslant \theta$, for the solution $z(t)$ we have (cf. (11))

$$\xi_i(t) > \frac{\delta_1[\xi_i(0)]^k}{\{4d_1[1 + \eta_i(0)]\}^k} \qquad \text{for } \frac{\xi_i(0)}{4d_1[1 + \eta_i(0)]} \leqslant t \leqslant \theta, \qquad \xi_i(0) > 0. \tag{13}$$

We set $\theta_1 = \min\{\theta, 1/4d_1\}$. In the definition of $v_0(t)$ we replace θ by θ_1. For $0 \leqslant t \leqslant \theta_1$ we have

$$\eta_i(t) - \eta_i(0) \geqslant -2d_1 t\,[1 + \eta_i(0)], \qquad 1 + \eta_i(t) \geqslant {}^1/_2[1 + \eta_i(0)].$$

From this it follows that (cf. (12) and (13))

$$\xi_i(t) > d_2 \xi_i^k(0)/[1 + \eta_i(t)]^k, \quad 0 \leqslant t \leqslant \theta_1,$$

where d_2 is a constant, $d_2 < 1$. From what has been said above it may be asserted that in applying the active control $v_0(t)$ on $(T_0, T_0 + \theta_1]$ we have the inequality

$$\xi_i(t) > \frac{d_2 \xi_i^k(T_0)}{[1 + \eta_i(t)]^k}, \qquad T_0 \leqslant t \leqslant T_0 + \theta_1, \tag{14}$$

where $0 < \xi_i(T_0) \leqslant 1$ for the given i.

Let us suppose that for given i the phase point $z(t)$ for all $t \in [T_1, T_2)$ is contained in the region Σ_i^2. Then

$$\xi_i(t) \geqslant \varepsilon > \frac{d_2 \varepsilon^k}{[1 + \eta_i(t)]^k}, \qquad T_1 \leqslant t \leqslant T_2. \tag{15}$$

In the method of avoiding encounter from the point $z_0 \in R^n \backslash M$ proposed above the control of the vector v consists of a sequence of active and passive parts (when $v_0(t) \equiv v_0$). Hence, if $0 < \xi_i(0) < \epsilon$ for given i, then (cf. (14) and (15))

$$\xi_i(t) > \frac{d_2 \xi_i^k(0)}{[1 + \eta_i(t)]^k}, \qquad 0 \leqslant t \leqslant \theta_1; \qquad \xi_i(t) > \frac{d_2 \varepsilon^k}{[1 + \eta_i(t)]^k}, \qquad t \geqslant \theta_1, \tag{16}$$

and if $\xi_i(0) \geqslant \epsilon$, then (cf. (14) and (15))

$$\xi_i(t) > d_2 \varepsilon^k/[1 + \eta_i(t)]^k \qquad t \geqslant 0. \tag{17}$$

The estimate (5) follows from (16) and (17). The proof of Theorem 1 is complete.

B. ASSUMPTION 3. Suppose that for the game (4): a) Assumption 1 is satisfied; b) in R^2 there exist vectors l_i such that the numbers c_1, $c_2 = \max_{i,u\in P,v\in Q}|\Phi_i(u, v) - l_i|$ and k_i satisfy the inequality $c_1 > 2^m(1 + k_1) \ldots (1 + k_m)c_2$.

THEOREM 2. *If Assumption 3 is satisfied, then evasion is possible in the game (4). Here the process of avoiding encounter with the set M from the point $z_0 \in R^n \backslash M$ can be controlled so that for all $t \geqslant 0$*

$$\xi_i(t) > \begin{cases} \dfrac{c\xi_i^k(0)}{[1+|z(t)|]^{k^2+k}} & \text{for } \xi_i(0) < \dfrac{\varepsilon}{[1+|z_0|]^k}, \\ \dfrac{c\varepsilon^k}{[1+|z(t)|]^{k^2+k}} & \text{for } \xi_i(0) \geqslant \dfrac{\varepsilon}{[1+|z_0|]^k}, \end{cases} \tag{18}$$

where c, ϵ are constants and $k = \max\{k_1, \ldots, k_m\}$.

PROOF. Let $u(t)$, $0 \leqslant t \leqslant 1$, be an arbitrary measurable function with values in P, and let ω be an arbitrary fixed point of Γ. Let each value $\widetilde{v}_\omega(t)$, $0 \leqslant t \leqslant 1$, be chosen as the lexicographic minimum among solutions $v \in Q$ of the equation $\Phi(u(t), v) = l - \omega$. In (6) we set $v(t) = \widetilde{v}_\omega(t)$, and we obtain (7). We represent that function $\varphi_i(t, z_0)$, $0 \leqslant t \leqslant 1$, in the form

$$\varphi_i(t, z_0) = b_{i0} + b_{i1}t + \ldots + b_{i,k_i}t^{k_i} - \psi_i(t, z_0)\frac{t^{k_i}}{k_i!}, \tag{19}$$

where

$$b_{i0} = F_i\pi_i z_0,\ b_{i1} = F_i\pi_i C z_0 + a_{i0}, \ldots, b_{i,k_i} = \frac{1}{k_i!}[F_i\pi_i C^{k_i} z_0 + (l + l_i)],$$

and the function $\psi_i(t, z_0)$, $0 \leqslant t \leqslant 1$, satisfies

$$|\psi_i(t, z_0)| \leqslant d_3 t[1 + |z_0|],$$

where d_3 is a constant. Hence (cf. (7) and (19))

$$F_i\pi_i z(t) = b_{i0} + b_{i1}t + \ldots + b_{i,k_i}t^{k_i} - \frac{t^{k_i}}{k_i!}[\omega + h_i^1(t) + h_i^2(t) + \psi_i(t, z_0)]. \tag{20}$$

Arguing as in the preceding section, we find that in partitioning the square Γ into $(2k + 2)^2$ equal squares for fixed z_0 there exists a square $\Gamma_1 \subset \Gamma$ with side $c_1/(1 + k_1)$ having no intersection with the curve

$$(b_{10} + b_{11}t + \ldots + b_{1,k_1}t^{k_1})k_1!/t^{k_1}, \quad 0 < t \leqslant 1.$$

Continuing this process as in the preceding section, we distinguish, as a result, a square $\Gamma_m \subset \Gamma_{m-1}$ with side

$$c_1/(1 + k_1)(2 + 2k_2) \ldots (2 + 2k_m),$$

having no intersection with the curve

$$(b_{m0} + b_{m1}t + \ldots + b_{m,k_m}t^{k_m})k_m!/t^{k_m}, \quad 0 < t \leqslant 1.$$

Let $\omega = \omega_0 = \omega_0(z_0)$ be the center of Γ_m. Altering the argument of subsection A in the appropriate way and using (20), we find that for all $t \in (0, \theta(z_0)]$ (cf. (10))

$$|F_i\pi_i z(t)| = |b_{i0} + b_{i1}t + \ldots + b_{i,k_i}t^{k_i} - \frac{t^{k_i}}{k_i!}[\omega_0 + h_i^1(t) + h_i^2(t) + \psi_i(t, z_0)]| > \delta_2 t^{k_i}, \quad (21)$$

where δ_2 is some constant, $\theta(z_0) = a/[1 + |z_0|]$, and a is a constant, $a \leqslant 1$. Using (21) and arguing as in subsection A, we find that for all $t \in [0, \theta(z_0)]$

$$\xi_i(t) > \delta_3 t^k, \quad (22)$$

where δ_3 is a constant, $\delta_3 \leqslant 1$, $k = \max\{k_1, \ldots, k_m\}$, and $\xi_i(0) > 0$.

It is obvious that for arbitrary measurable functions $u(t)$ and $v(t)$ defined on $[0, 1]$ the corresponding solution $z(t)$, $0 \leqslant t \leqslant 1$, of the equation $\dot{z} = Cz + f(u(t), v(t))$, $z(0) = z_0$, satisfies

$$|z(t) - z_0| \leqslant bt(1 + |z_0|), \qquad |z(t)| \leqslant b(1 + |z_0|), \quad (23)$$

where b is a constant, $b \geqslant 1$. From this we obtain in the obvious way the inequality $2[1 + |z(t)|] \geqslant 1 + |z_0|$, $0 \leqslant t \leqslant h$, where h is a constant, $h \leqslant 1$. It may be assumed with no loss of generality that $a \leqslant h$. We denote the constant $\delta_3 a^k/2^k$ by ϵ and in R^n we consider the hypersurfaces Σ^i, $i = 1, \ldots, m$, defined by

$$\xi_i = \frac{\varepsilon}{[1 + |z|]^k}, \qquad i = 1, 2, \ldots, m. \quad (24)$$

We denote the region bounded by Σ^i and containing M_i by Σ_1^i, and the region $R^n \backslash (\Sigma^i \cup \Sigma_1^i)$ by Σ_2^i. The construction of the control of evasion $v_0(t)$ is similar to that of subsection A using the repelling surfaces $\bigcup_1^m \Sigma^i$. A difference from subsection A is that the length of the active section of control now depends on the position of the point $z(t)$ at the initial moment τ_0 after which active control is exercised on $(\tau_0, \tau_0 + \theta(z(\tau_0))]$.

We now deduce the estimate for the distance $\xi_i(t)$. Suppose that $z_0 \in \Sigma_1^i \backslash M_i$ for a given i. We have from (23)

$$\xi_i(t) \geqslant \frac{\xi_i(0)}{2} \qquad \text{for } 0 \leqslant t \leqslant \frac{\xi_i(0)}{2b[1 + |z_0|]}. \quad (25)$$

On applying the control $\widetilde{v}_{\omega_0}(t)$, $0 \leqslant t \leqslant \theta(z_0)$, and using (22), we obtain

$$\xi_i(t) > \frac{\delta_3 \xi_i^k(0)}{2^{2k} b^k [1 + |z(t)|]^k} \qquad \text{for } \frac{\xi_i(0)}{2b[1 + |z_0|]} \leqslant t \leqslant \theta(z_0),\ \xi_i(0) > 0. \quad (26)$$

Thus, for all $t \in [0, \theta(z_0)]$ and $\xi_i(0) > 0$ (cf. (25) and (26))

$$\xi_i(t) > \frac{\delta_4 \xi_i^k(0)}{[1 + |z(t)|]^k}, \quad (27)$$

where δ_4 is a constant, $\delta_4 \leqslant 1$.

Suppose now that $z(t_0) \in \Sigma^i$, $t_0 \geqslant 0$, for a given i. Then (cf. (24))

$$\xi_i(t) \geqslant \frac{\xi_i(t_0)}{2} \geqslant \frac{\varepsilon}{2^{k+1}[1 + |z(t)|]^k} \qquad \text{for } t_0 \leqslant t \leqslant t_0 + \frac{\xi_i(t_0)}{2b[1 + |z(t_0)|]}. \quad (28)$$

On applying the control $\widetilde{v}_{\omega_0}(t-t_0)$, $t_0 \leqslant t \leqslant t_0 + \theta(z(t_0))$, $\omega_0 = \omega_0(z(t_0))$, we get

$$\xi_i(t) > \frac{\delta_3 \xi_i^k(t_0)}{2^k b^k [1+|z(t_0)|]^k} \geqslant \frac{\delta_3 \varepsilon^k}{2^{k^2+2k} b^k [1+|z(t)|]^{k^2+k}} \quad \text{for}$$

$$\frac{\xi_i(t_0)}{2b[1+|z(t_0)|]} \leqslant t \leqslant t_0 + \theta(z(t_0)). \tag{29}$$

Hence (cf. (28) and (29)) for all $t \in [t_0, t_0 + \theta(z(t_0))]$

$$\xi_i(t) > \frac{\delta_5 \varepsilon^k}{[1+|z(t)|]^{k^2+k}}, \tag{30}$$

where δ_5 is a constant, $\delta_5 \leqslant 1$.

Let us assume that for given i we have $z(t) \in \Sigma_2^i$ for all $t \in [t_1, t_2]$. Then

$$\xi_i(t) > \frac{\varepsilon}{[1+|z(t)|]^k}, \qquad t_1 \leqslant t \leqslant t_2. \tag{31}$$

Coarsening (27), (30) and (31), and using the inequality $\epsilon/[1+|z_0|]^k \leqslant \epsilon$, we obtain (18). The proof of Theorem 2 is complete.

C. ASSUMPTION 4. Suppose that $f(u, v) = -u + v$ in the game (4), and there exist natural numbers k_i, two-dimensional subspaces $W_i \subset L_i$, linear mappings F_i: $W_i \longrightarrow R^2$, and linear mappings Φ: $Q \longrightarrow R^2$ and Φ_i: $P \times Q \longrightarrow R^2$ such that:

a) F_i is an injective mapping;

b) each set $F_i \pi_i C^j(-P+Q)$, $j = 0, 1, \ldots, k_i - 2$, consists of a single point;

c) $F_i \pi_i C^{k_i - 1}(-u+v) = \Phi(v) + \Phi_i(u, v)$ for all $u \in P$ and $v \in Q$, and the set $R = \Phi(Q)$ contains an interior point.

We note that Assumption 4 is more restrictive than Assumption 1 (if Assumption 4 is satisfied, so is Assumption 1); however, it is simpler in form and more easily verified. It is obvious that if Assumption 4 is satisfied the set $R = \Phi(Q)$ is compact, and there exist a largest square $\Gamma \subset R^2$ with center at 0 and a vector $l \in R^2$ such that $l + \Gamma \subset R$; let $2c_1$ be the side of Γ.

THEOREM 3. *If Assumptions* 4 *and* 2 *are satisfied, then evasion is possible in the game* (4). *Here the process of avoiding encounter with the set M from the point* $z_0 \in R^n \backslash M$ *can be controlled so that for all* $t \geqslant 0$ *estimate* (5) *holds, while to construct the evasion control* $v(t)$ *it suffices to use only information on the values of* $z(s)$ *for* $s \in [0, t]$, *and* $v(t)$ *can be chosen piecewise constant.*

PROOF. For a vector $\omega \in \Gamma$ the function $\widetilde{v}_\omega(t) \equiv \widetilde{v}_\omega$, $0 \leqslant t \leqslant 1$, is chosen from the condition

$$\Phi(\tilde{v}_\omega) = l - \omega. \tag{32}$$

The remainder of the proof coincides with the proof of Theorem 1 (see subsection A). It is clear that in the method of controlling the vector v described in subsection A only values of $z(s)$, $s \in [0, t]$, are used for finding the value $v_0(t)$; the fact that the evasion control $V(z(s), 0 \leqslant s \leqslant t)$ is piecewise constant is obvious (cf. (32)). This completes the proof of Theorem 3.

ASSUMPTION 5. Suppose that for the game (4): a) Assumption 4 is satisfied; and b) in R^2 there exist vectors l_i such that the numbers

$$c_1,\ c_2 = \max_{i,\, u \in P,\, v \in Q} |\Phi_i(u, v) - l_i|$$

and the numbers k_i satisfy the inequality

$$c_1 > 2^m (1 + k_1) \ldots (1 + k_m) c_2.$$

THEOREM 4. *If Assumption 5 is satisfied, then evasion is possible in the game* (4). *Here the process of avoiding encounter with the set M from the point $z_0 \in R^n \backslash M$ can be controlled in such a way that for all $t \geqslant 0$ the estimate* (18) *holds, and to construct the evasion control $v(t)$ it suffices to use information on $z(s)$ for $s \in [0, t]$, while $v(t)$ can be chosen piecewise constant.*

PROOF. The rule (32) is also used in the proof of Theorem 4. We now argue just as in the proof of Theorem 2 in subsection B. Obviously, the control of evasion $v_0(t)$ so constructed has the form $V(z(s), 0 \leqslant s \leqslant t)$ and is piecewise constant. This complete the proof of Theorem 4.

REMARK 2. The conditions of Theorems 1–4 are naturally called *conditions of maneuvering superiority*.

REMARK 3. As is evident from the proofs of Theorems 1–4, Theorem 3 is a particular case of Theorem 1, and Theorem 4 is a particular case of Theorem 2. However, in Theorems 3 and 4 only information on the values of $z(s)$, $0 \leqslant s \leqslant t$, is used to construct the control. Further, finding the constants n_i in the formulations of Theorems 1 and 3 is a difficult analytical problem, while for Theorems 2 and 4 the corresponding constants are the numbers k_i. Therefore, the conditions of Theorems 2 and 4 are more easily verified than the conditions of Theorems 1 and 3 (see the examples in the next subsection).

D. EXAMPLES. I. The points $x_1, \ldots, x_m$ pursue the point y (the vectors $x_1, \ldots, x_m, y \in R^\nu$). The pursuit is considered finished if for at least one value of the index i we have $x_i = y$. The movements of the points $x_1, \ldots, x_m$ and y are described by the differential equations

$$\begin{aligned} x_i^{(p_i)} + a_{i1} x_i^{(p_i-1)} + \ldots + a_{i,p_i} x_i &= u_i, \qquad u_i \in P_i, \\ y^{(q)} + b_1 y^{(q-1)} + \ldots + b_q y &= v, \qquad v \in Q, \end{aligned} \tag{33}$$

where a_{ij} and b_l are linear mappings of R^ν into itself, $j = 1, \ldots, p_i$, $l = 1, \ldots, q$, and P_i and Q are nonempty compact subsets of R^ν.

We have the following assertion.

ASSERTION 1. Suppose that there exist a two-dimensional subspace $W \subset R^\nu$ and a ball $V \subset W$ such that the orthogonal projection πQ of the set Q onto W contains the ball V: $\pi Q \supset V$. Suppose, moreover, at least one of the following two conditions is satisfied: a) $q < p_i$ for all $i = 1, \ldots, m$; b) $q \leqslant p_i$ for all $i = 1, \ldots, m$, and the constants r_i, the radius r of the ball V, and the number q satisfy $r > 2^{m+1/2}(1 + q)^m \max\{r_1, \ldots, r_m\}$, where r_i is the minimum of the radii of two-dimensional balls in W containing πP_i. Then all the hypotheses of Theorem 4 are satisfied, i.e., evasion is possible in the game (33).

The assertion is proved by direct calculation (we set $k_i = q$, $c_1 = r/\sqrt{2}$, and we take as l_i the center of the two-dimensional ball of minimum radius r_i in W which contains πP_i).

II. We shall consider particular cases of the game (33).

The following example is an analogue of the control example of L. S. Pontrjagin (cf. [1], [2] and [15]):

$$\begin{aligned} &\ddot{x}_i + \alpha_i \dot{x}_i = \rho_i u_i, \quad |u_i| \leqslant 1, \quad x_i \in R^\nu, \quad \nu \geqslant 2, \\ &\ddot{y} + \beta \dot{y} = \sigma v, \quad |v| \leqslant 1, \quad y \in R^\nu, \end{aligned} \tag{34}$$

where α_i, β, ρ_i, and σ are nonnegative numbers. It is clear that for the game (34) the radii r_i are equal to ρ_i, and $q = 2$. It is not difficult to show that all conditions of the assertion are satisfied if

$$\nu \geqslant 2, \qquad \sigma > 2^{m+1/2} \cdot 3^m \max\{\rho_1, \rho_2, \ldots, \rho_m\}. \tag{35}$$

Theorem 3 is also applicable to this example if

$$\nu \geqslant 2, \qquad \sigma > 2^{m+1/2} \cdot 5^m \max\{\rho_1, \rho_2, \ldots, \rho_m\}, \tag{36}$$

or, as is easily seen, $k_i = 2$ and $n_i \leqslant 4$.

We point out that although Theorem 4 is this example guarantees the possibility of evasion under fewer assumptions than Theorem 3 (cf. (35) and (36)), the lower bound for the distance $\xi_i(t)$ in the first case is poorer than in the second.

The following example is an analogue of the problem of the "boy and the crocodile" (cf. [1], [2] and [15]):

$$\begin{aligned} &\ddot{x}_i = u_i, \quad |u_i| \leqslant 1, \quad x_i \in R^\nu, \quad \nu \geqslant 2, \\ &\dot{y} = v, \quad |v| \leqslant 1, \quad y \in R^\nu. \end{aligned} \tag{37}$$

It can be shown that for the game (37) all the conditions of Theorems 1–4 are satisfied with $k_i = n_i = 1$, $l = l_i = 0$, $c_1 = 1/\sqrt{2}$ and $c_2 = 0$.

REMARK 4. If the hypotheses of Theorem 1 of [15] are satisfied, then all the hypotheses of Theorem 1 of the present paper are satisfied with $c_2 = 0$.

REMARK 5. Suppose that $m = 1$. If the conditions for evasion of [5] are satisfied, then all the hypotheses of Theorem 1 of the present work are satisfied with $c_2 = 0$.

§2. Directional advantage

A. Let $\psi_1, \ldots, \psi_{2m}$ be arbitrary vectors of unit length in R^n with $\psi_{2i-1} = -\psi_{2i}$. From this system of vectors we form the sets $K_1, K_2, \ldots$ in the following manner: each set K_j contains only m vectors and only one from each pair ψ_{2i-1}, ψ_{2i}. It is clear that the set $\{K_1, K_2, \ldots\}$ is finite; we denote the number of elements in it by r.

ASSUMPTION 6. Suppose that for the game (4) there exist natural numbers k_i, subspaces $W_i \subset L_i$, and vectors $l_i \in W_i$ such that: a) dim $W_i \geqslant k_i + 2$; b) each set

$\pi_i C^j f(P, Q)$, $j = 0, 1, \ldots, k_i - 2$, consists of a single point; and c) at least one of the numbers N_1 or N_2, where

$$N_1 = \min_{\substack{\psi_1, \psi_2 \in S_1 \\ \psi_3, \psi_4 \in S_2 \\ \cdots\cdots \\ \psi_{2m-1}, \psi_{2m} \in S_m}} \max_{K_1, K_2, \ldots, K_r} \max_{v \in Q} \min_{u \in P} \min_i (\psi_i^j, \pi_i C^{k_i-1} f(u, v) - l_i),$$

$$N_2 = \min_{\substack{\psi_1, \psi_2 \in S_1 \\ \psi_3, \psi_4 \in S_2 \\ \cdots\cdots \\ \psi_{2m-1}, \psi_{2m} \in S_m}} \max_{K_1, K_2, \ldots, K_r} \min_{u \in P} \max_{v \in Q} \min_i (\psi_i^j, \pi_i C^{k_i-1} f(u, v) - l_i),$$

is positive (the vectors $\psi_1^j, \ldots, \psi_m^j$ form the set K_j).

THEOREM 5. *If Assumption 6 is satisfied, then evasion is possible in the game* (4). *Here the process of avoiding encounter with the set M from the point* $z_0 \in R^n \backslash M$ *can be controlled in such a way that for all* $t \geqslant 0$ *the estimate* (18) *holds. In the case where* $N_1 > 0$ *only values of* $z(s)$ *for* $s \in [0, t]$ *are used to find* $v(t)$.

PROOF. Let

$$b_{i0}(z_0) = \pi_i z_0, \quad b_{i1}(z_0) = \pi_i C z_0 + \pi_i f(P, Q),$$
$$\ldots, b_{i, k_i - 1}(z_0) = \pi_i C^{k_i-1} z_0 + \pi_i C^{k_i-2} f(P, Q), \quad b_{i, k_i}(z_0) = \pi_i C^{k_i} z_0 + l_i.$$

According to condition a) of Assumption 6, dim $W_i \geqslant k_i + 2$. Therefore, there exists a vector $\varphi_i(z_0) \in S_i$ for which $(b_{ij}(z_0), \varphi_i(z_0)) = 0$ for all $j = 1, \ldots, k_i$. It is clear that the vector $-\varphi_i(z_0) \in S_i$ also satisfies this condition. From the system of vectors $\psi_1 = \varphi_1(z_0)$, $\psi_2 = -\varphi_1(z_0)$; $\ldots$; $\psi_{2m-1} = \varphi_m(z_0)$, $\psi_{2m} = -\varphi_m(z_0)$ we form the sets $\bar{K}_1 = K_1(z_0), \ldots, \bar{K}_r = K_r(z_0)$. From the definition of N_1 it follows that

$$N_1 \leqslant \max_{\{\bar{K}_1, \bar{K}_2, \ldots, \bar{K}_r\}} \max_{v \in Q} \min_{u \in P} \min_i (\psi_i^j, \pi_i C^{k_i-1} f(u, v) - l_i),$$

where the vectors $\psi_1^j, \ldots, \psi_m^j$ form the sets $\bar{K}_j$, $j = 1, \ldots, r$. It is obvious that the function $(\psi_i^j, \pi_i C^{k_i-1} f(u, v) - l_i)$ is continuous in (u, v) jointly. Therefore, there exist a set $\bar{K}_{j_1}$ and a vector $v_0(z_0) \in Q$ such that

$$N_1 \leqslant \min_i (\psi_i^{j_1}, \pi_i C^{k_i-1} f(u, v_0(z_0)) - l_i), \tag{38}$$

where u is an arbitrary point of P.

It is obvious that there exists a set $\bar{K}_{j_2}$ such that

$$N_2 \leqslant \max_{\{\bar{K}_1, \bar{K}_2, \ldots, \bar{K}_r\}} \min_{u \in P} \max_{v \in Q} \min_i (\psi_i^j, \pi_i C^{k_i-1} f(u, v) - l_i)$$
$$= \min_{u \in P} \max_{v \in Q} \min_i (\psi_i^{j_2}, \pi_i C^{k_i-1} f(u, v) - l_i).$$

It is clear that

$$\min_{u \in P} \max_{v \in Q} \min_i (\psi_i^{j_2}, \pi_i C^{k_i-1} f(u, v) - l_i) \leqslant \max_{v \in Q} \min_i (\psi_i^{j_2}, \pi_i C^{k_i-1} f(u, v) - l_i),$$

where u is an arbitrary point of P. Let $u(s)$, $0 \leqslant s \leqslant 1$, be an arbitrary measurable function with values in P.

We consider the following equation for $v \in Q$:

$$\max_{v\in Q}\min_{i}(\psi_i^{j_2},\ \pi_i C^{k_i-1}f(u(s),\ v)-l_i)=\min_{i}(\psi_i^{j_2},\ \pi_i C^{k_i-1}f(u(s),\ v)-l_i). \tag{39}$$

It is obvious that the function

$$\min_{1\leqslant i\leqslant m}(\psi_i^{j_2},\ \pi_i C^{k_i-1}f(u,\ v)-l_i)$$

is continuous in (u, v) jointly. Therefore, a solution of (39) exists. The smallest solution of (39) in the lexicographic sense we denote by $v_0(s)$, $0 \leqslant s \leqslant 1$. Arguing as in [5], we can demonstrate the measurability of the function $v_0(s)$, $0 \leqslant s \leqslant 1$, on the segment [0, 1]. We note that $v_0(s) = v(z_0, u(s))$.

There are two possible cases: a) $N_1 > 0$; b) $N_1 \leqslant 0, N_2 > 0$. We denote by N, j, and $\bar{v}(s)$, $0 \leqslant s \leqslant 1$, respectively the numbers N_1, j_1 and the vector $v_0(z_0)$ in the case where $N_1 > 0$ and the numbers N_2, j_2 and the function $v_0(s)$, $0 \leqslant s \leqslant 1$, in the case where $N_1 \leqslant 0$, $N_2 > 0$.

We have (cf. (18) and (39))

$$N\leqslant\min_{1\leqslant i\leqslant m}(\psi_i^{j},\ \pi_i C^{k_i-1}f(u(s),\ \bar{v}(s))-l_i)\leqslant(\psi_i^{j},\ \pi_i C^{k_i-1}f(u(s),\ \bar{v}(s))-l_i) \tag{40}$$

for all $s \in [0, 1]$ and $i = 1, \ldots, m$.

For the solution $z(t)$, $0 \leqslant t \leqslant 1$, corresponding to the functions $u(t)$, $\bar{v}(t)$, $0 \leqslant t \leqslant 1$, and the initial condition $z(0) = z_0$ we have (cf. [5])

$$\pi_i z(t)=\pi_i e^{tC}z_0+a_{i0}t+\ldots+a_{i,\,k_i-2}\frac{t^{k_i-1}}{(k_i-1)!}+l_i\frac{t^{k_i}}{k_i!}$$
$$+\int_0^t\frac{(t-s)^{k_i-1}}{(k_i-1)!}\{[\pi_i C^{k_i-1}f(u(s),\ \bar{v}(s))-l_i]+g_i(t,\ s)\}\,ds,$$

where $|g_i(t, s)| \leqslant d_2(t-s) \leqslant d_2 t$, $0 \leqslant t \leqslant 1$, d_2 a constant.

Obviously, for $0 \leqslant t \leqslant 1$

$$\pi_i e^{tC}z_0=\pi_i z_0+t\pi_i Cz_0+\ldots+\frac{t^{k_i}}{k_i!}\pi C^{k_i}z_0+\varphi_i(t,\ z_0)\frac{t^{k_i}}{k_i!},$$

where $|\varphi_i(t, z_0)| \leqslant d_3 t(1 + |z_0|)$, d_3 a constant. Furthermore, let

$$\theta=\min\left\{1,\ \frac{N}{2d_2},\ \frac{N}{4d_3(1+|z_0|)}\right\}.$$

Then $|g_i(t, s)| \leqslant d_2\theta \leqslant N/2$ for $t \in [0, \theta]$. Hence (cf. (40))

$$(\pi_i z(t),\ \psi_i^j)\geqslant\sum_{\beta=0}^{k_i}(b_{i\beta}(z_0),\ \psi_i^j)\frac{t^\beta}{\beta!}+\frac{t^{k_i}}{2k_i!}[N-2|\varphi_i(t,\ z_0)|]\geqslant\frac{t^k}{4k!}N \tag{41}$$

for all $t \in [0, \theta]$ and $i = 1, \ldots, m$ (in (41), $k = \max\{k_1, \ldots, k_m\}$). It is clear that $(\pi_i z(t), \psi_i^j) \leqslant |\pi_i z(t)| \leqslant \xi_i(t)$. Therefore (cf. (41))

$$\xi_i(t)\geqslant\frac{t^k}{4k!}N,\qquad 0\leqslant t\leqslant\theta. \tag{42}$$

By enlarging the constant d_3, it may be assumed that

$$\theta = \frac{a}{1+|z_0|}, \tag{43}$$

where a is a constant, $a \leqslant 1$. The rest of the argument coincides completely with the argument used in §1 in the proof of Theorem 2. This completes the proof of Theorem 5.

B. EXAMPLES. I. We consider the differential game (33) of §1. We have the following assertion.

ASSERTION 2. Suppose that for the game (33): a) $q \leqslant p_i$, b) $\nu \geqslant q + 2$, and c) we have the inequality

$$\min_{\substack{\psi_1, \psi_2 \in S \\ \cdots\cdots \\ \psi_{2m-1}, \psi_{2m} \in S}} \max_{K_1, K_2, \ldots, K_r} \max_{v \in Q} \min_i (\psi_i^j, v) > \max_{\substack{\psi_1, \psi_2 \in S \\ \cdots\cdots \\ \psi_{2m-1}, \psi_{2m} \in S}} \min_{K_1, K_2, \ldots, K_r} \max_{u \in P} \max_i (\psi_i^j, u_i),$$

where S is the unit sphere in R^ν with center at the point 0. Then all the hypotheses of Theorem 5 are satisfied.

II. We now consider the game (34). It is obvious that conditions a) and b) of Assertion 2 are satisfied if $\nu \geqslant 4$. We now consider condition c) of Assertion 2. It is obvious that

$$\max_{\substack{\psi_1, \psi_2 \in S \\ \cdots\cdots \\ \psi_{2m-1}, \psi_{2m} \in S}} \min_{K_1, K_2, \ldots, K_r} \max_{u \in P} \max_i (\psi_i^j, \rho_i u_i) \leqslant \max\{\rho_1, \rho_2, \ldots, \rho_m\}.$$

On the other hand, it can be shown that

$$\min_{\substack{\psi_1, \psi_2 \in S \\ \cdots\cdots \\ \psi_{2m-1}, \psi_{2m} \in S}} \max_{K_1, K_2, \ldots, K_r} \max_{|v| \leqslant 1} \min_i (\psi_i^j, \sigma v) = \sigma\mu(\nu, m),$$

where

$$\mu(\nu, m) = \min_{\substack{\psi_1 \in S \\ \cdots \\ \psi_m \in S}} \max_{|v| \leqslant 1} \min_{i=1, \ldots, m} |(\psi_i, v)|,$$

and $\psi_1, \ldots, \psi_m$ are arbitrary vectors of the sphere S. Further, it is possible to show that condition c) of Assertion 2 is satisfied if $\sigma\mu(\nu, m) > \max\{\rho_1, \ldots, \rho_m\}$. Considering the special nature of (34), the constant 4 in the inequality $\nu \geqslant 4$ can be reduced to 3.

Thus, evasion is possible in the game (34) if

$$1)\ \nu \geqslant 3; \qquad 2)\ \sigma\mu(\nu, m) > \max\{\rho_1, \rho_2, \ldots, \rho_m\}. \tag{44}$$

It is possible to show that all the conditions of Assertion 2 are satisfied for the game (37) with $\nu \geqslant 3$ (using the basic ideas of the proof of Theorem 5 and the specific features of (37), it is not hard to prove the possibility of evasion in the game (37) for $\nu = 2$ as well).

§3. l-evasion

A. ASSUMPTION 7. Suppose that for the game (4) there exist natural numbers k_i, subspaces $W_i \subset L_i$, $\dim W_i \geqslant 1$, a constant θ, and vectors $l_i \in W_i$ such that: a) each set $\pi_i C^j f(P, Q)$, $j = 0, 1, \ldots, k_i - 2$, consists of a single point; b) for each point $z_0 \in R^n \backslash M$ there is a subspace $W_i(z_0) \subset W_i$, $\dim W_i(z_0) < \dim W_i$, such that

$$G_i(z_0, t) = \pi_i e^{tC} z_0 + a_{i0} t + \ldots + a_{i, k_i-2} \frac{t^{k_i-1}}{(k_i-1)!} + l_i \frac{t^{k_i}}{k_i!} \in W_i(z_0)$$

for all $t \in [0, 2\theta]$, where $a_{ij} = \pi_i C^j f(P, Q)$; and c) at least one of the numbers N_1 or N_2 defined as in §2 is positive.

THEOREM 6. *Let Assumption 7 be satisfied for the game* (4). *Then evasion is possible in the game* (4). *Here the process of avoiding encounter with the set M from the point $z_0 \in R^n \backslash M$ can be controlled in such a way that*

$$\xi_i(t) \geqslant \begin{cases} c_i t^{k_i}, & 0 \leqslant t \leqslant \theta_1, \\ l_i^0, & t \geqslant \theta_1, \end{cases} \tag{45}$$

where c_i, θ_1 and l_i^0 are constants. In the case where $N_1 > 0$ only values of $z(s)$, $s \in [0, t]$, are used to find the value of $v(t)$.

PROOF (cf. [10]). According to condition b) of Assumption 7 the curve

$$G_i(t, z_0) = \pi_i e^{tC} z_0 + a_{i0} t + \ldots + a_{i, k_i-2} \frac{t^{k_i-1}}{(k_i-1)!} + l_i \frac{t^{k_i}}{k_i!}, \quad 0 \leqslant t \leqslant 2\theta,$$

is contained in the subspace $W_i(z_0)$. Since $\dim W_i(z_0) < \dim W_i$, it is obvious that there exists a vector $\varphi_i^0(z_0) \in S_i$ orthogonal to the subspace $W_i(z_0)$. Hence for all $t \in [0, 2\theta]$

$$(G_i(t, z_0), \varphi_i^0(z_0)) = 0. \tag{46}$$

From the system of vectors $\overline{\psi}_1^0 = \varphi_1^0(z_0)$, $\overline{\psi}_2^0 = -\varphi_1^0(z_0), \ldots, \overline{\psi}_{2m-1}^0 = \varphi_m^0(z_0)$, $\overline{\psi}_{2m}^0 = -\varphi_m^0(z_0)$ we form the sets $K_1^0, \ldots, K_r^0$, where $K_j^0 = K_j(z_0)$ (cf. §2). Arguing as in the preceding section, we can prove the existence of a set $K^0 \in \{K_1^0, \ldots, K_r^0\}$ and construct a measurable function $\overline{v}_0(t)$, $0 \leqslant t \leqslant 2\theta$ (here either $\overline{v}_0(t)$ is the constant vector $v_0(z_0)$ or $\overline{v}_0(t) = v(z_0, u(t))$), such that for all $s \in [0, 2\theta]$ and $i = 1, \ldots, m$,

$$N \leqslant (\psi_i^0, \pi_i C^{k_i-1} f(u(s), \overline{v}_0(s)) - l_i), \tag{47}$$

where the vectors $\psi_1^0, \ldots, \psi_m^0$ form the set K^0, $N = N_1$ for $N_1 > 0$ and $N = N_2$ if $N_1 \leqslant 0$, $N_2 > 0$. For the solution $z(t)$, $0 \leqslant t \leqslant 1$, corresponding to the functions $u(t)$, $\overline{v}_0(t)$, $0 \leqslant t \leqslant 1$, and the initial condition $z(0) = z_0 \notin M$ we have (cf. §2 and [5])

$$\pi_i z(t) = \pi_i e^{tC} z_0 + a_{i0} t + \ldots + a_{i, k_i-2} \frac{t^{k_i-1}}{(k_i-1)!} + l_i \frac{t^{k_i}}{k_i!}$$
$$+ \int_0^t \frac{(t-s)^{k_i-1}}{(k_i-1)!} \{[\pi_i C^{k_i-1} f(u(s), \overline{v}_0(s)) - l_i] + g_i(t, s)\} \, ds,$$

where $|g_i(t, s)| \leqslant d \cdot (t - s) \leqslant d \cdot t$, $0 \leqslant t \leqslant 2\theta$, d a constant. Recalling (46) and (47), as in §2 we obtain

$$(\pi_i z(t), \psi_i^0) \geqslant \frac{t^{k_i}}{2k_i!} N, \quad 0 \leqslant t \leqslant 2\theta_1, \quad \theta_1 = \min\left\{\theta, \frac{N}{4d}\right\}. \tag{48}$$

Let $I(\gamma) = [\gamma\theta_1, (\gamma + 1)\theta_1)$, $\gamma = 0, 1, \ldots$. We construct functions $\bar{v}_\gamma(t)$, $t \in I(\gamma)$, inductively in the following manner. For $\gamma = 0$ the function $\bar{v}_0(t)$ has been constructed above (cf. (47)). Let γ be any natural number, and let $z_\gamma = z(\gamma\theta_1)$. For the point z_γ we find vectors $\varphi_i^\gamma(z_\gamma)$ from the condition that $(G_i(t, z_\gamma), \varphi_i^\gamma(z_\gamma)) = 0$ for all $t \in [\gamma\theta_1, (\gamma + 2)\theta_1]$, $i = 1, \ldots, m$. Further, we set $\bar{\psi}_1^\gamma = \varphi_1^\gamma(z_\gamma)$, $\bar{\psi}_2^\gamma = -\varphi_1^\gamma(z_\gamma), \ldots,$ $\bar{\psi}_{2m}^\gamma = -\varphi_m^\gamma(z_\gamma)$, and from the vectors $\bar{\psi}_1^\gamma, \ldots, \bar{\psi}_{2m}^\gamma$ we construct sets $K_1^\gamma, \ldots, K_r^\gamma$. As in the case $\gamma = 0$ we find a set $K^\gamma \in \{K_1^\gamma, \ldots, K_r^\gamma\}$ and a measurable function $\bar{v}_\gamma(s)$, $\gamma\theta_1 \leqslant s \leqslant (\gamma + 2)\theta_1$, with values in Q such that (cf. (47))

$$N \leqslant (\psi_i^\gamma, \ \pi_i C^{k_i - 1} f(u(s), \bar{v}_\gamma(s)) - l_i)$$

for all $s \in [\gamma\theta_1, (\gamma + 2)\theta_1]$, where $\psi_1^\gamma, \ldots, \psi_m^\gamma$ form the set K^γ. Noting the inequality (cf. §2)

$$(\psi_i^\gamma, \ \pi_i C^{k_i-1} f(u(s), \ \bar{v}_\gamma(s)) - l_i) + (\psi_i^\gamma, \ g_i(t, s)) \geqslant N - |g_i(t, s)| \geqslant N - d\theta_1 \geqslant 3N/4$$

$(0 \leqslant s \leqslant t \leqslant \theta_1, \gamma = 0)$, from the obvious equality

$$\pi_i z(t) = \pi_i e^{(t - \gamma\theta_1)C} z_\gamma + \int_{\gamma\theta_1}^{t} \pi_i e^{(t-s)C} f(u(s), \ \bar{v}_\gamma(s))\, ds, \qquad t \in I(\gamma), \qquad \gamma \geqslant 0, \tag{49}$$

we obtain for $t \in I(\gamma)$

$$(\pi_i z(t), \ \psi_i^\gamma) \geqslant \frac{3(t - \gamma\theta_1)^{k_i}}{4k_i!} N. \tag{50}$$

Hence

$$\xi_i(t) \geqslant \frac{3(t - \gamma\theta_1)^{k_i}}{4k_i!} N, \qquad t \in I(\gamma), \qquad \gamma \geqslant 0. \tag{51}$$

The possibility of evasion in the game (4) follows easily from (51). Indeed, it is obvious that $\xi_i(t) \neq 0$ for all $t \in (\gamma\theta_1, (\gamma + 1)\theta_1)$, $\gamma \geqslant 0$. Further, $\xi_i(0) \neq 0$, and $\xi_i(\gamma\theta_1) \geqslant 3\theta_1^{k_i} N/4k_i!$ for $\gamma \geqslant 1$. Hence $\xi_i(t) \neq 0$ for all $i = 1, \ldots, m$ and $t \in [0, \infty)$.

We now deduce (45). From the obvious equality

$$z_\gamma = e^{\theta_1 C} z_{\gamma-1} + \int_0^{\theta_1} e^{sC} f(u(\gamma\theta_1 - s), \ \bar{v}_{\gamma-1}(\gamma\theta_1 - s))\, ds, \qquad \gamma \geqslant 1, \tag{52}$$

it follows (cf. (49)) that for $\gamma\theta_1 \leqslant t \leqslant (\gamma + 1)\theta_1$

$$\pi_i z(t) = \pi_i e^{(t - \gamma\theta_1 + \theta_1)C} z_{\gamma-1} + \int_0^{\theta_1} \pi_i e^{(t - \gamma\theta_1 + s)C} f(u(\gamma\theta_1 - s), \ \bar{v}_{\gamma-1}(\gamma\theta_1 - s))\, ds$$
$$+ \int_{\gamma\theta_1}^{t} \pi_i e^{(t-s)C} f(u(s), \ \bar{v}_\gamma(s))\, ds, \qquad \gamma \geqslant 1. \tag{53}$$

Elementary calculations show that

$$(\pi_i z(t), \psi_i^{\gamma-1}) \geqslant \frac{3N}{4k_i!}\left[(t - \gamma\theta_1 + \theta_1)^{k_i} - (t - \gamma\theta_1)^{k_i}\right] - H_i(t - \gamma\theta_1), \tag{54}$$

where $i = 1, \ldots, m$, $t \in I(\gamma)$, $\gamma \geq 1$ and

$$H_i = \max_{\substack{r \in [0, \theta_1] \\ u \in P \\ v \in Q}} \left| \pi_i e^{rC} f(u, v) - \sum_{j=0}^{k_i - 2} a_{ij} \frac{r^j}{j!} - l_i \frac{r^{k_i - 1}}{(k_i - 1)!} \right|.$$

Let $t \in I(\gamma)$, $\gamma \geq 0$ and

$$\alpha_i(t, \gamma) = \max \left\{ 0, \frac{3N}{4k_i!} [(t - \gamma\theta_1 + \theta_1)^{k_i} - (t - \gamma\theta_1)^{k_i}] - H_i (t - \gamma\theta_1) \right\}. \quad (55)$$

From (54) and (55) we obtain

$$|(\pi_i z(t), \psi_i^{\gamma-1})| \geq \alpha_i(t, \gamma), \quad \gamma \geq 1. \quad (56)$$

It is clear that the length $|\pi_i z(t)|$ of the vector $\pi_i z(t)$ is not less than the lengths $|(\pi_i z(t), \psi_i^\gamma)|$ and $|(\pi_i z(t), \psi_i^{\gamma-1})|$ of the projections $(\pi_i z(t), \psi_i^\gamma)$ and $(\pi_i z(t), \psi_i^{\gamma-1})$ of the vector $\pi_i z(t)$ onto the unit vectors ψ_i^γ and $\psi_i^{\gamma-1}$. Hence

$$|\pi_i z(t)| \geq 1/2 [|(\pi_i z(t), \psi_i^\gamma)| + |(\pi_i z(t), \psi_i^{\gamma-1})|].$$

From (51) and (56) we have

$$\xi_i(t) \geq \frac{1}{2} \left[\frac{3(t - \gamma\theta_1)^{k_i}}{4k_i!} N + \alpha_i(t, \gamma) \right], \quad t \in I(\gamma), \quad \gamma \geq 1. \quad (57)$$

Let

$$c_i = \frac{3N}{4k_i!}, \quad l_i^0(\gamma) = \min_{t \in [\gamma\theta_1, (\gamma+1)\theta_1]} \frac{1}{2} [c_i (t - \gamma\theta_1)^{k_i} + \alpha_i(t, \gamma)], \quad \gamma \geq 1. \quad (58)$$

It is not hard to show that $l_i^0(\gamma) = l_i^0$, where $l_i^0 = l_i^0(0)$, and

$$l_i^0(0) = \min_{t \in [0, \theta_1]} \frac{1}{2} [c_i t^{k_i} + \alpha_i(t, 0)]. \quad (59)$$

The positivity of the numbers l_i^0 follows from the definition of the numbers $l_i^0(0)$ (cf. (59)). Thus (cf. (51) and (57)–(59))

$$\xi_i(t) \geq c_i t^{k_i}, \quad 0 \leq t \leq \theta_1; \qquad \xi_i(t) \geq l_i^0, \quad t \geq \theta_1.$$

This completes the proof of Theorem 6.

B. EXAMPLES. As examples we consider the games (34) and (37). We have the following assertion.

ASSERTION 3. All the conditions of Theorem 6 are satisfied for the game (34) with $\nu \geq 3$ and $\sigma\mu(\nu, m) > \max\{\rho_1, \ldots, \rho_m\}$, where $\mu(\nu, m)$ is the function introduced in §2, and for the game (37) for arbitrary $\nu \geq 3$.

The assertion is proved by direct calculation. For the examples (34) and (37) it is not hard to compute the constants c_i, k_i, θ_1 and l_i^0 (cf. [10], [14] and [16]).

Using the basic ideas of the proof of Theorem 6 and the specific features of (37), we can show that in game (37) evasion is also possible for $\nu = 2$.

§4. Differential games with phase restrictions

A. In this section a differential game with phase restrictions is considered which is described by the equations

$$\dot{x}_i = A_i x_i + B_i y + f_i(u, v), \qquad \dot{y} = \beta y + f_{m+1}(u, v), \tag{60}$$

where $x_i \in R^{\nu_i}$, $y \in R^\nu$, A_i and B_i are constant matrices, u is the control vector of pursuit and v the control vector of evasion, $u \in P$, $v \in Q$, where P and Q are nonempty compact subsets of the spaces R^p and R^q respectively, $f_j(u, v)$, $j = 1, \ldots, m+1$, are functions defined on $P \times Q$ and continuous there in (u, v) jointly, and β is some real number. The terminal set and the phase restrictions are defined in the following manner.

Let N_i be some subset of the set of natural numbers. We denote the elements of N_i by $n_i^1, \ldots, n_i^{s_i}$. Let $a = (a^1, \ldots, a^k)^T \in R^k$, $k \geq s_i$. The symbol $\{a\}_{N_i}$ denotes the vector of the space R^{s_i} with components $a^{n_i^1}, \ldots, a^{n_i^{s_i}}$. The game (60) is considered finished if for at least one value of i

$$\{x\}_{N_i} = \{y\}_{N_i}. \tag{61}$$

Let $q_k \in R^\nu$, $k = 1, \ldots, s$, be fixed vectors. It is assumed that during the game phase restrictions prescribed by the following inequalities are satisfied:

$$(q_k, y) \geqslant 0, \quad k = 1, 2, \ldots, s. \tag{62}$$

By (61) and (62) the sets N_i and vectors q_k define the terminal set (the rule for stopping the game (60)) and the phase restrictions imposed on the positions of the phase point y; they thus enter into the description (rules) of the game (60). We say that, starting from the initial position x_{i0}, y_0 satisfying the conditions $\{x_{i0}\}_{N_i} \neq \{y_0\}_{N_i}$ and $(q_k, y_0) \geq 0$ for all i and k, it is possible to avoid encounter if for any measurable functions $u(t)$, $t \geq 0$, $u(t) \in P$, it is possible to construct a measurable variation $v(t)$ of the vector v such that the solution $x_1(t), \ldots, x_m(t), y(t)$ of the equations

$$\begin{aligned} \dot{x}_i &= A_i x_i + B_i y + f_i(u(t), v(t)), & x_i(0) &= x_{i0}, \\ \dot{y} &= \beta y + f_{m+1}(u(t), v(t)), & y(0) &= y_0, \end{aligned} \tag{63}$$

for all $t \geq 0$ satisfies the conditions

$$(q_k, y(t)) \geqslant 0, \quad k = 1, 2, \ldots, s; \qquad \{x_i(t)\}_{N_i} \neq \{y(t)\}_{N_i}, \quad i = 1, 2, \ldots, m,$$

where in finding the values $v(t)$ it is permitted to use only values of $u(t)$ and $x_i(s)$, $y(s)$ for $s \in [0, t]$. In the case where from any initial position x_{i0}, y_0 ($\{x_{i0}\}_{N_i} \neq \{y_0\}_{N_i}$, $(q_k, y_0) \geq 0$) it is possible to avoid encounter we say that *evasion is possible in the game* (60).

We now formulate the assumption under which Theorem 7 below is proved.

ASSUMPTION 8. Suppose that for the game (60): a) the set

$$Q_1 = \{\omega \in R : (q_k, \omega) \geqslant 0, \quad k = 1, 2, \ldots, s\}, \qquad \text{where } R = \bigcap_{u \in P} f_{m+1}(u, Q),$$

contains an interior point; b) there exist two-element subsets N_i' of the sets N_i such that each of the sets

$$\{f_i(P,\ Q)\}_{N_i'} = \{y \in R^2 : y = \{f_i(u,\ v)\}_{N_i'},\quad u \in P,\ v \in Q\}$$

contains only one point.

THEOREM 7. *If Assumption* 8 *is satisfied, then evasion is possible in the game* (60). *Here the process of avoiding encounter can be controlled in such a way that the distance* $\xi_i(t)$ *between the points* $\{y(t)\}_{N_i}$ *and* $\{x_i(t)\}_{N_i}$ *for all* $t \geqslant 0$ *satisfies the estimate* (71).

PROOF. By Assumption 8 the set Q_1 contains an interior point. Therefore, in Q_1 it is possible to inscribe a ball of positive radius. Let $\bar{S}$ be some ball of positive radius contained in Q_1, let ω be an arbitrary point of $\bar{S}$, and let $u(t)$, $0 \leqslant t \leqslant 1$, be an arbitrary measurable function with values in the set P. We consider the equation

$$f_{m+1}(u(t),\ v) = \omega, \qquad v \in Q. \tag{64}$$

The smallest solution of (64) in the lexicographic sense we denote by $\tilde{v}_\omega(t)$; the function $\tilde{v}_\omega(t)$, $0 \leqslant t \leqslant 1$, is measurable on $[0,\ 1]$.

The solution $y(t)$, $0 \leqslant t \leqslant 1$, of the equation

$$\dot{y} = \beta y + f_{m+1}(u(t),\ \tilde{v}_\omega(t)), \qquad y(0) = y_0, \qquad (q_k,\ y_0) \geqslant 0, \qquad k = 1, 2, \ldots, s,$$

satisfies $(q_k,\ y(t)) \geqslant 0$ for all $k = 1, \ldots, s$ and $t \in [0,\ 1]$, since (cf. (64))

$$y(t) = e^{\beta t} y_0 + \int_0^t e^{\beta(t-\tau)} f_{m+1}(u(\tau),\ \tilde{v}_\omega(\tau))\, d\tau = e^{\beta t} y_0 + \int_0^t e^{\beta(t-\tau)} \omega\, d\tau = e^{\beta t} y_0 + e_1(t)\, \omega,$$

where $e_1(t) = (e^{\beta t} - 1)/\beta$ if $\beta \neq 0$ and $e_1(t) = t$ if $\beta = 0$, while $(q_k,\ y(t)) = e^{\beta t}(q_k,\ y_0) + e_1(t)(q_k,\ \omega) \geqslant 0$ because $(q_k,\ y_0) \geqslant 0$ for $\omega \in \bar{S} \subset Q_1$. Thus, if the function $\tilde{v}_\omega(t)$, $0 \leqslant t \leqslant 1$, is chosen according to (64), then the point $y(t)$, $0 \leqslant t \leqslant 1$, satisfies the required phase restrictions (62). Henceforth the initial point y_0 satisfies the restrictions (62), and the measurable function $\tilde{v}_\omega(t)$, $0 \leqslant t \leqslant 1$, satisfies (64).

For the solution $x_1(t), \ldots, x_m(t)$, $0 \leqslant t \leqslant 1$, of the equations

$$\dot{x}_i = A_i x_i + B_i y(t) + f_i(u(t),\ \tilde{v}_\omega(t)), \qquad x_i(0) = x_{i0},$$

we have

$$x_i(t) = e^{A_i t} x_{i0} + \int_0^t e^{A_i(t-\tau)} B_i e^{\beta\tau} y_0\, d\tau + \int_0^t e^{A_i(t-\tau)} [B_1 e_1(\tau)\, \omega + f_i(u(\tau),\ \tilde{v}_\omega(\tau))]\, d\tau.$$

Therefore,

$$\{x_i(t)\}_{N_i'} - \{y(t)\}_{N_i'} = \varphi_i(t,\ z_0) - e_1(t)\, \{\omega\}_{N_i'} - g_i'(t), \tag{65}$$

where

$$\varphi_i(t,\ z_0) = \left\{ e^{A_i t} x_{i0} + \int_0^t e^{A_i(t-\tau)} B_i e^{\beta\tau} y_0\, d\tau + t f_i(P,\ Q) \right\}_{N_i'} - e^{\beta t}\, \{y_0\}_{N_i'}, \qquad z_0 = (x_{10},\ x_{20}, \ldots, x_{m0},\ \ y_0)^T,$$

$$-g_i'(t) = \left\{ \int_0^t (e^{A_i(t-\tau)} [B_1 e_1(\tau)\, \omega + f_i(u(\tau),\ \tilde{v}_\omega(\tau))] - f_i(u(\tau),\ \tilde{v}_\omega(\tau)))\, d\tau \right\}_{N_i'}.$$

It is not hard to show that

$$|g_i'(t)| \leqslant c_1 t^2, \qquad 0 \leqslant t \leqslant 1, \tag{66}$$

where c_1 is a constant. Let $g_i(t) = e_1^{-1}(t)g_i'(t)$, $0 \leqslant t \leqslant 1$.

From (66), noting that $|e_1(t)| \leqslant c_2 t$, $0 \leqslant t \leqslant 1$, c_2 constant, we obtain

$$|g_i(t)| \leqslant c_3 t, \qquad 0 \leqslant t \leqslant 1, \tag{67}$$

where c_3 is a constant. From (65) and (67) we have

$$\{x_i(t)\}_{N_i'} - \{y(t)\}_{N_i'} = \varphi_i(t, z_0) - e_1(t)[\{\omega\}_{N_i'} + g_i(t)]. \tag{68}$$

Each two-element set N_i' generates a linear mapping of R^ν into R^2; namely, to each element $a = (a^1, \ldots, a^\nu)^T$ there corresponds the point $\{a\}_{N_i'} \in R^2$. We denote this mapping by G_i. By Assumption 8 the set $G_i\overline{S}$ is a disk. We fix a point z_0. From the lemma of L. S. Pontrjagin on the evasive maneuver (cf. [2], Russian §4.B, English §4.2) it follows that in the disk $G_1\overline{S}$ there exists a square Γ_1 with nonempty interior such that for all $t \in (0, \theta_1]$ and $\{\omega\}_{N_i'} = \bar{\omega}_1 \in \Gamma_1$

$$|\varphi_1(t, z_0) - e_1(t)[\bar{\omega}_1 + g_1(t)]| > \delta_1 t,$$

where θ_1 and δ_1 are constants. It is clear that $V_1 = G_1^{-1}\Gamma_1 \cap \overline{S}$ is a body. By the lemma cited above, in the region G_2V_1 there exists a square Γ_2 with nonempty interior such that for all $t \in (0, \theta_2]$ and $\{\omega\}_{N_2'} = \bar{\omega}_2 \in \Gamma_2$

$$|\varphi_2(t, z_0) - e_1(t)[\bar{\omega}_2 + g_2(t)]| > \delta_2 t,$$

where θ_2 and δ_2 are constants. Reasoning in a similar way, we conclude that in G_mV_{m-1} there exists a square Γ_m such that for all $t \in (0, \theta_m]$ and $\{\omega\}_{N_m'} = \bar{\omega}_m \in \Gamma_m$

$$|\varphi_m(t, z_0) - e_1(t)[\bar{\omega}_m + g_m(t)]| > \delta_m t,$$

where θ_m and δ_m are constants. Let $\theta = \min\{1, \theta_1, \ldots, \theta_m\}$, $\delta = \min\{\delta_1, \ldots, \delta_m\}$, and let $\omega_0 \in \overline{S}$ be one of the preimages of the point $\bar{\omega}_m$ under the mapping G_m: $R^\nu \to R^2$. From the construction described above it is clear that $G_i\omega_0 \in \Gamma_i$. This means that for all $i \in (0, \theta]$, $\omega = \omega_0 = \omega_0(z_0)$,

$$|\{x_i(t)\}_{N_i'} - \{y(t)\}_{N_i'}| = |\varphi_i(t, z_0) - e_1(t)[\{\omega_0\}_{N_i'} + g_i(t)]| > \delta t. \tag{69}$$

By the obvious inequality

$$|\{x_i(t)\}_{N_i} - \{y(t)\}_{N_i}| \geqslant |\{x_i(t)\}_{N_i'} - \{y(t)\}_{N'}|$$

(69) can be rewritten as (recall that $|\{x_i(0)\}_{N_i} - \{y(0)\}_{N_i}| \neq 0$)

$$|\{x_i(t)\}_{N_i} - \{y(t)\}_{N_i}| > \delta t, \qquad i = 1, 2, \ldots, m, \qquad t \in [0, \theta]. \tag{70}$$

On the basis of (70), we shall prove the possibility of evasion in the game (60). In the spaces $R^{s_1}, \ldots, R^{s_m}$ we consider the spheres $\Sigma^1 \subset R^{s_1}, \ldots, \Sigma^m \subset R^{s_m}$ defined respectively by the equations

$$|\{a_1\}_{N_1}| = \varepsilon, \qquad |\{a_2\}_{N_2}| = \varepsilon, \ldots, |\{a_m\}_{N_m}| = \varepsilon,$$

where $a_i \in R^{s_i}$ and $\epsilon = \min\{1, \delta\theta\}$. We denote the region contained inside Σ^i by Σ^i_1, and the region $R^{s_i}\backslash(\Sigma^i \cup \Sigma^i_1)$ by Σ^i_2. The rest of the argument is analogous to that used in §1A.

We now estimate the distance $\xi_i(t) = |\{x_i(t)\}_{N_i} - \{y(t)\}_{N_i}|$. To this end we apply a method of evasion analogous to that of §1A (the alterations required for the specific features of the present section are obvious).

We introduce operations of addition and subtraction of vectors from arithmetic spaces of different dimensions in the following manner. Let $a = (a^1, \ldots, a^k)^T \in R^k$ $(k \geqslant 1)$ and $b = (b^1, \ldots, b^l)^T \in R^l$ $(l \geqslant 1)$. We then set $\pm a \pm b = (\pm a^1 \pm b^1, \ldots, \pm a^m \pm b^m, c^{m+1}, \ldots, c^n)$ where $m = \min(k, l)$; $c^i = \pm a^i$ $(i = m+1, \ldots, k)$ if $k > m$, $c^i = \pm b^i$ $(i = m+1, \ldots, l)$ if $l > m$, and there are no components c^i if $k = l$. We define the operations of projection π_i and π^i as follows. Let $z = (z^1, \ldots, z^{\nu_i})^T \in R^{\nu_i}$. Then $\pi_i z = (c^1, \ldots, c^{\nu_i})^T$ where $c^{n^j_i} = z^{n^j_i}$, $j = 1, \ldots, s_i$, $c^l = 0$, $l \neq n^j_i$, and $\pi^i z = z - \pi_i z$. In this notation, $z = \pi_i z + \pi^i z$ and $|\{z\}_{N_i}| = |\pi_i z|$. We set $\eta_i(t) = |\pi^i[x_i(t) - y(t)]|$.

Arguing as in §1A in deriving (5), we find that for all $t \geqslant 0$

$$\xi_i(t) > \begin{cases} \dfrac{c\xi_i(0)}{1+\eta_i(t)}, & \text{if } \xi_i(0) < \varepsilon, \\ \dfrac{c\varepsilon}{1+\eta_i(t)}, & \text{if } \xi_i(0) \geqslant \varepsilon, \end{cases} \tag{71}$$

where c and ϵ are constants. This completes the proof of Theorem 7.

REMARK 6. It is possible to remove the condition that the sets $\{f_i(P, Q)\}_{N'_i}$ consist of one point by introducing (in a way similar to that of §1) constant vectors l_i and requiring the appropriate inequalities for certain constants characterizing the advantage of the evader over the pursuers.

REMARK 7. Differential games with phase restrictions have previously been considered in [12].

B. EXAMPLE. We consider the following analogue of the well-known "cornered rat" game (cf. [12] and [17]):

$$\dot{x}^1_i = x^2_i, \quad \dot{x}^2_i = u_i, \quad |u_i| \leqslant 1; \quad \dot{y} = v, \quad |v| \leqslant 1;$$
$$x^1_i, \quad x^2_i, \quad y \in R^\nu, \quad \nu \geqslant 2, \quad i = 1, 2, \ldots, m; \quad (q_1, y) \geqslant 0, \quad (q_2, y) \geqslant 0,$$

where q_1 and q_2 are arbitrary nonzero vectors with directions which are not opposite. The game is considered finished if for at least one value of i we get $x^1_i = y$. Assumption 8 is satisfied, since the set Q_1, as the intersection of the semiballs $\{v\colon |v| \leqslant 1, (q_1, v) \geqslant 0\}$ and $\{v\colon |v| \leqslant 1, (q_2, v) \geqslant 0\}$ defined by vectors not having opposed directions obviously contains an interior point and the sets $\{f_i(P, Q)\}_{N_i}$ coincide with the point 0. Thus (cf. Theorem 7), in this game with phase restrictions evasion is possible.

§5. The problem of damped oscillation of a pendulum

A. In this section we consider the nonlinear differential game (3) under the following assumption.

ASSUMPTION 9. Suppose that for the game (3): a) there exists a constant L such that for all $z_1, z_2 \in R^n$, $u \in P$ and $v \in Q$

$$|F(z_1, u, v) - F(z_2, u, v)| \leqslant L|z_1 - z_2|;$$

b) the terminal set M is no more than countable, $M = \{m_1, m_2, \ldots\}$, and there exists a constant ϵ such that for arbitrary m_k and m_j $(k \neq j)$ the Euclidean distance $\rho(m_k, m_j)$ between the points m_k and m_j satisfies the inequality $\rho(m_k, m_j) \geqslant \epsilon$; and c) we have the inclusion $0 \in \bigcap_{j \geqslant 1} [R_1^j \cup R_2^j]$ where

$$R_1^j = \operatorname{co}\{\psi \in S: \max_{v \in Q} \min_{u \in P} (\psi, F(m_j, u, v)) > 0\},$$

$$R_2^j = \operatorname{co}\{\psi \in S: \min_{u \in P} \max_{v \in Q} (\psi, F(m_j, u, v)) > 0\},$$

S is the unit sphere in R^n with center at the point 0, and $\operatorname{co}\{X\}$ denotes the convex hull of the set X.

THEOREM 8. *If Assumption 9 is satisfied, then evasion is possible in the game* (3). *Here the process of avoiding encounter with the set M from the point $z_0 \in R^n \backslash M$ can be controlled in such a way that the distance $\xi_j(t)$ from the point $z(t)$ to the point m_j, $j = 1, 2, \ldots$, for all $t \geqslant 0$ satisfies the inequality*

$$\xi_j(t) > \begin{cases} \dfrac{c_j \xi_j(0)}{1 + |m_j|}, & \text{if } \xi_j(0) < \varepsilon_j, \\ \dfrac{c_j \varepsilon_j}{1 + |m_j|}, & \text{if } \xi_j(0) \geqslant \varepsilon_j, \end{cases} \tag{72}$$

where c_j and ϵ_j are positive numbers depending only on the game (3) *and on the point m_j.*

PROOF. From the obvious inequalities

$$|F(z, u, v)| \leqslant |F(z, u, v) - F(0, u, v)| + |F(0, u, v)| \leqslant L|z| + L_1 \leqslant L_2(1 + |z|),$$

where

$$L_1 = \max_{u \in P, v \in Q} |F(0, u, v)|, \quad L_2 = \max\{L, L_1\},$$

it follows that for any measurable functions $u(t)$ and $v(t)$ defined on $[0, \infty)$ with values in the sets P and Q respectively and for any initial position z_0 each solution $z(t)$ of the equation $\dot{z} = F(z, u(t), v(t))$, $z(0) = z_0$, is defined for all $t \geqslant 0$. Further, the uniqueness of the solution $z(t)$ also follows from Assumption 9. We fix an index j. By Assumption 9, $0 \in R_1^j \cup R_2^j$. There are two possible cases: 1) $0 \in R_1^j$; 2) $0 \in R_2^j$. In the case where $0 \in R_1^j \cap R_2^j$ we assume that case 1) applies. Suppose that $0 \in R_1^j$. According to a well-known theorem of Carathéodory, there exist vectors $\psi_1^1, \ldots, \psi_{n+1}^1 \in S$ and numbers $\lambda_1^1, \ldots, \lambda_{n+1}^1$ such that $\lambda_k^1 \geqslant 0$, $k = 1, \ldots, n+1$, $\Sigma_1^{n+1} \lambda_k^1 = 1$, $\lambda_1^1 \psi_1^1 + \ldots + \lambda_{n+1}^1 \psi_{n+1}^1 = 0$ and for all $k = 1, \ldots, n+1$ (cf. [10])

$$N_j^1 = \max_{v \in Q} \min_{u \in P} (\psi_k^1, F(m_j, u, v)) > 0.$$

It is clear that for any initial position z_0 there is a vector $\psi_k^1 = \psi_k^1(z_0)$ such that $(z_0 - m_j, \psi_k^1) \geqslant 0$.

As shown in §2, there is a vector $v_0(z_0) \in Q$ which satisfies

$$(\psi_k^1(z_0),\ F(m_j,\ u(s),\ v_0(z_0))) \geqslant N_j^1 \tag{73}$$

for any measurable $u(s) \in P$, $s \in [0, 1]$. In case 2) there exist a vector $\psi_k^2 = \psi_k^2(z_0)$ and a measurable function $v_0(s)$, $0 \leqslant s \leqslant 1$, with values in Q such that (cf. §2)

$$(\psi_k^2(z_0),\ F(m_j,\ u(s),\ v_0(s))) \geqslant N_j^2 = \min_{u \in P} \max_{v \in Q} (\psi_k^2(z_0),\ F(m_j,\ u,\ v)), \tag{74}$$

where $(\psi_k^2(z_0), z_0 - m_j) \geqslant 0$ and ψ_k^2 is one of the vectors ψ_l^2 which satisfy the condition

$$\sum_{l=1}^{n+1} \lambda_l^2 \psi_l^2 = 0,\ \lambda_l^2 \geqslant 0,\ \sum_{l=1}^{n+1} \lambda_l^2 = 1.$$

We recall that z_0 and $u(s)$ are used to find the value of $v_0(s)$ (see §2).

We denote by N_j and $\psi_k(z_0)$ the number N_j^1 and the vector $\psi_k^1(z_0)$ if $0 \in R_1^j$, and the number N_j^2 and the vector $\psi_k^2(z_0)$ if $0 \in R_2^j$. Thus, in both cases there exists a measurable function $\bar{v}(s) \in Q$, $0 \leqslant s \leqslant 1$, such that for all $s \in [0, 1]$ (cf. (73) and (74))

$$(\psi_k(z_0),\ F(m_j,\ u(s),\ \bar{v}(s))) \geqslant N_j. \tag{75}$$

Let

$$H_j = \min\left\{\frac{N_j}{2L_2},\ \frac{\varepsilon}{2}\right\},\qquad |z_0 - m_i| \leqslant \min\left\{1,\ \frac{H_i}{2e^{L_2}}\right\},$$
$$\theta_j = \min\left\{1,\ \frac{H_j}{2e^{L_2}(1 + |m_j|)}\right\}. \tag{76}$$

For the solution $z(t)$, $0 \leqslant t \leqslant \theta_j$, of the equation $\dot{z} = F(z, u(t), \bar{v}(t))$, $z(0) = z_0$, we have

$$|z(t) - m_j| \leqslant |z_0 - m_j| + \int_0^t L_2[1 + |m_j| + |z(s) - m_j|]\, ds.$$

By Gronwall's lemma

$$|z(t) - m_j| \leqslant |z_0 - m_j| e^{L_2 t} + (1 + |m_j|)(e^{L_2 t} - 1).$$

But $e^{L_2 t} \leqslant e^{L_2}$ and $e^{L_2 t} - 1 \leqslant te^{L_2}$, $0 \leqslant t \leqslant 1$. Therefore (cf. (76)), for $0 \leqslant t \leqslant \theta_j$

$$|z(t) - m_j| \leqslant |z_0 - m_j| e^{L_2} + (1 + |m_j|) e^{L_2}\theta_j \leqslant H_j,$$
$$|z(t) - m_j| \leqslant \varepsilon/2. \tag{77}$$

On the other hand, for $0 \leqslant t \leqslant \theta_j$

$$(z(t) - m_j,\ \psi_k(z_0)) \geqslant \int_0^t (\psi_k(z_0),\ F(m_j,\ u(s),\ \bar{v}(s)))\, ds - \int_0^t |F(z(s),\ u(s),\ \bar{v}(s))$$
$$- F(m_j,\ u(s),\ \bar{v}(s))|\, ds \geqslant N_j t - \int_0^t L_2 |z(s) - m_j|\, ds$$
$$\geqslant N_j t - L_2 H_j t \geqslant N_j t - L_2 \frac{N_j}{2L_2} t = \frac{N_j}{2} t.$$

Thus,

$$(z(t)-m_j,\ \psi_k(z_0))\geqslant\frac{N_j}{2}t,\qquad 0\leqslant t\leqslant\theta_j. \tag{78}$$

Let

$$\varepsilon_j=\min\left\{1,\ \frac{N_j}{3}\theta_j,\ \frac{H_j}{2e^{L_2}}\right\}. \tag{79}$$

In R^n we consider the spheres Σ^j (cf. §1):

$$\xi_j=\varepsilon_j,\qquad j=1,\ 2,\ldots,$$

where $\xi_j=|z-m_j|$. By the inequality $\epsilon_j\leqslant H_j/2e^{L_2}\leqslant\epsilon/4e^{L_2}\leqslant\epsilon/4$ (cf. (76)) the spheres $\Sigma^1,\Sigma^2,\ldots$ are mutually nonintersecting. Therefore, any point $z\in R^n$ can satisfy at most one of the inequalities $\xi_j\leqslant\epsilon_j$.

Let $z_0\in R^n\backslash M$. There are two possible cases: 1) $\xi_j(0)=|z_0-m_j|\leqslant\epsilon_j$ for some $j=j_0$; 2) $\xi_j(0)>\epsilon_j$ for all $j=1,2,\ldots$.

In case 1) application of the control $\bar{v}(t)$, $0\leqslant t\leqslant\theta_{j_0}$, guarantees that $z(t)$ does not coincide with m_{j_0} for all $t\in[0,\theta_{j_0}]$, since (cf. (78)) $\xi_{j_0}(t)\geqslant(z(t)-m_{j_0},\psi_k(z_0))\geqslant N_{j_0}t/2>0$ for $t\in(0,\theta_{j_0}]$. For $t=\theta_{j_0}$ we have (cf. (79))

$$\xi_{j_0}(\theta_{j_0})\geqslant\frac{N_{j_0}\theta_{j_0}}{2}>\frac{N_{j_0}\theta_{j_0}}{3}\geqslant\varepsilon_{j_0},$$

i.e., the phase point $z(t)$ for $t=\theta_{j_0}$ lies in the region $\xi_{j_0}>\epsilon_{j_0}$. On the other hand, on $[0,\theta_{j_0}]$ (cf. (77)) we have $\xi_{j_0}(t)\leqslant\epsilon/2$.

Thus, in applying the control $\bar{v}(t)$, $0\leqslant t\leqslant\theta_{j_0}$, the phase point $z(t)$ does not enter an $\epsilon/4$ neighborhood of the points m_j, $j\neq j_0$. The remainder of the argument is analogous to that of §1 (with obvious alterations). Thus, the possibility of evasion in the game (3) has been demonstrated.

We now deduce (72). It is assumed that

$$0<|z_0-m_j|\leqslant\varepsilon_j.$$

We have

$$|z(t)-z_0|\leqslant\int_0^t L_2(1+|z_0|+|z(s)-z_0|)\,ds.$$

Hence for $0\leqslant t\leqslant 1$

$$|z(t)-z_0|\leqslant(1+|z_0|)e^{L_2t}\leqslant(1+|m_j|+|z_0-m_j|)e^{L_2t}\leqslant(2+|m_j|)e^{L_2t}.$$

But $|\xi_j(t)-\xi_j(0)|\leqslant|z(t)-z_0|$. Therefore,

$$\xi_j(t)\geqslant\xi_j(0)-(2+|m_j|)e^{L_2t}\geqslant\frac{\xi_j(0)}{2},\qquad 0\leqslant t\leqslant\frac{\xi_j(0)}{2(2+|m_j|)e^{L_2}}. \tag{80}$$

On applying the control $v_0(t)$, $0\leqslant t\leqslant\theta_j$ (cf. (78)), we get

$$\xi_j(t)\geqslant\frac{N_j}{2}t\geqslant\frac{N_j}{2}\frac{\xi_j(0)}{2(2+|m_j|)e^{L_2}},\qquad\frac{\xi_j(0)}{2(2+|m_j|)e^{L_2}}\leqslant t\leqslant\theta_j. \tag{81}$$

From these inequalities we obtain

$$\xi_j(t) > \frac{c_j \xi_j(0)}{1+|m_j|}, \qquad 0 \leqslant t \leqslant \theta_j. \tag{82}$$

Suppose that on $[\theta_j, t_0)$ the phase point $z(t)$ for given j lies outside the ball $\xi_j \leqslant \epsilon_j$, while $z(t_0) \in \Sigma^j$. Suppose that active evasion control is applied on the segment $[t_0, t_0 + \theta_j]$; then

$$\begin{aligned} \xi_j(t) > \varepsilon_j \geqslant \frac{c_j \varepsilon_j}{1+|m_j|}, \qquad & \theta_j \leqslant t \leqslant t_0; \\ \xi_j(t) > \frac{c_j \varepsilon_j}{1+|m_j|}, \qquad & t_0 \leqslant t \leqslant t_0 + \theta_j. \end{aligned} \tag{83}$$

With our method of evasion the inequality $\xi_j(t) > c_j \xi_j(0)/[1 + |m_j|]$ for $0 < \xi_j(0) < \epsilon_j$ and $0 \leqslant t < \infty$ follows in an obvious way for the given j from (82) and (83). If $\xi_j(0) \geqslant \epsilon_j$ for a given j, then arguing as above we obtain $\xi_j(t) > c_j \epsilon_j/[1 + |m_j|]$, $0 \leqslant t < \infty$. The estimate (72) follows from these inequalities. This completes the proof of Theorem 8.

B. EXAMPLES. I. We consider the problem of the damped oscillation of a mathematical pendulum (cf. [4]). The equations of motion in appropriate coordinates are

$$\dot{z}_1 = z_2, \qquad \dot{z}_2 = -a \sin z_1 + bu + bv, \tag{84}$$

where $z_1, z_2 \in R^1$, a and b are positive numbers (in the notation of [4], $a = -g/d$ and $b = 1/md^2$), $|u| \leqslant \beta$, $|v| \leqslant \alpha$, and the terminal set $M = \{m_{j_1} = 2\pi j, m_{j_2} = 0\}$, $j = 0, 1, \ldots$.

We shall show that if

$$\alpha > \beta, \tag{85}$$

then Assumption 9 is satisfied for the game (84). Indeed, conditions a) and b) of Assumption 9 are obviously satisfied, and it is possible to set $L = a$ and $\epsilon = 2\pi$. Further, let $I = \{z: z_1 = 0, |z_2| \leqslant 1\} \subset R^2$, $\psi = (0, \chi)^T$, $\chi \in R^1$, $|\chi| = 1$, $F_1(z, u, v) = z_2$, $F_2(z, u, v) = -a \sin z_1 + bu + bv$ and $m_j = (2\pi j, 0)^T$ (cf. (84)).

We have $(\psi, F(m_j, u, v)) = \chi(bu + bv)$. Therefore, $I \subset R_1^j$ and $I \subset R_2^j$, since (cf. (85))

$$\max_{|v| \leqslant \alpha} \min_{|u| \leqslant \beta} (\psi, F(m_j, u, v)) = \min_{|u| \leqslant \beta} \max_{|v| \leqslant \alpha} (\psi, F(m_j, u, v)) = b(\alpha - \beta) > 0.$$

Since $\bigcap_{j \geqslant 0} [R_1^j \cup R_2^j] \supset I \ni 0$, it follows that condition c) of Assumption 9 is satisfied. Thus, if $\alpha > \beta$ all the hypotheses of Theorem 8 are satisfied for the game (84).

II. Suppose that in the game (3) $F(z, u, v) = Cz - u + v$, $n = 2$, $P = \{u: u_1^2 + u_2^2 \leqslant 1\}$, Q is an equilateral triangle with center at the point 0 and side length 2, and $M = \{0\}$ (cf. [13]). It is easy to show that Assumption 9 is satisfied for this game. However, the conditions of the papers [1]–[3], [5]–[9], [11], [12] and [14]–[16] are not satisfied for this game. We point out that in the case where in Assumption 9 the terminal set consists of a single point the conditions of Theorem 8 are close to the necessary conditions (cf. [13]).

BIBLIOGRAPHY

1. L. S. Pontrjagin and E. F. Miščenko, *Avoidance of contact in linear differential games,* Differencial'nye Uravnenija **7** (1971), 436–445; English transl. in Differential Equations **7** (1971).

2. L. S. Pontrjagin, *A linear differential evasion game,* Trudy Mat. Inst. Steklov. **112** (1971), 30–63; English transl. in Proc. Steklov Inst. Math. **112** (1971).

3. ————, *On the evasion process in differential games,* Appl. Math. Optim. **1** (1974), 5–19.

4. N. N. Krasovskiĭ and A. I. Subbotin, *Positional differential games,* "Nauka", Moscow, 1974; English transl., *Closed-loop differential games,* Springer-Verlag, Berlin and New York (to appear).

5. E. F. Miščenko and N. Satimov, *The evasion problem in differential games with nonlinear controls,* Differencial'nye Uravenija **9** (1973), 1792–1797; English transl. in Differential Equations **9** (1973).

6. R. V. Gamkrelidze and G. L. Kharatishvili [Haratišvili], *A differential game of evasion with nonlinear control,* SIAM J. Control **12** (1974), 332–349.

7. M. S. Nikol'skiĭ, *On a linear escape problem,* Dokl. Akad. Nauk SSSR **218** (1974), 1024–1027; English transl. in Soviet Math. Dokl. **15** (1974).

8. ————, *On a quasilinear escape problem,* Dokl. Akad. Nauk SSSR **221** (1975), 539–542; English transl. in Soviet Math. Dokl. **16** (1975).

9. N. Satimov, *On the escape problem in differential games with nonlinear controls,* Dokl. Akad. Nauk SSSR **216** (1974), 744–747; English transl. in Soviet Math. Dokl. **15** (1974).

10. ————, *On a way to avoid contact in differential games,* Mat. Sb. **99(141)** (1976), 380–393; English transl. in Math. USSR Sb. **28** (1976).

11. B. N. Pšeničnyĭ, *The escape problem,* Kibernetika (Kiev) **1975**, no. 4, 120–127; English transl. in Cybernetics **11** (1975).

12. A. A. Čikriĭ, *The evasion problem in nonlinear differential games,* Kibernetika (Kiev) **1975**, no. 3, 65–68; English transl. in Cybernetics **11** (1975).

13. A. Azamov, *The evasion problem in two-dimensional differential games,* Dokl. Akad. Nauk UzSSR **1974**, no. 8, 3–4. (Russian).

14. P. B. Gusjatnikov, *On the l-evasion of contact in a linear differential game,* Prikl. Mat. Meh. **38** (1974), 417–421; English transl. in J. Appl. Math. Mech. **38** (1974).

15. E. F. Miščenko and N. Satimov, *The avoidance of contact in n-person differential games,* Dokl. Akad. Nauk SSSR **224** (1975), 285–288; English transl. in Soviet Math. Dokl. **16** (1975).

16. P. B. Gusjatnikov, *l-evasion in linear differential games,* Differencial'nye Uravnenija **12** (1976), 446–455; English transl. in Differential Equations **12** (1976).

17. Rufus Isaacs, *Differential games,* Wiley, New York, 1965.

Translated by J. R. SCHULENBERGER

Труды Мат. Инст. Стеклов.
143 (1977)

Proceedings of the Steklov
Institute of Mathematics
1980, Issue 1

SEQUENCES OF NORMS OF FOURIER SUMS OF BOUNDED FUNCTIONS

UDC 517.512

K. I. OSKOLKOV

ABSTRACT. This paper investigates sequences of norms of the Fourier sums $\{\|S_n(f)\|\}_0^\infty$ of a given bounded function f. It is shown that if $\lambda = \{\lambda_\nu\}_0^\infty$ is a sequence of nonnegative numbers,

$$\|\lambda\|_1 = \sum_{\nu=0}^{\infty} \lambda_\nu < \infty, \qquad \|\lambda\|_2 = \left(\sum_{\nu=0}^{\infty} \lambda_\nu^2\right)^{1/2}, \qquad d(\lambda) = \frac{\|\lambda\|_2}{\|\lambda\|_1},$$

and if $f \in L^\infty$, $\|f\| \leqslant 1$, then

$$\sum_{\nu=0}^{\infty} \lambda_\nu \|S_\nu(f)\| \leqslant \frac{4}{\pi^2} \sum_{\nu=0}^{\infty} \lambda_\nu \log c(1+\nu d(\lambda)).$$

A similar inequality is applied to estimate

$$\mathscr{L}(\mathscr{N}) = \sup_{f, \|f\| \leqslant 1} \varlimsup_{k \to \infty} \frac{\|S_{n_k}(f)\|}{L_{n_k}},$$

where $\mathscr{N} = \{n_k\}$ is an increasing sequence of integers and L_n are the Lebesgue constants.

Bibliography: 9 titles.

§1. Introduction. On the principle of localization

Let $f(x)$ be a 2π-periodic summable function, and $S_n(f, x)$ the Fourier sums of this function ($n = 0, 1, \dots$). According to the Riemann-Lebesgue principle of localization (cf. [1], p. 55), for each $\epsilon > 0$ we have, uniformly in x,

$$S_n(f, x) = \frac{1}{\pi} \int_{-\varepsilon}^{\varepsilon} f(t+x) \frac{\sin nt}{t} \, dt + o(1) \ (n \to \infty). \tag{1.1}$$

Note that here ϵ is an arbitrary positive number, *independent of n*.

In the present paper we investigate properties of the sequence of norms[1] $\{\|S_n(f)\|\}_0^\infty$ of the Fourier sums of a given $f \in L^\infty$ as $n \to \infty$. (The main results are stated in §2 and proved in §3.) In particular, the following result will be proved; it is, in a certain sense, a variant of the principle of localization (1.1) with an ϵ depending on n.

1980 *Mathematics Subject Classification*. Primary 42A16, 42A20; Secondary 42A24.

[1] L^∞ is the space of 2π-periodic measurable and almost everywhere bounded functions $f(x)$ with norm $\|f\| = \operatorname{vrai\,sup}_x |f(x)|$. In this paper c denotes positive absolute constants, generally different at different occurrences; $[y]$ is the integral part of the positive number y.

THEOREM 1. *Let* $f \in L^\infty$, $\|f\| \leqslant 1$; $n_1, \ldots, n_P$ *are different positive integers. Then for every set of points* $x_1, \ldots, x_P$ *we have*

$$\frac{1}{P}\sum_{j=1}^{P}\left| S_{n_j}(f, x_j) - \frac{1}{\pi}\int_{-P^{-1/2}}^{P^{-1/2}} f(t+x_j)\frac{\sin n_j t}{t}\,dt\right| \leqslant c. \tag{1.2}$$

The inequality (1.2), whose proof depends on the interference between Dirichlet kernels of different orders (cf. also §2, Theorem 2, below) means that for any points $x_1, \ldots, x_P$ the average of the Fourier sums $S_{n_j}(f, x_j)$, $1 \leqslant j \leqslant P$, can be estimated with error $O(1)$ by a Dirichlet integral over intervals of length $\epsilon_P = O(P^{-1/2})$.

Let L_n be the Lebesgue constants, i.e.

$$L_n = \sup_{f,\,\|f\|\leqslant 1} \|S_n(f)\| \quad (n = 0, 1, \ldots),$$

and let $\mathcal{N} = \{n_k\}_0^\infty$ be an increasing sequence of integers. Estimates of the form (1.2) (cf. §3, Lemma 3), which represent Fourier sums localized over intervals of short length, are used to obtain upper bounds for

$$\underline{\mathcal{S}}(\mathcal{N}) = \sup_{f,\,\|f\|\leqslant 1} \varliminf_{k\to\infty} \frac{\|S_{n_k}(f)\|}{L_{n_k}}.$$

The number $\underline{\mathcal{S}}(\mathcal{N})$ describes the accuracy of the estimate

$$\|S_n(f)\| \leqslant L_n \qquad (f \in L^\infty,\ \|f\| \leqslant 1;\ n = 0, 1, \ldots) \tag{1.3}$$

for a given function f, simultaneously for all sufficiently large $n \in \mathcal{N}$. This distinguishes $\underline{\mathcal{S}}(\mathcal{N})$ from the number

$$\overline{\mathcal{S}}(\mathcal{N}) = \sup_{f,\,\|f\|\leqslant 1} \varlimsup_{k\to\infty} \frac{\|S_{n_k}(f)\|}{L_{n_k}},$$

which also describes the accuracy of (1.3), but only, generally speaking, for a sufficiently sparse subsequence of $\mathcal{N}$.

The exact value of $\overline{\mathcal{S}}(\mathcal{N})$ was determined by Bohr and Neder [2], [3]. In fact, it was shown in [3] that for every sequence $\mathcal{N} = \{n_k\}$ there is a function f such that $\|f\| = 1$ and

$$\varlimsup_{k\to\infty} \frac{S_{n_k}(f, 0)}{L_{n_k}} = 1. \tag{1.4}$$

It follows, in particular, that $\overline{\mathcal{S}}(\mathcal{N})$ is the same for all sequences $\mathcal{N}$ and equals 1.

The problem of determining $\underline{\mathcal{S}}(\mathcal{N})$ arose in connection with results of Men'šov [4]. Men'šov showed that there is a continuous 2π-periodic function for which every subsequence of the Fourier sums diverges at some point and, in particular, is not uniformly convergent. Busko [5] proved that there is a function $f \in L^\infty$ such that $\|S_n(f)\| \geqslant \log n$ for $n = 2, 3, \ldots$. It follows from this result that the number

$$\underline{\mathcal{S}} = \sup_{f,\,\|f\|\leqslant 1} \varliminf_{n\to\infty} \frac{\|S_n(f)\|}{L_n}$$

is positive, and therefore $\underline{\mathscr{S}}(\mathscr{N})$ is positive for every sequence $\mathscr{N}$, since $\underline{\mathscr{S}}(\mathscr{N}) \geqslant \underline{\mathscr{S}}$. It was shown in [6] that $\underline{\mathscr{S}} = ½$. From this and (1.4) it follows that

$$\frac{1}{2} = \underline{\mathscr{S}} \leqslant \underline{\mathscr{S}}(\mathscr{N}) \leqslant \bar{\mathscr{S}}(\mathscr{N}) = 1 \tag{1.5}$$

for every sequence $\mathscr{N}$.

Theorem 3 (§2) contains a necessary and sufficient condition for the equation $\underline{\mathscr{S}}(\mathscr{N}) = 1$ $(= \bar{\mathscr{S}}(\mathscr{N}))$ and also the exact value of $\underline{\mathscr{S}}(\mathscr{N})$ for the sequences $\mathscr{N} = \mathscr{N}^{(\kappa)} = \{[k^\kappa] + 1\}_{k=0}^\infty$, where κ is a given number, $\kappa \geqslant 1$. The principal difficulty is the proof of the upper bound (3.2) for $\underline{\mathscr{S}}(\mathscr{N}^{(\kappa)})$, which is based on Lemma 3. The essence of the proof is the construction of a special sequence of decompositions of the interval $I_0 = [-\pi, \pi)$, depending on the distribution on I_0 of the points x_ν, where $\|S_\nu(f)\| = |S_\nu(f, x_\nu)|$, and on repeated use of the interference property of the Dirichlet kernels, mentioned above. As a consequence of the construction we obtain (3.14) which provides a localized representation for Fourier sums by Dirichlet integrals over a sequence of intervals of decreasing length.

For this sequence of local Dirichlet integrals we use arguments that are close in certain respects to Carleson's method [7].

The proof of the necessity of (2.5) and the lower bound of $\underline{\mathscr{S}}(\mathscr{N})$ depend on an estimate obtained in [6]. That paper also contains a more complete survey of results about the sequences of Fourier sums of given functions; here we mention only the results that we need.

§2. Statements of results

THEOREM 2. *Let $\lambda = \{\lambda_\nu\}_0^\infty$ be a sequence of nonnegative numbers,*

$$\|\lambda\|_1 = \sum_{\nu=0}^\infty \lambda_\nu < \infty, \qquad \|\lambda\|_2 = \left(\sum_{\nu=0}^\infty \lambda_\nu^2\right)^{1/2}, \qquad d(\lambda) = \frac{\|\lambda\|_2}{\|\lambda\|_1}. \tag{2.1}$$

Then the following estimate holds for every $f \in L^\infty$, $\|f\| \leqslant 1$:

$$\sum_{\nu=0}^\infty \lambda_\nu \|S_\nu(f)\| \leqslant \frac{4}{\pi^2} \sum_{\nu=0}^\infty \lambda_\nu \log c(1 + \nu d(\lambda)). \tag{2.2}$$

In comparison with the trivial estimate

$$\sum_{\nu=0}^\infty \lambda_\nu \|S_\nu(f)\| \leqslant \frac{4}{\pi^2} \sum_{\nu=0}^\infty \lambda_\nu \log c(1 + \nu) \qquad (f \in L^\infty,\ \|f\| \leqslant 1),$$

which follows from the asymptotic formula for the Lebesgue constants (cf. [1], p. 67),

$$L_\nu = \frac{4}{\pi^2} \log \nu + O(1) \quad (\nu \to \infty), \tag{2.3}$$

the estimate (2.2) contains the smaller factor $d(\lambda) \leqslant 1$.

It is interesting to compare (2.2) with the following inequality in which the norm is outside the summation:

$$\left\| \sum_{\nu=0}^\infty \lambda_\nu |S_\nu(f, x)| \right\| \leqslant \frac{4}{\pi^2} \sum_{\nu=0}^\infty \lambda_\nu \log c(1 + \nu d^2(\lambda)) \qquad (f \in L^\infty,\ \|f\| \leqslant 1). \tag{2.4}$$

This differs from (2.2) by having the factor $d(\lambda)$ squared. For various choices of the sequence λ in (2.4) one can obtain results of the type of strong summability of Fourier series (cf., for example, [8]). We shall omit the proof of (2.4) since it differs little from the reasoning used in problems of strong summability.

THEOREM 3. A. *Let $\mathcal{N} = \{n_k\}_0^\infty$ be an increasing sequence of integers. A necessary and sufficient condition that*

$$\underline{\mathcal{S}}(\mathcal{N}) = \sup_{f, \|f\| \leqslant 1} \varliminf_{k\to\infty} \frac{\| S_{n_k}(f) \|}{L_{n_k}} = 1 \ (= \bar{\mathcal{S}}(\mathcal{N})), \tag{2.5}$$

is that the sequence $\mathcal{N}$ satisfies

$$\log k = o(\log n_k) \qquad (k \to \infty). \tag{2.6}$$

B. *Let $\mathcal{N} = \mathcal{N}^{(\kappa)} = \{[k^\kappa] + 1\}_{k=0}^\infty$, where $\kappa \geqslant 1$. Then*

$$\underline{\mathcal{S}}(\mathcal{N}^{(\varkappa)}) = \begin{cases} 1/2, & \text{if} \quad 1 \leqslant \varkappa \leqslant 2, \\ 1 - 1/\varkappa, & \text{if} \quad \varkappa > 2. \end{cases} \tag{2.7}$$

Theorem 2 can be used to obtain precise estimates for the sequences of norms of certain linear means of Fourier series, for example the de la Vallée Poussin means. Let $q = \{q_n\}_0^\infty$ be a sequence of integers,

$$V_{n, q_n}(f) = \frac{1}{q_n} \sum_{\nu=n}^{n+q_n-1} S_\nu(f) \tag{2.8}$$

the de la Vallée Poussin means, and

$$U_{n, q_n} = \sup_{f, \|f\| \leqslant 1} \| V_{n, q_n}(f) \|$$

the Lebesgue constants corresponding to these means. Nikol′skiĭ [9] showed that

$$\left| U_{n, q_n} - \frac{4}{\pi^2} \log \frac{n}{q_n} \right| \leqslant c. \tag{2.9}$$

We consider an analog of $\underline{\mathcal{S}}$ for the sequence (2.8) of means, i.e.

$$\mathcal{U}(q) = \sup_{f \in L^\infty, \|f\| \leqslant 1} \varliminf_{n\to\infty} \frac{\| V_{n, q_n}(f) \|}{U_{n, q_n}}.$$

THEOREM 4. *If the sequence $q = \{q_n\}_0^\infty$ of integers satisfies the conditions*

$$q_N = O\left(\frac{N}{\log N}\right), \qquad \max_{N \leqslant s, r \leqslant 2N} \frac{q_s}{q_r} = O(1) \ (N \to \infty), \tag{2.10}$$

then

$$\mathcal{U}(q) = \max_{f, \|f\|=1} \varliminf_{n\to\infty} \frac{\| V_{n, q_n}(f) \|}{U_{n, q_n}} = \varliminf_{n\to\infty} \min\left(1, \frac{1}{2} \frac{\log n}{\log (n/q_n)}\right). \tag{2.11}$$

Condition (2.10) is satisfied if, for example, $q_n = [n^\alpha]$ with $0 \leqslant \alpha < 1$; in this case (2.11) yields

$$\mathscr{U}(q) = \min\left(1, \frac{1}{2(1-a)}\right).$$

§3. Proofs

We begin by proving Theorems 1 and 2, and we also establish the following statements for every $f \in L^\infty$ with $\|f\| \leqslant 1$:

a) *If* $\mathscr{N} = \{n_k\}_0^\infty$ *is an arbitrary increasing sequence of integers, then*

$$\varlimsup_{n\to\infty} \frac{\|S_{n_k}(f)\|}{L_{n_k}} \leqslant \varlimsup_{k\to\infty} \frac{\log n_k k^{-1/2}}{\log n_k}. \tag{3.1}$$

b) *If* $n_k = [k^\kappa] + 1$ $(k = 0, 1, 2, \dots)$, *where* $\kappa \geqslant 1$, *then*

$$\varlimsup_{k\to\infty} \frac{\|S_{n_k}(f)\|}{L_{n_k}} \leqslant \begin{cases} 1/2, & \textit{if} \quad 1 \leqslant \varkappa < 2; \\ 1 - 1/\varkappa, & \textit{if} \quad \varkappa \geqslant 2. \end{cases} \tag{3.2}$$

c) *If* (2.10) *is satisfied for the sequence* $q = \{q_n\}_1^\infty$ *of integers, then*

$$\varlimsup_{n\to\infty} \frac{\|V_{n,q_n}(f)\|}{U_{n,q_n}} \leqslant \frac{1}{2} \varlimsup_{n\to\infty} \frac{\log n}{\log(n/q_n)}. \tag{3.3}$$

The necessity of (2.6) in Theorem 3 follows from (3.1), and the upper bounds for $\mathscr{S}(\mathscr{N}^{(\kappa)})$ in Theorem 3 and for $\mathscr{U}(q)$ in Theorem 4 follow from (3.2) and (3.3).

We let I_0 denote the half-open interval $[-\pi, \pi)$; by I (or i) we denote a half-open interval $[a, b)$ contained in I_0; by $|I|$ we mean the length of I and by $\eta(I)$, the center of I. For each $I \subset I_0$ we denote by I' the half-open interval concentric with I and of length $|I'| = 2|I|$.

The proofs of Theorems 1 and 2 and of (3.1)–(3.3) depend on the following lemmas.

LEMMA 1. *Let* $f \in L^\infty$, $|f| \leqslant 1$, $\operatorname{vrai\,sup}_{u\in I_0'} |F(u)| \leqslant 1$ *and* $I \subset I_0$. *Then for* $x \in I$ *and* $\nu = 0, 1, \dots$ *we have the inequalities*

$$\left| S_\nu(f, x) - \frac{1}{\pi} \int_{I_0'} \frac{\sin \nu(u-x)}{u-x} f(u)\, du \right| \leqslant c, \tag{3.4}$$

$$\left| \int_{I_0'\setminus I'} \frac{\sin \nu(u-x)}{u-x} F(u)\, du \right| \leqslant \left| \int_{I_0'\setminus I'} \frac{\sin \nu(u-x)}{u-\eta(I)} F(u)\, du \right| + c, \tag{3.5}$$

$$\frac{1}{\pi} \left| \int_{I'} \frac{\sin \nu(u-x)}{u-x} F(u)\, du \right| \leqslant \frac{4}{\pi^2} \log c(1 + \nu|I|). \tag{3.6}$$

We omit the proof of this lemma since (3.4) is well known, and (3.5) and (3.6) are established, for example, in [6].

LEMMA 2. *Let* $\lambda = \{\lambda_\nu\}_0^\infty$, $\lambda_\nu \geqslant 0$, *and (cf.* (2.1))

$$N = N(\lambda) = [d^{-1}(\lambda)].$$

Let $\mathscr{I} = \mathscr{I}(\lambda)$ *be a partition of* I_0 *into* N *mutually disjoint half-open intervals* I

of equal length, i.e. $\mathcal{J} = \{I\}$, *where*

$$\bigcup_{I\in\mathcal{J}} I = I_0, \quad |I| = |I_0|/N; \quad I_1 \cap I_2 = \varnothing \quad (I_1, I_2 \in \mathcal{J}, I_1 \neq I_2).$$

If vrai $\sup_{u\in I_0'}|F(u)| \leqslant 1$ *and* $\{x_\nu\}_0^\infty$ *is an arbitrary sequence*, $x_\nu \in I_0$ $(\nu \geqslant 0)$, *then*

$$\sum_{I\in\mathcal{J}} \sum_{\nu:\, x_\nu\in I} \lambda_\nu \left| \int_{I_0'\setminus I'} \frac{\sin\nu(u-x_\nu)}{u-x_\nu} F(u)\, du \right| \leqslant c\|\lambda\|_1, \tag{3.7}$$

where the sum $\Sigma_{I\in\mathcal{J}}$ *extends over all* $I \in \mathcal{J}$.

PROOF. For each $I \in \mathcal{J}$, put

$$\varepsilon_\nu(I) = \operatorname{sgn} \int_{I_0'\setminus I'} \frac{\sin\nu(u-x_\nu)}{u-\eta(I)} F(u)\, du \quad (\nu \geqslant 0),$$

$$\tau(I; u) = \sum_{\nu:\, x_\nu\in I} \varepsilon_\nu(I)\lambda_\nu \sin\nu(u-x_\nu).$$

In view of (3.5), in order to prove (3.7) it is enough to verify that

$$\sum_{I\in\mathcal{J}} \int_{I_0'\setminus I'} \frac{F(u)}{u-\eta(I)} \tau(I; u)\, du \leqslant c\|\lambda\|_1. \tag{3.8}$$

Using Cauchy's inequality and Parseval's equation, we obtain

$$\int_{I_0'\setminus I'} \frac{F(u)}{u-\eta(I)} \tau(I; u)\, du \leqslant \left\{ \int_{I_0'\setminus I'} \left(\frac{F(u)}{u-\eta(I)} \right)^2 du \right\}^{1/2} \left\{ \int_{I_0'} \tau^2(I; t)\, dt \right\}^{1/2}$$
$$\leqslant c|I|^{-1/2} \Big(\sum_{\nu:\, x_\nu\in I} \lambda_\nu^2 \Big)^{1/2} \leqslant cN^{1/2} \Big(\sum_{\nu:\, x_\nu\in I} \lambda_\nu^2 \Big)^{1/2},$$

and therefore, applying Cauchy's inequality again, we see that

$$\sum_{I\in\mathcal{J}} \int_{I_0'\setminus I'} \frac{F(u)}{u-\eta(I)} \tau(I; u)\, du \leqslant cN^{1/2} \sum_{I\in\mathcal{J}} \Big(\sum_{\nu:\, x_\nu\in I} \lambda_\nu^2 \Big)^{1/2}$$
$$\leqslant cN \Big(\sum_{I\in\mathcal{J}} \sum_{\nu:\, x_\nu\in I} \lambda_\nu^2 \Big)^{1/2} = cN\|\lambda\|_2 \leqslant c\|\lambda\|_1.$$

This establishes (3.8), and simultaneously Lemma 2.

LEMMA 3. *Let* $0 < \delta < 2\pi$, *let* P *be a positive integer, let* $\xi = \{\xi_j\}_1^P$ *be a finite set of points*, $\xi \subset I_0$, *and let* $m_1, \ldots, m_P$ *be integers such that*

$$m_1 \geqslant \frac{4\pi}{\delta}, \quad m_{j+1} - m_j \geqslant \frac{4\pi}{\delta} \quad (j = 1, \ldots, P-1). \tag{3.9}$$

For each $I \subset I_0$ *let* $P(I)$ *be the number of integers* j, $1 \leqslant j \leqslant P$, *for which* $\xi_j \in I$:

$$P(I) = \sum_{j:\, \xi_j\in I} 1, \tag{3.10}$$

and put

$$N(I) = N(\xi, \delta, I) = \begin{cases} \max(1, [P^{1/2}(I)]), & \text{if } |I| \geqslant \delta; \\ 1, & \text{if } |I| < \delta. \end{cases} \tag{3.11}$$

For $l = 0, 1, \ldots$ *we define inductively a partition* $\mathcal{J}_l = \{I\}$ *of the half-open interval* I_0 *into a finite set of mutually disjoint half-open intervals* I *in the following way.*

Put $\mathcal{J}_0 = \{I_0\}$ *and suppose that the partition* $\mathcal{J}_l = \{I\}$ *has been constructed for a given* $l \geqslant 0$; $\mathcal{J}_{l+1}$ *is obtained from* $\mathcal{J}_l$ *by partitioning each* $I \in \mathcal{J}_l$ *into* $N(I)$ *half-open intervals* i *of the same length* $|i| = |I|N^{-1}(I)$.

Then for $l = 0, 1, \ldots$ *we have the estimates*

$$|I| \geqslant \delta P^{-1/2} \quad (I \in \mathcal{J}_l); \tag{3.12}$$

$$\sum_{I \in \mathcal{J}_l} P(I) \log |I| \leqslant -\min\left\{(1-2^{-l}) P \log P,\ P \log \frac{1}{\delta}\right\} + c(l+1)P, \tag{3.13}$$

and

$$\sum_{I \in \mathcal{J}_l} \sum_{j : \xi_j \in I} \left| S_{m_j}(f, \xi_j) - \frac{1}{\pi} \int_{I'} \frac{\sin m_j (t - \xi_j)}{t - \xi_j} f(t)\, dt \right| \leqslant c(l+1)P \tag{3.14}$$

if $f \in L^\infty$ *and* $\|f\| \leqslant 1$.

PROOF. (3.12) is a simple corollary of (3.11). Let us prove (3.13). Put

$$\mathcal{Q} = \mathcal{Q}_l = \{I : I \in \mathcal{J}_{l-1},\ |I| \geqslant \delta\}, \quad Q = \sum_{I \in \mathcal{Q}} P(I);$$

$$\mathcal{R} = \mathcal{R}_l = \{I : I \in \mathcal{J}_{l-1},\ |I| < \delta\}, \quad R = \sum_{I \in \mathcal{R}} P(I).$$

For each $I \in \mathcal{J}_{l-1}$ put

$$Q(I) = \begin{cases} P(I), & \text{if} \quad I \in \mathcal{Q}, \\ 0, & \text{if} \quad I \in \mathcal{R}; \end{cases}$$

if $I \in \mathcal{J}_s$, with $0 \leqslant s \leqslant l-2$, we have

$$Q(I) = \sum_{i \in I,\ i \in \mathcal{Q}} P(i).$$

By the definition of $\mathcal{J}_l$

$$\sum_{I \in \mathcal{J}_l} P(I) \log |I| = \sum_{I \in \mathcal{J}_{l-1}} P(I) \log \frac{|I|}{N(I)} = \sum_{I \in \mathcal{Q}} P(I) \log \frac{|I|}{N(I)} + \sum_{I \in \mathcal{R}} P(I) \log \frac{I}{N(I)}. \tag{3.15}$$

If $I \in \mathcal{R}$, then $|I| < \delta$ and consequently $N(I) = 1$ by (3.11). Hence

$$\sum_{I \in \mathcal{R}} P(I) \log \frac{|I|}{N(I)} \leqslant \sum_{I \in \mathcal{R}} P(I) \log \delta = R \log \delta. \tag{3.16}$$

Let r and s be integers, $0 \leqslant s \leqslant r$. For each $i \in \mathcal{J}_r$ let $\omega_s(i)$ denote the interval I of the partition $\mathcal{J}_s$ for which $i \subset I$. If $I \in \mathcal{Q}$, then $|I| \geqslant \delta$, and it follows from (3.11) that

$$N(\omega_s(I)) \geqslant \frac{1}{2} P^{1/2}(\omega_s(I)) \quad (0 \leqslant s \leqslant l-1),$$

$$\frac{|I|}{N(I)} = 2\pi \left(\prod_{s=0}^{l-1} N(\omega_s(I)) \right)^{-1} \leqslant 2^l \cdot 2\pi \left(\prod_{s=0}^{l-1} P(\omega_s(I)) \right)^{-1/2}.$$

Therefore

$$\sum_{I\in Q} P(I)\log\frac{|I|}{N(I)}\leqslant clQ-\frac{\tau_l}{2},\tag{3.17}$$

where

$$\tau_l=\sum_{I\in Q}P(I)\sum_{s=0}^{l-1}\log P(\omega_s(I))=\sum_{I\in Q}P(I)\left[\log P(I)+\sum_{s=0}^{l-2}\log P(\omega_s(I))\right]$$

$$=\sum_{I\in\mathcal{J}_{l-1}}Q(I)\left[\log Q(I)+\sum_{s=0}^{l-2}\log P(\omega_s(I))\right].\tag{3.18}$$

We shall show by induction that

$$\tau_l\geqslant\sum_{I\in\mathcal{J}_{l-r}}Q(I)\left[2(1-2^{-r})\log Q(I)+\sum_{s=0}^{l-r-1}\log P(\omega_s(I))\right]\tag{3.19}$$

for $r=1,\dots,l$. For $r=1$, (3.19) coincides with (3.18). Supposing that (3.19) has been established for $r=q$, where $1\leqslant q\leqslant l-1$, we establish it for $r=q+1$. We have

$$\sum_{I\in\mathcal{J}_{l-q}}Q(I)\left[2(1-2^{-q})\log Q(I)+\sum_{s=0}^{l-q-1}\log P(\omega_s(I))\right]$$

$$=\sum_{I\in\mathcal{J}_{l-q-1}}\ \sum_{i\subset I,\ i\in\mathcal{J}_{l-q}}Q(i)\left[2(1-2^{-q})\log Q(i)+\sum_{s=0}^{l-q-1}\log P(\omega_s(i))\right].\tag{3.20}$$

For arbitrary $I\in\mathcal{J}_{l-q-1}$ and for $s=0,\dots,l-q-1$ we have

$$\omega_s(i)=\omega_s(I)\qquad(i\subset I,\ i\in\mathcal{J}_{l-q}),$$

and the sum $\Sigma_{i\in I,i\in\mathcal{J}_{l-q}}$ contains, by the definition of $\mathcal{J}$, at most $N(I)=\max([P^{1/2}(I)],1)$ terms. Hence, using the convexity of $\varphi(u)=u\log u$ $(u\geqslant 0)$, Jensen's inequality, and the equation

$$\sum_{i\subset I,\ i\in\mathcal{J}_{l-q}}Q(i)=Q(I)\quad(I\in\mathcal{J}_{l-q-1})$$

we obtain from the induction hypothesis (cf. (3.19) for $r=q$) that

$$\tau_l\geqslant\sum_{I\in\mathcal{J}_{l-q-1}}Q(I)\left\{2(1-2^{-q})\log\frac{Q(I)}{N(I)}+\sum_{s=0}^{l-q-1}\log P(\omega_s(I))\right\}$$

$$=\sum_{I\in\mathcal{J}_{l-q-1}}Q(I)\left\{2(1-2^{-q})\log Q(I)-2(1-2^{-q})\log[P^{1/2}(I)]\right.$$

$$\left.+\log P(I)+\sum_{s=0}^{l-q-2}\log P(\omega_s(I))\right\}$$

$$\geqslant\sum_{I\in\mathcal{J}_{l-q-1}}Q(I)\left\{2(1-2^{-q})\log Q(I)+2^{-q}\log Q(I)+\sum_{s=0}^{l-q-2}\log P(\omega_s(I))\right\}$$

$$=\sum_{I\in\mathcal{J}_{l-q-1}}Q(I)\left\{2(1-2^{-q-1})\log Q(I)+\sum_{s=0}^{l-q-2}\log P(\omega_s(I)),\right.$$

from which (3.19) follows for $r=q+1$.

Let $r = l$. Then it follows from (3.19) and (3.16)–(3.18) that

$$\sum_{I\in\mathcal{J}_l} P(I)\log|I| \leqslant R\log\delta-(1-2^{-l})Q\log Q+cR. \tag{3.21}$$

A simple calculation shows that

$$\max_{R,\,Q\geqslant 0,\,R+Q=P}(R\log\delta-(1-2^{-l})Q\log Q)$$
$$\leqslant -\min\left\{(1-2^{-l})P\log P,\ P\log\frac{1}{\delta}\right\}+cP,$$

and hence (3.13) follows from (3.21).

We now prove (3.14) by induction on l. For $l = 0$ (3.14) coincides with (3.4). Suppose that (3.14) holds for some $l \geqslant 0$; we deduce it for $l + 1$. It is enough to establish that, for each $I \in \mathcal{J}_l$,

$$\sum_{i\subset I,\ i\in\mathcal{J}_{l+1}}\ \sum_{j:\xi_j\in i}\left|\int_{I'\setminus i'}\frac{\sin m_j(t-\xi_j)}{t-\xi_j}f(t)\,dt\right|\leqslant cP(I). \tag{3.22}$$

In proving (3.22) we may suppose that $|I| \geqslant \delta$ and $P(I) \geqslant 1$ (cf. (3.11)). Put $p_j = [m_j|I|/2\pi]$, and for $i \in I$ let

$$A_j(i)=\int_{I'\setminus i'}\frac{\sin m_j(t-\xi_j)}{t-\xi_j}f(t)\,dt,$$

$$B_j(i)=\int_{I'\setminus i'}\frac{\sin p_j 2\pi|I|^{-1}(t-\xi_j)}{t-\xi_j}f(t)\,dt.$$

Since $|I'| = 2|I|$ and

$$|p_j 2\pi|I|^{-1}-m_j|\leqslant 2\pi|I|^{-1},$$

we have (cf. (3.10))

$$\sum_{i\subset I,\ i\in\mathcal{J}_{l+1}}\ \sum_{j:\xi_j\in i}|A_j(i)-B_j(i)|\leqslant cP(I). \tag{3.23}$$

We now use Lemma 2 to show that

$$\sum_{i\subset I,\ i\in\mathcal{J}_{l+1}}\ \sum_{j:\xi_j\in i}|B_j(i)|\leqslant cP(I). \tag{3.24}$$

For this purpose we note that by (3.9)

$$p_1\geqslant 1,\qquad p_{j+1}-p_j\geqslant 1\qquad (j=1,\ldots,P).$$

Hence, putting

$$x_\nu=\begin{cases}2\pi|I|^{-1}(\xi_j-\eta(I)), & \text{if }\ \nu=p_j,\ \text{ where }\ \xi_j\in I;\\ 0 & \text{for other values of }\ \nu,\end{cases}$$

$$\lambda_\nu=\begin{cases}1, & \text{if }\ \nu=p_j,\ \text{where }\ \xi_j\in I;\\ 0 & \text{for other values of }\ \nu\end{cases}$$

and making the change of variable $u = 2\pi|I|^{-1}(t - \eta(I))$, we obtain

$$\sum_{i\subset I,\ i\in\mathcal{J}_{l+1}}\ \sum_{j:\xi_j\in i}|B_j(i)|=\sum_{i\in\mathcal{J}}\ \sum_{\nu:x_\nu\in i}\lambda_\nu\left|\ \int\limits_{I_0'\setminus i'}\frac{\sin\nu(u-x_\nu)}{u-x_\nu}F(u)\,du\right|, \tag{3.25}$$

where

$$F(u)=f\left(u\frac{|I|}{2\pi}+\eta(I)\right)\qquad(u\in I_0'),$$

and $\mathcal{J}$ is the decomposition of the half-open interval I_0 described in the hypotheses of Lemma 2. This is easily seen if we note that

$$\|\lambda\|_1=\sum_{j:\xi_j\in I}1=P(I),\qquad\|\lambda\|_2=\Big(\sum_{j:\xi_j\in I}1\Big)^{1/2}=P^{1/2}(I), \tag{3.26}$$

and therefore by (3.11)

$$N(I)=[P^{1/2}(I)]=[d^{-1}(\lambda)]=N(\lambda).$$

Then (3.22) follows from (3.25), (3.7) and (3.23). This completes the proof of Lemma 3.

We can now prove Theorems 1 and 2, and (3.1)–(3.3). To prove Theorem 1 put

$$\lambda_\nu=\begin{cases}1, & \text{if}\quad \nu=n_j,\quad 1\leqslant j\leqslant P,\\ 0 & \text{for other values of}\ \ \nu,\end{cases}$$

and use (3.4) and Lemma 2. Since

$$\|\lambda\|_1=P,\qquad\|\lambda\|_2=P^{1/2},\qquad d(\lambda)=P^{-1/2},$$

it follows from (3.7) that

$$\sum_{I\subset\mathcal{J}}\ \sum_{j:x_j\in I}\left|S_{n_j}(f,x_j)-\frac{1}{\pi}\int\limits_{I'}\frac{\sin n_j(t-x_j)}{t-x_j}f(t)\,dt\right|\leqslant cP. \tag{3.27}$$

We note that $|I| = 2\pi N^{-1}(\lambda)\leqslant cP^{-1/2}$, and therefore

$$\left|\int\limits_{I'}\frac{\sin n_j(t-x_j)}{t-x_j}f(t)\,dt-\int\limits_{-P^{-1/2}}^{P^{-1/2}}f(t+x_j)\frac{\sin n_j t}{t}\,dt\right|\leqslant c\qquad(x_j\in I).$$

From this and (3.27) we obtain (1.2). This completes the proof of Theorem 1.

Theorem 2 is a corollary of Lemma 2, (3.4) and (3.6). In fact, for each $\nu=0,1,\dots$ let x_ν denote the point of I_0 at which the norm $\|S_\nu(f)\|$ is attained, i.e.

$$\|S_\nu(f)\|=|S_\nu(f,x_\nu)|,\qquad x_\nu\in I_0\qquad(\nu=0,1,\dots). \tag{3.28}$$

Then by (3.4) and (3.7),

$$\sum_{\nu=0}^{\infty}\lambda_\nu\|S_\nu(f)\|\leqslant\frac{1}{\pi}\sum_{I\in\mathcal{J}}\ \sum_{\nu:x_\nu\in I}\lambda_\nu\left|\int\limits_{I'}\frac{\sin\nu(u-x_\nu)}{u-x_\nu}f(u)\,du\right|+c\|\lambda\|_1,$$

whence by noticing that $|I| = 2\pi[d^{-1}(\lambda)]^{-1}$ and $d(\lambda)\leqslant 1$, and using (3.6), we obtain (2.2).

We now deduce (3.1) from (2.2). Let P be an integer and

$$\lambda_\nu=\lambda_\nu^{(P)}=\begin{cases}\dfrac{1}{PL_{n_k}}, & \text{if}\quad \nu=n_k,\ \ P\leqslant k\leqslant 2P-1;\\ 0 & \text{for other values of}\ \ \nu.\end{cases}$$

Using (2.3) we see that for all sufficiently large P

$$\|\lambda\|_1=\frac{1}{P}\sum_{k=P}^{2P-1}\frac{1}{L_{n_k}}\geqslant\frac{\pi^2}{8}\frac{1}{\log n_{2P}},$$

$$\|\lambda\|_2=\frac{1}{P}\left(\sum_{k=P}^{2P-1}\frac{1}{L_{n_k}^2}\right)^{1/2}\leqslant P^{-1/2},$$

so that

$$d(\lambda)\leqslant cP^{-1/2}\log n_{2P}. \tag{3.29}$$

Since $P^{-1/2}\leqslant d(\lambda)\leqslant 1$, we find from (2.2), (2.3) and (3.29) that

$$\frac{1}{P}\sum_{k=P}^{2P-1}\frac{\|S_{n_k}(f)\|}{L_{n_k}}\leqslant\frac{1}{P}\sum_{k=P}^{2P-1}\frac{\log n_k d(\lambda)}{\log n_k}+o(1)$$

$$\leqslant 1+\frac{\log d(\lambda)}{\log n_{2P}}+o(1)\leqslant\frac{\log\left(n_{2P}P^{-1/2}\right)}{\log n_{2P}}+o(1)\qquad(P\to\infty).$$

From this, (3.1) follows, i.e. we have shown that (2.6) is necessary for (2.5).

We now prove (3.3). We estimate the average of the norms $\|V_{n,q_n}(f)\|$ for $P\leqslant n\leqslant 2P-1$, i.e.

$$W_P=W_P(q,\ f)=\frac{1}{P}\sum_{n=P}^{2P-1}\|V_{n,q_n}(f)\|.$$

For each integer $\nu\geqslant 0$ let $\mathcal{M}_\nu$ be the set of integers n for which

$$P\leqslant n\leqslant 2P-1,\qquad n\leqslant\nu\leqslant n+q_n-1, \tag{3.30}$$

and put

$$\lambda_\nu=\lambda_\nu^{(P)}=\sum_{n\in\mathcal{M}_\nu}\frac{1}{Pq_n}. \tag{3.31}$$

Then (cf. (2.8))

$$W_P=\frac{1}{P}\sum_{n=P}^{2P-1}\frac{1}{q_n}\left\|\sum_{\nu=n}^{n+q_n-1}S_\nu(f)\right\| \tag{3.32}$$

$$\leqslant\frac{1}{P}\sum_{n=P}^{2P-1}\frac{1}{q_n}\sum_{\nu=n}^{n+q_n-1}\|S_\nu(f)\|=\sum_{\nu=0}^{\infty}\lambda_\nu\|S_\nu(f)\|.$$

Using (2.10), we find that

$$d(\lambda^{(P)})=O(P^{-1/2})\qquad(P\to\infty). \tag{3.33}$$

In fact, $\|\lambda\|_1=1$. Let $M_\nu=\Sigma_{n\in\mathcal{M}_\nu}1$; this is the number of integers n belonging to $\mathcal{M}_\nu$ $(\nu\geqslant 0)$. Then by (3.30) we have $M_\nu\leqslant n_2-n_1+1$, where

$$n_1=n_1(\nu)=\min\{n:P\leqslant n\leqslant 2P-1,\ \nu\leqslant n+q_n-1\},$$

$$n_2=n_2(\nu)=\max\{n:P\leqslant n\leqslant 2P-1,\ \nu\geqslant n\}.$$

In particular, $\nu \leqslant n_1 + q_{n_1} - 1$ and $\nu \geqslant n_2$; hence $M_\nu \leqslant \max_{P \leqslant s \leqslant 2P-1} q_s$. Hence it follows from (3.30) and the second condition (2.10) that

$$\max_\nu \lambda_\nu^{(P)} = O(P^{-1}) \quad (P \to \infty).$$

Hence (3.33) holds, and therefore, by (2.2) and (2.10),

$$W_P \leqslant \frac{1}{2} \frac{4}{\pi^2} \log P + O(1) \quad (P \to \infty),$$

whence by (2.9), the first condition (2.10), and (3.32) we obtain (3.3).

We now prove (3.2). First let $1 \leqslant \kappa \leqslant 2$. We take $\gamma = 16$, $k_j = [(\gamma(j + P - 1)P)^{1/\kappa}]$, $m_j = n_{k_j}$, $\xi_j = x_{m_j}$ $(1 \leqslant j \leqslant P)$, and

$$\delta = \frac{4\pi}{P}, \qquad l = \left[\frac{\log\log P}{\log 2}\right] + 1, \tag{3.34}$$

where P is a sufficiently large integer. Then

$$m_j = [[(\gamma(j + P - 1))^{1/\varkappa}]^\varkappa] + 1 \leqslant \gamma(j + P - 1)P + 1,$$

$$m_j \geqslant ((\gamma(j + P - 1)P)^{1/\varkappa} - 1)^\varkappa \geqslant \gamma(j + P - 1)P - \varkappa(\gamma(j + P - 1)P)^{\frac{\varkappa-1}{\varkappa}}$$
$$\geqslant \gamma(j + P - 1)P - 2(2\gamma P^2)^{1/2},$$

and hence for $1 \leqslant j \leqslant P$

$$4P^2 \leqslant \gamma P^2 - 2^{3/2}\gamma^{1/2}P \leqslant m_j \leqslant \gamma(2P - 1)P + 1 \leqslant 32P^2, \tag{3.35}$$

and if $1 \leqslant j \leqslant P - 1$ we have

$$m_{j+1} - m_j \geqslant \gamma P - 2^{3/2}\gamma^{1/2}P - 1 \geqslant P.$$

It follows that, for the integers m_j and the selected value of δ, hypothesis (3.9) of Lemma 2 is satisfied. Hence by using (3.28), (3.14) and (3.6) we obtain

$$\sum_{j=1}^{P} \|S_{m_j}(f)\| \leqslant \frac{4}{\pi^2} \sum_{I \in \mathcal{F}_l} \sum_{j : \xi_j \in I} \log(1 + m_j |I|) + c(l + 1)P.$$

From this estimate, using the fact that $|I| \geqslant \delta P^{-1/2} = 4\pi P^{-3/2}$ by (3.12) and that, in particular (cf. (3.35)), $m_j|I| \geqslant 1$, we obtain

$$\sum_{j=1}^{P} \|S_{m_j}(f)\| \leqslant \frac{4}{\pi^2} \sum_{I \in \mathcal{F}_l} \sum_{j : \xi_j \in I} \log m_j |I| + c(l + 1)P$$
$$= \frac{4}{\pi^2} \sum_{j=1}^{P} \log m_j + \frac{4}{\pi^2} \sum_{I \in \mathcal{F}_l} P(I) \log |I| + c(l + 1)P.$$

Hence by using (3.35) and (3.13), and (2.3), we conclude that

$$\frac{1}{P} \sum_{j=1}^{P} \|S_{m_j}(f)\| \leqslant \frac{4}{\pi^2} [\log P^2 - (1 - 2^{-l}) \log P] + c(l + 1)$$
$$= \frac{1}{2} \frac{4}{\pi^2} \log P^2 + O(\log\log P) = \frac{1}{2} \frac{1}{P} \sum_{j=1}^{P} L_{m_j} + O(\log\log P).$$

From this, inequality (3.2) follows for $1 \leqslant \kappa \leqslant 2$. Now let $\kappa > 2$. Then for $P \leqslant k \leqslant 2P-1$

$$P^{\varkappa} \leqslant n_k \leqslant 2^{\varkappa} P^{\varkappa} + 1, \tag{3.36}$$

$$n_{k+1} - n_k = [(k+1)^{\varkappa}] - [k^{\varkappa}] \geqslant \varkappa k^{\varkappa-1} - 1 \geqslant k \geqslant P. \tag{3.37}$$

Hence by choosing δ and l as in (3.34) we find from (3.36), (3.13) and (2.3) that

$$\frac{1}{P} \sum_{k=1}^{P} \| S_{n_k}(f) \| \leqslant \frac{4}{\pi^2} \{\log P^{\varkappa} - (1 - 2^{-l}) \log P\} + c(l+1)$$

$$= \frac{1}{P}\left(1 - \frac{1}{\varkappa}\right) \frac{4}{\pi^2} \sum_{k=1}^{P} L_{n_k} + O(\log\log P).$$

This proves (3.2) for $\kappa > 2$.

We now complete the proofs of Theorems 3 and 4. It follows from (3.1) and (3.2) that Theorem 3 will be established if we establish for each increasing sequence $\mathscr{N} = \{n_k\}_0^\infty$ the existence of a function $f \in L^\infty$ such that $\|f\| \leqslant 1$ and

$$\varliminf_{k\to\infty} \frac{\| S_{n_k}(f) \|}{L_{n_k}} \geqslant \varliminf_{k\to\infty} \frac{\log (n_k k^{-1})}{\log n_k}. \tag{3.38}$$

In fact, it follows from (3.38) that (2.6) is a sufficient condition for (2.5) to hold.

If $\mathscr{N} = \{n_k\}_0^\infty = \{[k^\kappa] + 1\}_0^\infty$, then (2.7) follows from (3.38) and (3.2) if $\kappa > 2$. If $1 \leqslant \kappa \leqslant 2$, however, (2.7) is a corollary of (3.2) and the inequality $\underline{\mathscr{S}}(\mathscr{N}) \geqslant \underline{\mathscr{S}} = 1/2$ (cf. (1.5)).

To construct an example of a function $f(x)$ with the property (3.38) we use the 2π-periodic functions $\varphi_{p,m}(t)$ (p and m integers, $p \leqslant m$) defined on $[-\pi, \pi)$ as follows:

$$\varphi_{p,m}(t) = \begin{cases} \operatorname{sgn} \sin m|t|, & \text{if} \quad |t| < \pi p/m; \\ 0, & \text{if} \quad \pi p/m \leqslant |t| \leqslant \pi. \end{cases}$$

The Fourier sums of these functions have the following properties (cf. [6], formula (47)):

$$\left| S_n(\varphi_{p,m}, 0) - \frac{4}{\pi^2} \log \min\left(p, \frac{m}{n-m}\right) \right| \leqslant c \qquad (m \leqslant n \leqslant 2m). \tag{3.39}$$

Let k be an integer, $k \geqslant k_0$, where k_0 is sufficiently large; $n = m = n_k$; $p = p_k = [n_k / k \log^3 k]$, $x_k = 1/\log k$ and $f_k(t) = \varphi_{p_k, n_k}(t - x_k)$. The number k_0 is to be so large that when $k \geqslant k_0$ the intervals

$$I_k = \left(x_k - \frac{\pi p_k}{n_k},\ x_k + \frac{\pi p_k}{n_k} \right),$$

support the functions f_k, belong to $[0, \pi)$ and are mutually disjoint. It follows from (3.39) that

$$\left| S_{n_k}(f_k, x_k) - \frac{4}{\pi^2} \log \frac{n_k}{k} \right| \leqslant c \log\log k. \tag{3.40}$$

Put

$$f(t) = \sum_{k=k_0}^{\infty} f_k(t).$$

Since the supports of the f_k are mutually disjoint, $\|f\| = 1$. We show that (3.38) holds for the Fourier sums of these functions. In fact, let $k \geqslant k_0$,

$$\Phi_k(t) = f(t) - f_k(t), \qquad \mathscr{G}_k = \Big\{ t : t \in \Big(\bigcup_{l=k_0}^{\infty} I_l \Big) \setminus I_k \,(\mathrm{mod}\, 2\pi) \Big\},$$

and let $\chi(t)$ be the characteristic function of G. Since $\Phi(t) = 0$ if $t \notin \mathscr{G}_k$, we have (cf. [6], Lemma 3)

$$|S_{n_k}(\Phi_k, x_k)| \leqslant c \int_{1/n_k}^{\pi} \frac{\chi(x_k + z) + \chi(x_k - z)}{z}\, dz + c.$$

Here the integral on the right-hand side is bounded in absolute value (cf. [6], formula (53) et seq.).

From this it follows by (3.40) and (2.3) that

$$\frac{|S_{n_k}(f, x_k)|}{L_{n_k}} \geqslant \frac{\log(n_k k^{-1})}{\log n_k} + o(1) \qquad (k \to \infty).$$

This establishes (3.38), and simultaneously Theorem 3.

To complete the proof of Theorem 4, we use the function f constructed in [6] to obtain a lower bound for $\underline{\mathscr{L}}$. In [6] (the proof of Theorem 1) we established the existence of a function $f \in L^\infty$ such that $\|f\| = 1$ and

$$\Big| S_\nu(f, x_k) - \frac{4}{\pi^2} \log \min\Big(\frac{k}{\log^3 k}, \frac{k^2}{\nu - k^2}\Big) \Big| \leqslant c$$

$$(k^2 \leqslant \nu \leqslant 2k^2, \ x_k = 1/\log k, \ k \geqslant k_0).$$

It follows that if $k^2 \leqslant n \leqslant (k+1)^2$ then

$$\|V_{n, q_n}(f)\| \geqslant \frac{1}{q_n} \Big| \sum_{\nu=n}^{n+q_n-1} S_\nu(f, x_k) \Big| \geqslant \frac{4}{\pi^2} \frac{1}{q_n} \sum_{\nu=n}^{n+q_n-1} \log \min\Big(\frac{k}{\log^3 k}, \frac{k^2}{\nu - k^2}\Big) - c$$

$$\geqslant \frac{4}{\pi^2} \log \min\Big(\frac{k}{\log^3 k}, \frac{k^2}{2k + Q_k}\Big) - c,$$

where $Q_k = \max_{k^2 \leqslant n < (k+1)^2} q_n$, and therefore by (2.10)

$$\|V_{n, q_n}(f)\| \geqslant \frac{4}{\pi^2} \min\Big(\frac{1}{2} \log n, \ \log \frac{n}{q_n}\Big) - c \log\log n \qquad (n \geqslant k_0^2).$$

Hence by using the first condition (2.10), and (2.9), we find that

$$\varliminf_{n \to \infty} \frac{\|V_{n, q_n}(f)\|}{U_{n, q_n}} \geqslant \varliminf_{n \to \infty} \min\Big(1, \ \frac{1}{2} \frac{\log n}{\log(n/q_n)}\Big].$$

Then (2.11) follows from this and (3.3). This completes the proof of the theorem.

The author thanks S. A. Teljakovskiĭ and V. N. Temljakov for helpful discussions of the results of the present paper.

BIBLIOGRAPHY

1. A. Zygmund, *Trigonometric series,* 2nd rev. ed., Vol. 1, Cambridge Univ. Press, New York, 1959.

2. Harold Bohr, *Über die Koeffizientensumme einer beschränkten Potenzreihe,* Nachr. Königl. Ges. Wiss. Göttingen Math.-Phys. Kl. **1916**, 276–291.

3. Ludwig Neder, *Über die Koeffizientensumme einer beschränkten Potenzreihe,* Math. Z. **11** (1921), 115–123.

4. D. E. Menchoff [Men′šov], *Sur les sommes partielles des séries de Fourier des fonctions continues,* Mat. Sb. **15** (**57**) (1944), 385–432.

5. Edmond Busko, *Fonctions continues et fonctions bornées non adhérentes dan $L^\infty(T)$ à la suite de leurs sommes partielles de Fourier,* Studia Math. **34** (1970), 319–337.

6. K. I. Oskolkov, *An estimate in the approximation of continuous functions by subsequences of Fourier sums,* Trudy Mat. Inst. Steklov. **134** (1975), 240–253; English transl. in Proc. Steklov Inst. Math. **134** (1975).

7. Lennart Carleson, *On convergence and growth of partial sums of Fourier series,* Acta Math. **116** (1966), 135–157.

8. G. Alexits and D. Králik, *Über die Approximation mit starken de la Vallée-Poussinschen Mitteln,* Acta Math. Acad. Sci. Hungar. **16** (1965), 43–49.

9. S. M. Nikol′skiĭ, *Sur certaines méthodes d'approximation au moyen de sommes trigonométriques,* Izv. Akad. Nauk SSSR Ser. Mat. **4** (1940), 509–520. (Russian; French summary)

Translated by R. P. BOAS

Труды Мат. Инст. Стеклов.
143 (1977)

Proceedings of the Steklov
Institute of Mathematics
1980, Issue 1

AN ANALYTIC METHOD FOR ESTIMATES OF THE CONCENTRATION FUNCTION

UDC 519.21

L. P. POSTNIKOVA AND A. A. JUDIN

ABSTRACT. The authors solve a problem posed by Kesten on the proof, by means of characteristic functions, of an estimate for the concentration function of the sum of n independent identically distributed random variables: $Q(S_n, \lambda) \ll \lambda Q(X, \lambda)/l\sqrt{n(1 - Q(X, l))}$, where l is any positive number not greater than 2λ.

Bibliography: 6 titles.

The concentration function $Q(X, \lambda)$ of a random variable X is defined by

$$Q(X, \lambda) = \sup_x \mathrm{P}(x \leqslant X \leqslant x + \lambda).$$

Consider the sum

$$S_n = \sum_{i=1}^{n} X_i \tag{1}$$

of independent identically distributed random variables. Kesten [1] found the estimate

$$Q(S_n, \lambda) \leqslant A \frac{\lambda Q(X, \lambda)}{l\sqrt{n(1 - Q(X, l))}}, \tag{2}$$

where A is a constant and l is a positive number not greater than 2λ; in fact, [1] contains a more general result.

To prove (2), Kesten developed the combinatorial method of Rogozin [2]. In this connection, Kesten [1] asks whether one can obtain (2) (and more general estimates) by means of characteristic functions. In the present paper we answer this question for identically distributed summands.

THEOREM. *The sum* (1) *satisfies the estimate*

$$Q(S_n, \lambda) \ll \frac{\lambda Q(X, \lambda)}{l\sqrt{n(1 - Q(X, l))}}, \tag{2'}$$

where $\ll$ *is Vinogradov's symbol.*

The proof of the theorem is based on several auxiliary lemmas.

LEMMA 1. *If X is a random variable and $f(t)$ and $Q(X, \lambda)$ are the characteristic function and the concentration function of X, then*

$$Q(X, \lambda) \leqslant \left(\frac{96}{95}\right)^2 \max\left(\lambda, \frac{1}{a}\right) \int_{-a}^{a} |f(t)|\, dt. \tag{3}$$

1980 *Mathematics Subject Classification.* Primary 60E10, 60G50.

For the proof, see [3], Chapter III, Lemma 3.

LEMMA 2. *Let A and B be nonempty measurable bounded sets of real numbers and put $A + B = \{x_1 + x_2 : x_1 \in A, x_2 \in B\}$. Then*

$$\operatorname{mes}(A+B) \geqslant \operatorname{mes} A + \operatorname{mes} B.$$

PROOF. We may assume, without loss of generality, that $\sup A = \inf B = 0$. If $x \in A$, $y \in B$, $x > -\epsilon$ and $y < \epsilon$, then $(A + y) \cup (B + x) \subset A + B$. Hence

$$\operatorname{mes}(A+B) \geqslant \operatorname{mes}[(A+y) \cup (B+x)] = \operatorname{mes} A + \operatorname{mes} B - \operatorname{mes}[(A+y) \cap (B+x)] \geqslant \operatorname{mes} A + \operatorname{mes} B - 2\varepsilon.$$

Since ϵ was arbitrary, we have $\operatorname{mes}(A + B) \geqslant \operatorname{mes} A + \operatorname{mes} B$.

We say that a set E of reals belongs to the class P_T, and write $E \in P_T$, when E is closed, symmetric relative to 0, contained in $(-T, T)$ and satisfies $(-\delta, \delta) \subset E$ for some $\delta > 0$. Define the density $\alpha_T(E)$ of a set $E \in P_T$ to be the number

$$\alpha_T(E) = \inf_{l \leqslant T} \frac{\operatorname{mes}(E \cap [-l, l])}{2l}.$$

LEMMA 3. *If $E_1, E_2 \in P_T$, then*

$$\alpha_T(E_1 + E_2) \geqslant \min(1, \alpha_T(E_1) + \alpha_T(E_2)). \tag{4}$$

PROOF. First assume that both E_1 and E_2 are the unions of finitely many disjoint closed intervals:

$$E_i = \bigcup_{j=1}^{m_i} [a_j^{(i)}, b_j^{(i)}] \qquad (i = 1, 2).$$

When N is a sufficiently large positive integer and $i = 1, 2$, define a set F_i^N of integers by

$$F_i^N = \left\{k \in \mathbf{Z} : \left[\frac{k - 1/2}{N}, \frac{k + 1/2}{N}\right] \subset E_i\right\}.$$

As $E_1, E_2 \in P_T$, for sufficiently large N we have $0 \in F_i^N$, $i = 1, 2$.

Put

$$A_i(N, m) = \sum_{\substack{0 \leqslant x \leqslant m \\ x \in F_i^N}} 1.$$

Let $l_j^{(i)}$ be the length of $[a_j^{(i)}, b_j^{(i)}]$. The number of intervals $[(k - ½)/N, (k + ½)/N]$ contained in $[a_j^{(i)}, b_j^{(i)}]$ is equal* to $l_j^{(i)}N - 2$. By definition, $A_i(N, m)$ is the number of intervals $[(k - ½)/N, (k + ½)/N]$ contained in $E_i \cap [-½/N, (m + ½)/N]$. Hence

$$A_i(N, m) \geqslant \left(\sum l_j^{(i)}\right) N - 2m_i;$$

the sum is taken over those j for which $[a_j^{(i)}, b_j^{(i)}] \subset [0, m/N]$. By definition,

$$\sum l_j^{(i)} \geqslant \alpha_T(E_i)\frac{m}{N},$$

whence

**Translator's note.* More exactly, the number is greater than $l_j^{(i)}N - 2$ and less than or equal to $l_j^{(i)}N$. The error is small for large N.

$$A_i(N,\ m) \geqslant \alpha_T(E_i)\,m - 2m_i = \left(\alpha_T(E_i) - \frac{2m_i}{m}\right)m.$$

Now we distinguish two cases. If $m \geqslant \delta_i N$ then

$$A_i(N,\ m) \geqslant \left(\alpha_T(E_i) - \frac{2m_i}{\delta_i N}\right)m;$$

if $m < \delta_i N$, then from $(-\delta_i, \delta_i) \subset E_i$ we get $A_i(N, m) \geqslant m$, and this together with $\alpha_T(E_i) \leqslant 1$ gives

$$A_i(N,\ m) \geqslant \left(\alpha_T(E_i) - \frac{2m_i}{\delta_i N}\right)m \tag{5}$$

for any $m = 1, 2, \ldots, [TN]$.

Put

$$G_+^{(i)} = \{z \in \mathbf{Z};\ z \geqslant 0,\ z \in F_i^N\} \qquad (i = 1,\ 2).$$

Taking (5) into account, we apply Dyson's theorem ([4], p. 17) to sets $G_+^{(1)}$ and $G_+^{(2)}$, and obtain

$$\Phi(m) = \sum_{\substack{1 \leqslant n \leqslant m,\ n = n_1 + n_2 \\ n_1 \in G_+^{(1)},\ n_2 \in G_+^{(2)}}} 1 \geqslant \min\left(1,\quad \alpha_T(E_1) + \alpha_T(E_2) - \frac{2m_1}{\delta_1 N} - \frac{2m_2}{\delta_2 N}\right)m$$

for any m, $1 \leqslant m \leqslant [TN]$. Denote by $\hat{F}_i^N$ the set of the numbers k/N, where $k \in F_i^N$, $i = 1, 2$. Since

$$E_1 + E_2 \supset \hat{F}_1^N + \hat{F}_2^N + \left[-\frac{1}{2N},\ \frac{1}{2N}\right],$$

it follows that

$$\frac{1}{2l}\operatorname{mes}(E_1 + E_2 \cap [-l,\ l]) \geqslant \frac{2\Phi([lN])\frac{1}{N}}{2l} = \frac{\Phi([lN])}{lN}.$$

As $N \longrightarrow \infty$, we get

$$\frac{\operatorname{mes}(E_1 + E_2 \cap [-l,\ l])}{2l} \geqslant \min(1,\ \alpha_T(E_1) + \alpha_T(E_2)). \tag{6}$$

Passing to the infimum over $l \leqslant T$, we obtain

$$\alpha_T(E_1 + E_2) \geqslant \min(1,\ \alpha_T(E_1) + \alpha_T(E_2)).$$

Now we prove the assertion for arbitrary closed sets E_1 and E_2. Thus let $E_i \in P_T$, $i = 1, 2$. Obviously $E_1 + E_2$ is closed and the set

$$\mathfrak{M} = (-T,\ T) \setminus (E_1 + E_2)$$

is open. Therefore $\mathfrak{M}$ is the union of countably many disjoint open intervals $\delta_1, \delta_2, \ldots$. Let $|\delta_i|$ be the length of δ_i. Assume that $|\delta_i| \geqslant |\delta_{i+1}|$, $i = 1, 2, \ldots$. We have $\Sigma_1^\infty |\delta_i| < \infty$, and therefore, by the Cauchy criterion, for each $\epsilon > 0$ we can find an n_0 such that

$$|\delta_{[n/2]}| + |\delta_{[n/2]+1}| + \ldots + |\delta_n| \leqslant \varepsilon$$

for $n \geqslant n_0$. Hence $[n/2]|\delta_n| \leqslant \epsilon$, and $n|\delta_n| \longrightarrow 0$ as $n \longrightarrow \infty$. It follows that

$$n|\delta_n| + \sum_{i=n}^{\infty} |\delta_i| < \varepsilon, \tag{7}$$

when $n \geqslant n_0(\epsilon)$. Put

$$A_{\delta_n} = E_1 + E_2 + \left[-\frac{|\delta_n|}{2}, \frac{|\delta_n|}{2}\right].$$

By adding $[-|\delta_n|/2, |\delta_n|/2]$ to $E_1 + E_2$, we cut a part of length $|\delta_n|$ from each of $\delta_1, \ldots, \delta_n$, and $\delta_{n+1}, \delta_{n+2}, \ldots$ are covered entirely. Hence, by (7),

$$\operatorname{mes} A_{\delta_n} < \operatorname{mes}(E_1 + E_2) + \varepsilon. \tag{8}$$

Obviously

$$\left(E_1 + \left[-\frac{|\delta_n|}{4}, \frac{|\delta_n|}{4}\right]\right) + \left(E_2 + \left[-\frac{|\delta_n|}{4}, \frac{|\delta_n|}{4}\right]\right) = A_{\delta_n} \tag{9}$$

and, for any $\delta > 0$ and $i = 1, 2$, the set $E_i + [-\delta, \delta]$ is the union of finitely many closed intervals. Substitute (6) and (8) into (9) to get

$$\frac{\operatorname{mes}(E_1 + E_2 \cap [-l, l])}{2l} > \frac{\operatorname{mes}(A_{\delta_n} \cap [-l, l]) - \varepsilon}{2l} \geqslant \min\left(1, \alpha_T\left(E_1 + \left[-\frac{|\delta_n|}{4}, \frac{|\delta_n|}{4}\right]\right)\right.$$

$$\left.+ \alpha_T\left(E_2 + \left[-\frac{|\delta_n|}{4}, \frac{|\delta_n|}{4}\right]\right)\right) - \frac{\varepsilon}{2l} \geqslant \min\left(1, \alpha_T(E_1) + \alpha_T(E_2)\right) - \frac{\varepsilon}{2l}.$$

As $\epsilon \longrightarrow 0$, we obtain

$$\frac{\operatorname{mes}(E_1 + E_2 \cap [-l, l])}{2l} \geqslant \min(1, \alpha_T(E_1) + \alpha_T(E_2)).$$

We conclude the proof by taking the infimum over l.

LEMMA 4. *If $f_1 \geqslant f_2 \geqslant \cdots \geqslant f_n \geqslant 0$ and $\alpha_1, \ldots, \alpha_n$ satisfy $\Sigma_1^k \alpha_i \geqslant k\delta$ for $1 \leqslant k \leqslant n$, then*

$$\sum_{i=1}^{n} \alpha_i f_i \geqslant \delta \sum_{i=1}^{n} f_i.$$

PROOF. Apply Abel's partial summation formula:

$$\sum_{i=1}^{n} f_i \alpha_i = \sum_{k=1}^{n} (\alpha_1 + \ldots + \alpha_k)(f_k - f_{k+1})$$

$$+ (\alpha_1 + \ldots + \alpha_n) f_n \geqslant \sum_{k=1}^{n-1} k\delta (f_k - f_{k+1}) + n\delta f_n$$

$$= \delta\left(\sum_{k=1}^{n-1} k(f_k - f_{k+1}) + nf_n\right) = \delta \sum_{k=1}^{n} f_k.$$

LEMMA 5. *Let $f(t)$ be the characteristic function of a random variable X. Put*

$$\omega = \int_{-a}^{a} |f(t)|^2\, dt, \quad m = \min_{t \in [-a, a]} |f(t)|, \quad \varphi^* = \arc\cos m.$$

If $0 < \varphi \leqslant \varphi^*/4$ *and* $E_\varphi = \{t \in [-a, a] : |f(t)| \geqslant \cos\varphi\}$, *then*

$$\mu(\varphi) = \operatorname{mes} E_\varphi \leqslant \frac{24\omega}{\varphi^*}\varphi. \tag{10}$$

PROOF. Put $G_y = [0, y]$ for $y \leqslant a$. There is a t_0 such that

$$\operatorname{mes}(E_\varphi \cap G_y + t_0) \geqslant \frac{1}{[2a/y] + 1}\mu(\varphi). \tag{11}$$

In fact, decompose $[-a, a]$ into $[2a/y]$ parts of length y and a remaining part. Let $\mu_i(\varphi)$, $i = 1, 2, \ldots, [2a/y] + 1$, be the measure of the intersection of the ith part with E_φ. Clearly

$$\sum \mu_i(\varphi) = \mu(\varphi). \tag{$*$}$$

We may assume that $\max_i \mu_i(\varphi)$ is attained at a part of length y. It follows from $(*)$ that

$$([2a/y] + 1)\max_i \mu_i(\varphi) \geqslant \mu(\varphi).$$

But the part where $\max \mu_i(\varphi)$ is attained is the shift of G_y by some t_0.

Obviously, (11) implies that

$$\operatorname{mes}(E_\varphi \cap G_y + t_0) \geqslant \frac{y}{2a + y}\mu(\varphi). \tag{11$'$}$$

Put $F_y(t_0) = E_\varphi \cap G_y + t_0$. Clearly, the set $F_y(t_0) - F_y(t_0)$ is contained in $[-y, y]$, and it is symmetric relative to 0 and closed. By Lemma 2 and (11$'$), we have

$$\operatorname{mes}(F_y(t_0) - F_y(t_0)) \geqslant 2 \operatorname{mes} F_y(t_0) \geqslant \frac{2y}{2a + y}\mu(\varphi)$$

and

$$\frac{\operatorname{mes}(F_y(t_0) - F_y(t_0))}{2y} \geqslant \frac{1}{2a + y}\mu(\varphi). \tag{12}$$

It follows from Lemma 1 in [5] that $F_y(t_0) - F_y(t_0) \subset E_{2\varphi}$. Hence

$$F_y(t_0) - F_y(t_0) \subset E_{2\varphi} \cap [-y, y].$$

This together with (12) gives

$$\frac{\operatorname{mes}(E_{2\varphi} \cap [-y, y])}{2y} \geqslant \frac{1}{2a + y}\mu(\varphi) \tag{13}$$

for any $y \leqslant a$. Therefore

$$\alpha_a(E_{2\varphi}) \geqslant \frac{1}{3a}\mu(\varphi). \tag{14}$$

We use the following standard notation:

$$kE_\theta = \begin{cases} E_\theta, & k=1; \\ nE_\theta + E_\theta, & k=n+1. \end{cases}$$

By Lemma 3,

$$\alpha_a(kE_{2\varphi}) \geqslant \min\left(1, \frac{k}{3a}\mu(\varphi)\right). \tag{15}$$

Again it follows from Lemma 1 in [5] that $kE_{2\varphi} \subset E_{2k\varphi}$, and therefore

$$\alpha_a(E_{2k\varphi}) \geqslant \min\left(1, \frac{k}{3a}\mu(\varphi)\right). \tag{16}$$

Assume that $\alpha_a(E_{2k'\varphi}) = 1$ for some $k = k'$; apply the definition of density to conclude that mes $E_{2k'\varphi} = 2a$. Since $|f(t)|$ is continuous, we have $|f(t)| \geqslant \cos 2k'\varphi$ for any $t \in [-a, a]$. It follows that $m \geqslant \cos 2k'\varphi$ and $\varphi^* \leqslant 2k'\varphi$. Hence, if $2k\varphi \leqslant \varphi^*$, then

$$\alpha_a(E_{2k\varphi}) \geqslant \frac{k}{3a}\mu(\varphi). \tag{17}$$

Putting $k_0 = [\varphi^*/4\varphi]$, we obtain $2(2k_0)\varphi \geqslant \varphi^*$ and $E_{4k_0\varphi} = [-a, a]$. Moreover,

$$\omega = \int_{-a}^{a} |f(t)|^2\,dt = \sum_{s=1}^{k_0-1} \int_{E_{2(s+1)\varphi}\setminus E_{2s\varphi}} |f(t)|^2\,dt$$

$$\geqslant \sum_{s=0}^{k_0-1} \cos^2 2(s+1)\varphi\,(\text{mes}\,E_{2(s+1)\varphi} - \text{mes}\,E_{2s\varphi}).$$

Put $\alpha_i = \text{mes}\,E_{2(i+1)\varphi} = \text{mes}\,E_{2i\varphi}$ for $i = 0, 1, \ldots$. Then

$$\sum_{i=0}^{k-1} \alpha_i = \text{mes}\,E_{2k\varphi}.$$

By (17),

$$\text{mes}\,E_{2k\varphi} \geqslant 2a\alpha_a(E_{2k\varphi}) \geqslant \frac{2k}{3}\mu(\varphi)$$

and, applying Lemma 4, we get

$$\omega \geqslant \frac{2}{3}\mu(\varphi)\sum_{s=0}^{k_0-1}\cos^2 2(s+1)\varphi = \frac{2}{3}\mu(\varphi)\sum_{s=1}^{k_0}\cos^2 2s\varphi$$

$$= \frac{\mu(\varphi)}{3}\sum_{s=1}^{k_0}(1+\cos 4s\varphi) = \frac{k_0\mu(\varphi)}{3} + \frac{\mu(\varphi)}{3}\sum_{s=1}^{k_0}\cos 4s\varphi.$$

Now $4k_0\varphi \leqslant \varphi^* \leqslant \pi/2$, and therefore $\omega \geqslant k_0\mu(\varphi)/3$. Hence $\mu(\varphi) \leqslant 3\omega/k_0$. For $\varphi \leqslant \varphi^*/4$ and $k_0 \geqslant 1$ this yields

$$\mu(\varphi) \leqslant \frac{6\omega}{\varphi^*/4\varphi} = 24\frac{\omega}{\varphi^*}\varphi,$$

and the proof is complete.

LEMMA 6. *Let $f(t)$ be the characteristic function of a random variable X, and let*

$$\omega = \int_{-1/\lambda}^{1/\lambda} |f(t)|^2\,dt, \qquad m = \min_{t\in[-1/\lambda,\,1/\lambda]} |f(t)| = \cos\varphi^*.$$

Then

$$\omega \leqslant 40\frac{Q(X,\lambda)}{\lambda} \tag{18}$$

and for $l \leqslant 2\lambda$

$$\varphi^{*2} \geqslant \frac{l^2}{6\pi^2\lambda^2}(1 - Q(X, l)). \tag{19}$$

PROOF. The first estimate is well known ([3], Chapter III, Lemma 4). We prove the second. Let X be the given random variable and $f(t)$ its characteristic function. Denote by $\widetilde{X}$ the symmetrized random variable $\widetilde{X} = X - X$, where the two variables on the right-hand side are independent. Denote by $\widetilde{F}(x)$ the distribution function of $\widetilde{X}$. It is well known that the characteristic function of $\widetilde{X}$ is equal to

$$|f(t)|^2 = \int_{-\infty}^{\infty} \cos xt\,d\widetilde{F}(x).$$

Use the definition of m to compute

$$\frac{2}{\lambda}(1 - m^2) \geqslant \int_{-1\lambda}^{1/\lambda}(1 - |f(t)|^2)\,dt = \int_{-1/\lambda}^{1/\lambda}\left(\int_{-\infty}^{\infty}(1 - \cos xt)\,d\widetilde{F}(x)\right)dt$$
$$= 2\int_{-1/\lambda}^{1/\lambda}\left(\int_{-\infty}^{\infty}\sin^2\frac{tx}{2}\,d\widetilde{F}(x)\right)dt.$$

Denote by $\|\alpha\|$ the distance from a real number α to the nearest integer. It is easy to see that $|\sin\alpha\pi| \geqslant 2\|\alpha\|$ and

$$\frac{2}{\lambda}(1 - m^2) \geqslant 8\int_{-1/\lambda}^{1/\lambda}\left(\int_{-\infty}^{\infty}\left\|\frac{tx}{2\pi}\right\|^2 d\widetilde{F}(x)\right)dt = 8\int_{-1/\lambda}^{1/\lambda}\left(\int_{|x|\leqslant\pi\lambda}\left\|\frac{tx}{2\pi}\right\|^2 d\widetilde{F}(x)\right)dt$$
$$+ 8\int_{|x|>\pi\lambda}\left(\int_{-1/\lambda}^{1/\lambda}\left\|\frac{xt}{2\pi}\right\|^2 dt\right)d\widetilde{F}(x) \geqslant 8\int_{-1/\lambda}^{1/\lambda}\left(\int_{|x|\leqslant\pi\lambda}\frac{t^2x^2}{4\pi^2}\,d\widetilde{F}(x)\right)dt$$
$$+ 8\int_{|x|>\pi\lambda}\left(2\int_0^{\pi/|x|}\left(\frac{xt}{2\pi}\right)^2 dt\right)\left[\frac{|x|}{\pi\lambda}\right]d\widetilde{F}(x)$$
$$\geqslant \frac{4}{3\pi^2}\frac{1}{\lambda^3}\int_{|x|\leqslant\pi\lambda} x^2\,d\widetilde{F}(x) + \frac{4\pi}{3}\int_{|x|>\pi\lambda}\frac{1}{|x|}\frac{1}{2}\frac{|x|}{\pi\lambda}\,d\widetilde{F}(x)$$

because $[\alpha] > \alpha/2$ for $\alpha \geqslant 1$. Hence

$$\frac{2}{\lambda}(1-m^2) \geqslant \frac{4}{3\pi^2}\frac{1}{\lambda^2}\int\limits_{|x|\leqslant\pi\lambda} x^2 d\tilde{F}(x) + \frac{2}{3}\int\limits_{|x|>\pi\lambda} d\tilde{F}(x) \geqslant \frac{4}{3\pi^2\lambda^3}\frac{l^2}{4}\int\limits_{l/2\leqslant|x|\leqslant\pi\lambda} d\tilde{F}(x)$$

$$+\frac{2}{3\lambda}\int\limits_{|x|>\pi\lambda} d\tilde{F}(x) \geqslant \min\left(\frac{l^2}{3\pi^2\lambda^3}, \frac{2}{3\lambda}\right)\int\limits_{|x|>l/2} d\tilde{F}(x)$$

$$\geqslant \min\left(\frac{l^2}{3\pi^2\lambda^3}, \frac{2}{3\lambda}\right)(1-Q(\tilde{X}, l)) \geqslant \min\left(\frac{l^2}{3\pi^2\lambda^3}, \frac{2}{3\lambda}\right)(1-Q(X, l)).$$

(We use the well-known property of the concentration function ([3], Chapter III, Lemma 1).) If $l \leqslant 2\lambda$, then $l^2/3\pi^2\lambda^3 \leqslant 2/3\lambda$, and therefore

$$\frac{2}{\lambda}(1-m^2) \geqslant \frac{l}{3\pi^2\lambda^3}(1-Q(X, l)). \tag{20}$$

Since $\cos m = \varphi^*$, we have

$$\varphi^{*2} \geqslant 2(1-m) \geqslant 1-m^2. \tag{21}$$

Lemma 6 follows by comparing (20) with (21).

We proceed to the proof of the theorem. By Lemma 1,

$$Q(S_n, \lambda) \leqslant \left(\frac{96}{95}\right)^2 \max\left(\lambda, \frac{1}{a}\right)\int_{-a}^{a} |f(t)|^n\, dt.$$

With $a = 1/\lambda$ we get

$$Q(S_n, \lambda) \ll \lambda \int_{-1/\lambda}^{1/\lambda} |f(t)|^n\, dt \ll \lambda \int_0^{1/\lambda} |f(t)|^n\, dt. \tag{22}$$

Lemma 5 yields

$$\mu(\varphi) = \operatorname{mes}\left(t \in \left[0, \frac{1}{\lambda}\right] : |f(t)| \geqslant \cos\varphi\right) \leqslant 12\frac{\omega}{\varphi^*}\varphi, \tag{23}$$

when $\varphi \leqslant \varphi^*/4$. Using Tchebycheff's inequality

$$\frac{1}{2}\omega = \int_0^{1/\lambda} |f(t)|^2\, dt \geqslant \mu(\varphi)\cos^2\varphi,$$

we get the estimate

$$\mu(\varphi) \leqslant \omega/2\cos^2\varphi, \tag{24}$$

valid for all φ, $0 \leqslant \varphi \leqslant \pi/2$.

Now we return to the integral (22). We will apply the rearrangement lemma ([6], Chapter X). Let $|f(t)|^*$ be the rearrangement of $|f(t)|$ in decreasing order on $[0, 1/\lambda]$. Then

$$Q(S_n, \lambda) \ll \lambda \int_0^{1/\lambda} |f(t)|^{*n}\, dt. \tag{25}$$

From (23) we get the estimate for the rearranged function:

$$|f(t)|^* \leqslant \cos\frac{\varphi^*}{12\omega}t, \quad 0 \leqslant t \leqslant 3\omega. \tag{23'}$$

For $3\omega < t < 1/\lambda$ we use (24) to obtain

$$|f(t)|^* \leqslant \sqrt{\frac{\omega}{2t}}. \tag{24'}$$

For the integral on the right-hand side of (25) we find that

$$J = \int_0^{1/\lambda} |f(t)|^{*^n}\, dt \leqslant \int_0^{3\omega} \left(\cos \frac{\varphi^*}{12\omega} t\right)^n dt + \int_{3\omega}^{1/\lambda} \left(\frac{\omega}{2t}\right)^{n/2} dt$$

$$\leqslant \int_0^{3\omega} e^{-\frac{n}{4}\left(\frac{\varphi^*}{12\omega}\right)^2 t^2} dt + \left(\frac{\omega}{2}\right)^{n/2} \frac{1}{(3\omega)^{n/2-1}} \leqslant \frac{1}{\sqrt{n}\varphi^*/\omega} + \frac{\omega}{6^{n/2}}.$$

Lemma 6 gives $\omega \leqslant (1/\lambda)Q(X, \lambda)$, and if $l \leqslant 2\lambda$ then

$$(\varphi^*)^2 \geqslant \frac{l^2}{\lambda^2}(1 - Q(X, l)).$$

Consequently,

$$J \leqslant \frac{Q(X, \lambda)}{l\sqrt{n}(1 - Q(X, l))} + \frac{Q(X, \lambda)}{\lambda 6^{n/2}} \leqslant \frac{Q(X, \lambda)}{l\sqrt{n}(1 - Q(X, l))}.$$

Hence

$$Q(S_n, \lambda) \leqslant \frac{\lambda Q(X, \lambda)}{l\sqrt{n}(1 - Q(X, l))},$$

where $l \leqslant 2\lambda$. The theorem is proved.

BIBLIOGRAPHY

1. Harry Kesten, *A sharper form of the Doeblin-Lévy-Kolmogorov-Rogozin inequality for concentration functions*, Math. Scand. **25** (1969), 133–144.

2. B. A. Rogozin, *An estimate of the concentration functions*, Teor. Verojatnost. i Primenen. **6** (1961), 103–105; English transl. in Theor. Probability Appl. **6** (1961).

3. V. V. Petrov, *Sums of independent random variables*, "Nauka", Moscow, 1972; English transl., Springer-Verlag, Berlin and New York, 1975.

4. H. Halberstam and K. F. Roth, *Sequences*. Vol. 1, Clarendon Press, Oxford, 1966.

5. L. P. Postnikova and A. A. Judin, *The concentration function*, Teor. Verojatnost. i Primenen. **22** (1977), 371–375; English transl. in Theor. Probability Appl. **22** (1977).

6. G. H. Hardy, J. E. Littlewood and G. Pólya, *Inequalities*, Cambridge Univ. Press, New York, 1934.

Translated by JAN K. PACHL

Труды Мат. Инст. Стеклов.
143 (1977)

Proceedings of the Steklov
Institute of Mathematics
1980, Issue 1

THE ARITHMETIC STRUCTURE OF INTEGRAL POLYNOMIALS AND CLASS NUMBERS

UDC 511

V. G. SPRINDŽUK

ABSTRACT. A new effective bound is obtained for solutions of Diophantine equations of hyperelliptic type: all integral solutions x, y of the equation $f(x) = Ay^m$ $(m \geqslant 3)$, where $f(x)$ is an integral polynomial of degree $n \geqslant 2$ with no multiple roots, satisfy an inequality of the form

$$\max(|x|, |y|) < \exp\{c\,|A|^{\alpha}\,|D_f|^{\beta}\,(\ln H)^{\gamma}\},$$

where D_f is the discriminant of $f(x)$, H is the height of $f(x)$, and c, α, β, and γ can be expressed explicitly in terms of n and m. This result leads to a characterization of families of algebraic number fields whose class number grows without bound as the discriminant grows.

Bibliography: 21 titles.

§1. Introduction

It was recently established that there is a close connection between bounds for solutions of Diophantine equations of classical type in two unknowns and class numbers of algebraic number fields (see [1]–[4]). The discovery of this connection was made possible when methods were found for clarifying the effect of all of the parameters in the equation on the magnitude of solutions.

In this article we consider an equation of hyperelliptic type

$$f(x) = Ay^m, \tag{1}$$

where $f(x)$ is an integral polynomial, and A and $m \geqslant 2$ are integers. We obtain a new estimate from above for integral solutions x, y of this equation, which we then use to give very simple sequences of algebraic number fields whose class number increases without bound.

Although equations of the form (1) have been an object of study for mathematicians of many generations (special cases were considered by Fermat, Euler and others [5]), the basic facts about their solutions were obtained comparatively recently. In the case

$$m = 2, \quad A = 1, \quad f(x) \text{ is a cubic polynomial with no multiple roots,} \tag{2}$$

Mordell [6], [7] proved that equation (1) has only finitely many solutions, by reducing the equation to a bounded number of Thue equations. This reduction can be accomplished using the reduction theory for binary cubic forms when $f(x) = x^3 - a$, and using the reduction theory for binary biquadratic forms when $f(x)$ is a cubic poly-

1980 *Mathematics Subject Classification.* Primary 12A50; Secondary 10B10, 10B15.

nomial of general type. Such a method cannot be extended to the general case

$$m = 2,\ A = 1,\ f(x) \text{ is a polynomial of degree } n \geqslant 3 \text{ having at least three simple roots.} \tag{3}$$

Using the ideal theory for algebraic number fields and theorems obtained by Thue's method on approximating algebraic numbers by algebraic numbers from a fixed field, Siegel [8] proved finiteness of the number of solutions of (1) in the case (3). Siegel's argument also makes it possible to prove finiteness of the number of solutions of equation (1) when

$$m \geqslant 3,\ A = 1,\ f(x) \text{ is a polynomial of degree } n \geqslant 2 \text{ having at least two simple roots} \tag{4}$$

Neither Mordell's argument nor Siegel's makes it possible to obtain a bound for the solutions of (1), since they are based on ineffective results on solutions of the Thue equation or its generalizations to relative fields.

By obtaining an effectivization of Thue's equation and its generalizations, Baker [9], [10] removed this deficiency. In particular, by supplementing Mordell's method with his own results on the Thue equation, Baker [11] obtained the following estimate for solutions of equation (1) in the case (2):

$$\max(|x|,\ |y|) < \exp(10^6 H)^{10^6}, \tag{5}$$

where H is the height of $f(x)$ (the maximum modulus of its coefficients). By supplementing Siegel's argument with his own results on the generalized Thue equation, Baker [12] obtained the following estimate for solutions of equation (1) in case (3):

$$\max(|x|,\ |y|) < \exp\exp\exp(n^{10n} H)^{n^2}, \tag{6}$$

and the following estimate in case (4):

$$\max(|x|,\ |y|) < \exp\exp\{(5m)^{10}(n^{10n} H)^{n^2}\}. \tag{7}$$

Comparing (5) with (6) and (7) reveals a sharp difference between Mordell's method, on which the reduction of (1) to Thue's equation in case (2) is based, and Siegel's method for reducing (1) to the generalized Thue equation in cases (3) and (4). The piling up of exponentials in (6) and (7) as opposed to (5) is caused by successively adjoining to the field of rational numbers **Q**, in which the coefficients and solutions of (1) lie, certain algebraic numbers connected with equation (1), in order to obtain a generalized Thue equation in a finite extension of **Q**.

In [2] and [4] the author described a new method for analyzing equation (1) when $m = 2$, which does not use the reduction of (1) to a generalized Thue equation, but rather is based on an investigation of linear dependence between three algebraic units (the "ternary" exponent equation). In this way the following assertion was proved [4] concerning equation (1) with $m = 2$.

Suppose that the leading coefficient of $f(x)$ is 1; α, α_1 and α_2 are three simple roots of the polynomial, f_α, f_{α_1} and f_{α_2} are the minimal polynomials, D_α, D_{α_1} and D_{α_2} are the discriminants of the numbers α, α_1 and α_2, respectively, R_α,

R_{α_1} and R_{α_2} are the resultants of the polynomials f' and f_α, f' and f_{α_1}, f' and f_{α_2}, respectively, d, d_1 and d_2 are the relative degrees of the field $\mathbf{F} = \mathbf{Q}(\alpha, \alpha_1, \alpha_2)$ over $\mathbf{Q}(\alpha)$, $\mathbf{Q}(\alpha_1)$ and $\mathbf{Q}(\alpha_2)$, respectively, and s is the degree of $\mathbf{F}$ over $\mathbf{Q}$. Then all integral solutions of (1) with $m = 2$ satisfy

$$\max(|x|,\ |y|) < \exp\{c_1 |A|^{8(d+d_1+d_2)+\varepsilon} B^{4+\varepsilon} (\ln H)^{1+\varepsilon}\}, \tag{8}$$

where

$$B = |R_\alpha|^{2d} |R_{\alpha_1}|^{2d_1} |R_{\alpha_2}|^{2d_2} |D_\alpha|^{4s+6d} |D_{\alpha_1}|^{4s+6d_1} |D_{\alpha_2}|^{4s+6d_2},$$

c_1 can be expressed effectively in terms of n and ϵ, and $\epsilon > 0$ is arbitrary.

As shown in [2]–[4], an analysis of how the bounds for the solutions of (1) depend on varying A (for fixed $f(x)$) is important in order to obtain new information on the magnitude of class numbers of algebraic number fields. Inequality (8) shows that in the case $m = 2$

$$\max(|x|,\ |y|) < \exp\{c_2 |A|^{8(d+d_1+d_2)+\varepsilon}\},$$

and if, in particular, $f(x)$ is a normal polynomial of at least third degree, then

$$\max(|x|,\ |y|) < \exp\{c_2 |A|^{24+\varepsilon}\}. \tag{9}$$

This last inequality makes it possible to prove that, for any natural number $k \geqslant 2$, the discriminant D_q and the class number h_q of the real quadratic field $\mathbf{Q}(\sqrt{q^2 - 1})$, where $q = 1, 2, \ldots$, satisfy the inequalities

$$D_q > c_3 (\ln q)^{\tau(2k)/24 - \delta(k) - \varepsilon}, \tag{10}$$

$$h_q > c_4 D_q^{1/2 - \frac{24}{\tau(2k) - \delta_1(k)} - \varepsilon}, \tag{11}$$

where $\tau(2k)$ is the number of divisors of $2k$, $c_3 > 0$ can be expressed effectively in terms of k and ϵ, $c_4 > 0$ can be expressed in terms of k and ϵ (ineffectively), $\epsilon > 0$ is arbitrary, $\delta_1(k) = 24\delta(k)$ and

$$\delta(k) = \begin{cases} 1/12 & \text{for } 2 \nmid k, \quad 3 \nmid k, \\ 3/32 & \text{for } 2 \mid k, \quad 3 \nmid k, \\ 7/48 & \text{for } 2 \mid k, \quad 3 \mid k, \\ 5/48 & \text{for } 2 \nmid k, \quad 3 \mid k. \end{cases} \tag{12}$$

If $\tau(2k) \geqslant 51$ for k not divisible by 6 and $\tau(2k) \geqslant 52$ for k divisible by 6, then (10) and (11) imply that h_q grows with q as some positive power of D_q. Then the class number in each genus also grows as some positive power of D_q, and one can choose an infinite sequence of q's such that the number of prime divisors of $q^{2k} - 1$ remains bounded and so the number of genuses is bounded.

In this case we apply the method in [4] to analyze equation (1) in the case $m \geqslant 3$, and we obtain a bound of the form (8) for the solutions of the equation (see §2, Theorem 1, below). In addition, using (9) and an analogous inequality for $m = 3$, we show that inequalities of the form (10), (11) hold for fields of the form $\mathbf{Q}(\sqrt[t]{q^{tk} - 1})$, $t = 3$, 4 or 6 (see §2, Theorems 2–4). Finally, we consider more general fields of the form $\mathbf{Q}(\sqrt[t]{f(q)})$ and also the fields of Ankeny-Brauer-Chowla [13] (see §2, Theorems 5–7).

§2. Basic results

The following Theorem 1 gives a bound for the solutions of (1) when $m \geqslant 3$ which is analogous to (8).

THEOREM 1. *Suppose that $m \geqslant 3$ is a natural number, $f(x)$ is an integral polynomial of degree $n \geqslant 2$ and height no greater than $H \geqslant 2$ with leading coefficient* 1, *which has two simple roots α and α_1; f and f_1 are the minimal polynomials, D_α and D_{α_1} are the discriminants of the numbers α and α_1, R_α and R_{α_1} are the resultants of the polynomials f' and f_α, f' and f_{α_1}, respectively*; $\mathbf{K}$, $\mathbf{K}_1$ *and* $\mathbf{F}$ *are the algebraic number fields* $\mathbf{Q}(\alpha)$, $\mathbf{Q}(\alpha_1)$ *and* $\mathbf{Q}(\alpha, \alpha_1)$, *respectively, and*

$$s = [\mathbf{K} : \mathbf{Q}], \qquad s_1 = [\mathbf{K}_1 : \mathbf{Q}], \qquad d = [\mathbf{F} : \mathbf{K}], \qquad d_1 = [\mathbf{F} : \mathbf{K}_1].$$

Then all integral solutions of (1) *satisfy*

$$\max(|x|, |y|) < \exp\{c_5 |A|^{2m(m-1)\varphi(m)(d+d_1)+\varepsilon} (B \ln H)^{1+\varepsilon}\}, \tag{13}$$

where

$$B = |D_\alpha D_{\alpha_1}|^{ss_1 m^3 (m-1)\varphi(m)} \, |D_\alpha^d D_{\alpha_1}^{d_1}|^{2m(m-1)\varphi(m)} \, |R_\alpha^d R_{\alpha_1}^{d_1}|^{2m(m-1)^2\varphi(m)},$$

c_5 can be expressed effectively in terms of n, m and ϵ, $\varphi(m)$ is the Euler function, and $\epsilon > 0$ is arbitrary.

The proof of this theorem, although it follows the method of argument in [4], differs from that argument because of certain simplifications connected with the special nature of the case $m \geqslant 3$. Inequality (13) is by no means the most that is possible from this method; even the proof of Theorem 1 given below (see §5) actually gives lower values for the exponents of $|A|$ and B on the right side of (13). Nevertheless, (13) is essentially stronger than (6), and it is very problematic to strengthen it in an essential way.

All of the other theorems in this article concern explicit constructions of algebraic number fields with large (unboundly increasing) class number. The following three theorems are analogous to the above assertion concerning the quadratic fields $\mathbf{Q}(\sqrt{q^{2k} - 1})$. In what follows $\tau(N)$ is the number of divisors of the natural number N.

THEOREM 2. *Let $k \geqslant 1$ be a natural number, and let D_q and h_q be the discriminant and class number of $\mathbf{Q}(\sqrt[3]{q^{3k} - 1})$, $q = 1, 2, \ldots$. Then*

$$|D_q| > c_6 (\ln q)^{\frac{\tau(3k)-1}{48} - \varepsilon}, \tag{14}$$

$$h_q > c_7 |D_q|^{1/2 - \frac{48}{\tau(3k)-1} - \varepsilon}, \tag{15}$$

where $c_6 > 0$ and $c_7 > 0$ can be expressed in terms of k and ϵ, and $\epsilon > 0$ is arbitrary.

THEOREM 3. *Let $k \geqslant 1$ be a natural number, and let D_q and h_q be the discriminant and class number of* $\mathbf{Q}(\sqrt[4]{q^{4k}-1})$, $q = 1, 2, \ldots$. *Then*

$$|D_q| > c_8 (\ln q)^{\tau(4k)/12-\delta_2(k)-\epsilon}, \tag{16}$$

$$h_q > c_9 D_q^{1/2 - \frac{24}{\tau(4k)-12\delta_2(k)} - \epsilon}, \tag{17}$$

where $\delta_2(k) = 7/24$ if $3|k$, $\delta_2(k) = 3/16$ if $3 \nmid k$, the numbers $c_8 > 0$ and $c_9 > 0$ can be expressed in terms of k and ϵ, and $\epsilon > 0$ is arbitrary.

THEOREM 4. *Let $k \geqslant 1$ be a natural number, and let D_q and h_q be the discriminant and class number of* $\mathbf{Q}(\sqrt[6]{q^{6k}-1})$, $q = 1, 2, \ldots$. *Then*

$$|D_q| > c_{10} (\ln q)^{\tau(6k)/8-7/16-\epsilon}, \tag{18}$$

$$h_q > c_{11} D_q^{1/2 - \frac{24}{\tau(6k)-3,5} - \epsilon}, \tag{19}$$

where $c_{10} > 0$ and $c_{11} > 0$ can be expressed in terms of k and ϵ, and $\epsilon > 0$ is arbitrary.

Inequalities (14) and (15) show that h_q grows without bound as q grows when $\tau(3k) \geqslant 98$. Analogously, (16) and (17) imply a similar assertion when $\tau(4k) \geqslant 51$ if $3 \nmid k$, and when $\tau(4k) \geqslant 52$ if $3 \mid k$. Inequalities (18) and (19) require the condition $\tau(6k) \geqslant 52$.

Theorems 3 and 4 are proved using (9), and Theorem 2 uses the analogous inequality which is implied by (13) in the case $m = 3$. One feature these theorems have in common is the need to assume that k has a sufficiently large number of divisors if we want to assert that h_q increases without bound. It is likely that such assumptions can be essentially weakened if the present method is refined. The ideal result would be that the class number of the fields $\mathbf{Q}(\sqrt[t]{q^t - 1})$, $q = 1, 2, \ldots$, grows without bound, but, since we are not able to prove such an assertion (which may even be false), we consider a subsequence of such fields, making the substitution $q \longrightarrow q^k$ with an arithmetic condition on k.

This suggests the idea of considering compositions of polynomials in more general problems. In this way we obtain the following assertions ($f * g(x)$ denotes the composition $f(g(x))$ of the polynomials $f(x)$ and $g(x)$).

THEOREM 5. *Let $f(x)$ be an arbitrary nonconstant integral polynomial, and let the integral polynomials $e(x)$, $a(x)$ and $b(x)$ be chosen so that*

$$e(x)\, a^2(x) - f(x)\, b^2(x) = \pm 1. \tag{20}$$

Then for any real number $B > 2$ there exists an integral polynomial $g(x)$, which can be determined effectively from $a(x)$, $b(x)$, $e(x)$, $f(x)$, and B, such that the real quadratic fields

$$\mathbf{Q}\left(\sqrt{[e * g(q)][f * g(q)]}\right) \qquad (q \geqslant q_0)$$

have discriminant D_q *and class number* h_q *satisfying*

$$D_q > c_{12}(\ln q)^B, \qquad h_q > c_{13} D_q^{1/2 - 1/B - \varepsilon} \qquad (q \geqslant q_0),$$

where $c_{12} > 0$, $c_{13} > 0$, *and* q_0 *are expressed in terms of the coefficients of* $a(x)$, $b(x)$, $e(x)$ *and* $f(x)$ *and the numbers* B *and* ϵ; $\epsilon > 0$ *is arbitrary.*

In particular, if $e(x) = 1$ in (20), i.e., the analog of Pell's equation with $f(x)$ is solvable in the ring of integral polynomials, then Theorem 5 concerns the fields $\mathbf{Q}(\sqrt{f * g(q)})$ with the polynomial $f(x)$ in a rather broad class (one can use any polynomials $f(x)$ of the form $u^2(x) + v(x)$, where $v(x)|u(x)$). This can be considered as a generalization of the analogous assertion in §1 concerning the fields $\mathbf{Q}(\sqrt{q^{2k} - 1})$. The following theorem generalizes Theorems 2–4 in a similar way.

THEOREM 6. *Let* $t = 3, 4$, *or* 6, *and let* $f(x)$ *be a nonconstant integral polynomial for which the equation*

$$a^t(x) - f(x) b^t(x) = 1 \tag{21}$$

is solvable in integral polynomials $a(x)$ *and* $b(x)$. *Then for any real number* $B > 4$ *there exists an integral polynomial* $g(x)$, *which can be determined effectively from the polynomials* $a(x)$, $b(x)$, $f(x)$ *and the number* B, *such that the fields* $\mathbf{Q}(\sqrt[t]{f * g(q)})$ *have discriminant* D_q *and class number* h_q *satisfying*

$$|D_q| > c_{14}(\ln q)^B, \qquad h_q > c_{15} D_q^{1/2 - e/B - \varepsilon} \qquad (q \geqslant q_1),$$

where $e = 1$ *for* $t = 3$, $e = 2$ *for* $t = 4$ *and* $e = 3$ *for* $t = 6$; *the numbers* $c_{14} > 0$, $c_{15} > 0$ *and* q_1 *are expressed in terms of the coefficients of the polynomials* $a(x)$, $b(x)$, $f(x)$ *and the numbers* B *and* ϵ; $\epsilon > 0$ *is arbitrary.*

Finally, our last theorem concerns Ankeny-Brauer-Chowla fields [13].

THEOREM 7. *Let* $n \geqslant 3$ *be an integer, and let* $a_1, \ldots, a_{n-1}$ *be integers satisfying*

$$|a_i - a_j| > 2 \qquad (i, j = 1, 2, \ldots, n-1;\ i \neq j).$$

Then for any $B > 2(n-1)$ *there exists an integral polynomial* $g(x)$ *such that the algebraic number field generated by a root of the polynomial*

$$(x - a_1)\ldots(x - a_{n-1})(x - g(q)) \pm 1 \qquad (q = 1, 2, \ldots)$$

has discriminant D_q *and class number* h_q *satisfying*

$$|D_q| > c_{16}(\ln q)^B, \qquad h_q > c_{17} D_q^{1/2 - (n-1)/B - \varepsilon} \qquad (q = 1, 2, \ldots),$$

where $c_{16} > 0$ *and* $c_{17} > 0$ *are expressed in terms of* $a_1, \ldots, a_{n-1}$ *and the numbers* B *and* ϵ; $\epsilon > 0$ *is arbitrary.*

It was earlier known that in the sequence of fields generated by a root of the polynomial

$$(x - a_1)\ldots(x - a_{n-1})(x - N) \pm 1 \qquad (N \geqslant N_0),$$

for "almost all" N the class number grows without bound as N grows [3]. Theorem 7 shows that this is also true when N runs through the values of a certain integral polynomial.

The inequalities for the class number in Theorems 2–7 are obtained using the Siegel-Brauer theorem [14]–[15]–more precisely, its "principal part"

$$hR > c(n, \varepsilon) |D|^{1/2-\varepsilon}, \qquad (*)$$

where h, R, D and n are the class number, the regulator, the discriminant and the degree of the algebraic number field $\mathbf{K}$. The number $c(n, \epsilon)$ in this inequality is ineffective, so that these inequalities for h_q are also ineffective. Recently Stark [16] showed that, if $\mathbf{K}$ does not contain a quadratic subfield, then there is an effective refinement of $(*)$, and in the general case when $\epsilon \geqslant 1/n$ and $n \geqslant 4$ the number $c(n, \epsilon)$ can be given effectively. This makes the inequality for h_q in Theorem 2 effective, and it gives somewhat weakened effective inequalities in Theorems 3, 4, 6 and 7.

As for the inequalities for the discriminant D_q, they are all effective.

§3. Notation

If α is an algebraic number, we let $\overline{|\alpha|}$ denote the maximum modulus of the conjugates of α, and we let $h(\alpha)$ denote the height of α, i.e., the maximum modulus of the coefficients of the minimal polynomial of α.

It is obvious that for algebraic α and β

$$\overline{|\alpha+\beta|} \leqslant \overline{|\alpha|} + \overline{|\beta|}, \qquad \overline{|\alpha\beta|} \leqslant \overline{|\alpha|}\,\overline{|\beta|}. \qquad (22)$$

It can be proved (see [1], §2) that

$$\max(\ln h(\alpha+\beta),\ \ln h(\alpha\beta)) < c_{18}(\ln h(\alpha) + \ln h(\beta)), \qquad (23)$$

where c_{18} can be expressed explicitly in terms of the degrees of α and β. In addition,

$$\frac{1}{n}\overline{|\alpha|} \leqslant h(\alpha) \leqslant 2^n a \max(1, \overline{|\alpha|}^n), \qquad (24)$$

where $a > 0$ is the leading coefficient of the minimal polynomial for α, and n is its degree.

In what follows we shall use (22)–(24) without special explanations.

For natural numbers N and $k \geqslant 2$, we let $A_k[N]$ denote the kth power free divisor of N. If $f(x)$ is an integral polynomial, then for q an integer $A_k[f(q)]$ is the kth power free divisor of $f(q)$ (taken to be positive regardless of the sign of $f(q)$).

For a polynomial $f(x)$ we let D_f or $D[f(x)]$ denote its discriminant. For a polynomial $f(x, y)$ in two variables x, y, we let $f(x \mid y)$ denote the polynomial in one variable x obtained by fixing the value of the other variable y. The notation $D(y) = D[f(x \mid y)]$ shows that $D(y)$ is the discriminant of $f(x \mid y)$.

We let $f * g(x)$ denote the composition $f(g(x))$ of two polynomials $f(x)$ and $g(x)$. If $h(x)$ is a third polynomial, then

$$f * g * h(x) = f * (g * h(x)),$$

and similarly for any number of polynomials.

In what follows $c_{19}, c_{20}, \ldots$ are numbers of the same type as $c_1, \ldots, c_{18}$ before, i.e., positive numbers whose explicit form is not important in our investigation.

§4. Algebraic units and regulators

Let **L** be an algebraic number field of degree l over Q, let l be the rank of the group of independent units of **L**, and let $R = R_{\mathbf{L}}$ be the regulator of the field **L**.

LEMMA 4.1. *Every unit* η *of infinite order in* **L** *satisfies*

$$\lceil \eta \rceil \geqslant c_{19} > 1,$$

where c_{19} *depends only on* l.

Assertions of this sort are well known (Kronecker's "lemma"). An explicit expression for c_{19} is given in [17] and [18].

The following lemma is due to Siegel [17].

LEMMA 4.2. *There exist independent units* $\eta_1, \ldots, \eta_r \in \mathbf{L}$ *such that*

$$\prod_{i=1}^{r} \ln \lceil \eta_i \rceil < c_{20} R, \tag{25}$$

where c_{20} *can be expressed explicitly in terms of* l.

PROOF. Let $\epsilon_1, \ldots, \epsilon_r$ be a basis of the group of units of infinite order in **L**. We consider the system of linear forms

$$L_j(\mathbf{x}) = \sum_{i=1}^{r} x_i \ln |\varepsilon_i^{(j)}| \qquad (j = 1, 2, \ldots, r), \tag{26}$$

where $\epsilon_1^{(j)}, \ldots, \epsilon_r^{(j)}$, corresponding to $\epsilon_1, \ldots, \epsilon_r$, are units of the conjugate field $L^{(j)}$, where the number of the $\mathbf{L}^{(j)}$ is such that the $\mathbf{L}^{(j)}$ are real for $j = 1, \ldots, s$ and complex for $j = s + 1, \ldots, s + t$, and $\mathbf{L}^{(j)}$ and $\mathbf{L}^{(j+t)}$, $j = s + 1, \ldots, s + t$, are complex conjugates.

Let λ_1 be the least real number λ for which the system of inequalities

$$\max_{1 \leqslant j \leqslant r} |L_j(\mathbf{x})| \leqslant \lambda \tag{27}$$

has a nontrivial integral solution $\mathbf{x}_1$, let λ_2 be the least λ for which this system has two linearly independent integral solutions $\mathbf{x}_{21}, \mathbf{x}_{22}$, let λ_3 be the least λ for which there are three linearly independent solutions $\mathbf{x}_{31}, \mathbf{x}_{32}, \mathbf{x}_{33}$, and so on. By Minkowski's theorem on successive minima, the product $\lambda_1 \cdots \lambda_r$ does not exceed the absolute value of the determinant of the system (27), i.e.,

$$\lambda_1 \ldots \lambda_r \leqslant 2^{-t+1} R.$$

One can obviously choose the indices $i_1, \ldots, i_r$ ($1 \leqslant i_h \leqslant h$, $h = 1, \ldots, r$) so that the vectors

$$\mathbf{y}_1 = \mathbf{x}_{1i_1}, \ \mathbf{y}_2 = \mathbf{x}_{2i_2}, \ \ldots, \ \mathbf{y}_r = \mathbf{x}_{ri_r}$$

are linearly independent, and, by the above,

$$\begin{gathered} \max_{1 \leqslant j \leqslant r} |L_j(\mathbf{y}_k)| \leqslant \lambda_k \qquad (k = 1, 2, \ldots, r), \\ \prod_{k=1}^{r} \max_{1 \leqslant j \leqslant r} |L_j(\mathbf{y}_k)| \leqslant 2^{-t+1} R. \end{gathered} \tag{28}$$

Setting $\mathbf{y}_k = (y_{k1}, \ldots, y_{kr})$, we introduce the units $\eta_k = \epsilon_1^{y_{k1}} \cdots \epsilon_k^{y_{kr}}$ $(k = 1, \ldots, r)$. If we substitute $\mathbf{y}_k$ in place of $\mathbf{x}$, equations (26) along with inequality (28) give

$$\prod_{k=1}^{r} \max_{1\leqslant j\leqslant r} |\ln|\eta_k^{(j)}|| \leqslant 2^{-t+1}R. \tag{29}$$

Since $r = s + t - 1$ and

$$\sum_{j=1}^{s} \ln|\eta_k^{(j)}| + 2\sum_{j=s+1}^{s+t} \ln|\eta_k^{(j)}| = 0, \tag{30}$$

we find that

$$\ln \overline{|\eta_k|} \leqslant c_{21} \max_{1\leqslant j\leqslant r} |\ln|\eta_k^{(j)}||,$$

which, by (29), gives (25).

LEMMA 4.3. *Let $U = U_{\mathbf{L}}$ be the group of units of* **L** *generated by the units $\eta_1, \ldots, \eta_r$ in Lemma 4.2, and let $\alpha \neq 0$ be a number in* **L**. *Then there exists a unit $\eta \in U$ such that*

$$\overline{|\alpha\eta|} < |\operatorname{Nm}(\alpha)|^{1/l} e^{c_{22}R},$$

where Nm(α) *is the absolute norm of α, and c_{22} can be expressed explicitly in terms of l.*

PROOF. We number the conjugate fields $\mathbf{L}^{(j)}$, $j = 1, \ldots, l$, just as in the proof of the preceding lemma, and we consider the system of equations

$$\sum_{i=1}^{r} x_i \ln|\eta_i^{(j)}| = \ln\frac{|\alpha^{(j)}|}{N^{1/l}} \qquad (j = 1, 2, \ldots, r),$$

where $N = \operatorname{Nm}(\alpha)$. Since the determinant of the system is nonzero (the units $\eta_1, \ldots, \eta_r$ are independent), it follows that the system has a solution $(x_{10}, \ldots, x_{r0})$. Replacing each x_{i0} by the nearest integer a_i, we obtain

$$\left|\sum_{i=1}^{r} a_i \ln|\eta_i^{(j)}| - \ln\frac{|\alpha^{(j)}|}{N^{1/l}}\right| \leqslant \frac{r}{2} \max_{1\leqslant i\leqslant r} |\ln|\eta_i^{(j)}||. \tag{31}$$

Since

$$\overline{|\eta_i|}^{-l+1} \leqslant |\eta_i^{(j)}| \leqslant \overline{|\eta_i|} \qquad (1\leqslant i\leqslant r,\ 1\leqslant j\leqslant l),$$

we have

$$|\ln|\eta_i^{(j)}|| \leqslant (l-1)\ln\overline{|\eta_i|}.$$

In addition, Lemma 4.1 and (25) show that

$$\ln\overline{|\eta_i|} < c_{20}c_{19}^{r+1}R \qquad (i = 1, 2, \ldots, r). \tag{32}$$

Consequently, the left side of (31) does not exceed

$$\frac{r}{2}(l-1)c_{20}c_{19}^{r+1}R = c_{23}R. \tag{33}$$

Because of our numbering of the conjugate fields $\mathbf{L}^{(j)}$, inequality (31) with the right side of (33) remains in force for $j = r + 1, \ldots, l - 1$. Taking (30) into account, we have

$$\left|\alpha^{(j)}(\eta_1^{(j)})^{-a_1}\ldots(\eta_r^{(j)})^{-a_r}\right| < N^{1/l}e^{c_{22}R},$$

where $c_{22} = lc_{23}$. We see that the unit $\eta = \eta_1^{-a_1}\cdots\eta_r^{-a_r}$ gives the required assertion.

LEMMA 4.4. *Let α, β and γ be nonzero numbers of* **L** *of height no greater than $T \geqslant 2$, and let U be the group of units in Lemma* 4.3. *Then all solutions of the equation*

$$\alpha x + \beta y = \gamma \tag{34}$$

in units x, y of the group U satisfy

$$\max(\overline{|x|}, \overline{|y|}) < \exp\{c_{24}R^{3+\epsilon}(\ln T)^{1+\epsilon}\},$$

where c_{24} can be effectively determined from l and ϵ; $\epsilon > 0$ is arbitrary.

The proof of this lemma is contained as an intermediate step in the argument in [4], pp. 101–103. The proof uses Siegel's lemma (Lemma 4.2) and Stark's theorem [19] on linear forms in the logarithms of algebraic numbers.

LEMMA 4.5. *The regulator R and the discriminant D of* **L** *are connected by the inequality*

$$R < c_{25}|D|^{1/2}(\ln|D|)^{l-1}, \tag{35}$$

where c_{25} can be expressed explicitly in terms of l.

This is a well-known inequality of Landau [20] (see also [17]).

Finally, we give one more lemma, which we only use in our comments on the proof of Theorem 1 (see §9), but which is of independent interest, and perhaps conceals new facts concerning class numbers.

LEMMA 4.6. *Let* **K** *be a subfield of* **L**. *Then the regulator*

$$R_{\mathbf{K}} < c_{26}R, \tag{36}$$

where c_{26} can be expressed explicitly in terms of l.

PROOF. Let $n = [\mathbf{K}\colon \mathbf{Q}]$ and $m = [\mathbf{L}\colon \mathbf{K}]$. We denote the conjugate fields of **K** by $\mathbf{K}^{(j)}$, $j = 1, \ldots, n$, and we denote the conjugates of **L** by $\mathbf{L}^{(j,k)}$, where for fixed j the fields $\mathbf{L}^{(j,k)}$ are all conjugate to $\mathbf{K}^{(j)}$, $k = 1, \ldots, m$. We introduce the analogous notation for elements of these fields. Then, for example, if $\eta \in \mathbf{L}$, we have

$$\mathrm{Nm}^{(j)}_{\mathbf{L}/\mathbf{K}}(\eta) = \eta^{(j,1)}\ldots\eta^{(j,m)} \qquad (j = 1, 2, \ldots, n).$$

Let $\eta_1, \ldots, \eta_r$ be a system of independent units of **L** satisfying

$$\prod_{i=1}^{r}\max_{(j,k)}|\ln|\eta_i^{(j,k)}|| \leqslant R. \tag{37}$$

The existence of such a system of units was established in the proof of Lemma 4.2 (inequality (29)).

We set

$$\theta_i^{(j)} = \mathrm{Nm}_{\mathbf{L}/\mathbf{K}}^{(j)}(\eta_i) \qquad (i=1, 2, \ldots, r;\ j=1, 2, \ldots, n).$$

The units $\theta_i = \mathrm{Nm}_{\mathbf{L}/\mathbf{K}}(\eta_i)$, $i = 1, \ldots, r$, lie in **K**, and the rank of this system of units coincides with the rank of the system of all units of infinite order in **K**. In fact, if $\epsilon_1, \ldots, \epsilon_q$ is a system of fundamental units for **K**, then for each unit ϵ_u there exist integers $a_u \neq 0$ and $a_{1u}, \ldots, a_{ru}$ satisfying

$$\varepsilon_u^{a_u} = \eta_1^{a_{1u}} \ldots \eta_r^{a_{ru}} \qquad (u=1, 2, \ldots, q).$$

Taking the norm from **L** to **K**, we find that the units $\epsilon_u^{ma_u}$ are expressed multiplicatively in terms of $\theta_1, \ldots, \theta_r$. If M is the least common multiple of $ma_1, \ldots, ma_q$, then the Mth power of each unit $\epsilon_1, \ldots, \epsilon_q$ is expressed multiplicatively in terms of $\theta_1, \ldots, \theta_r$. This shows that the rank of the system of units $\theta_1, \ldots, \theta_r$ cannot be less than q, and hence the rank of the matrix

$$(\ln|\theta_i^{(j)}|)_{\substack{i=1, 2, \ldots, r\\ j=1, 2, \ldots, n}}$$

is equal to q. Let

$$\det(\ln|\theta_i^{(j)}|) \neq 0 \qquad (i, j=1, 2, \ldots, q).$$

We then find that

$$R_{\mathbf{K}} \leqslant |\det(\ln|\theta_i^{(j)}|)| = |\det(\ln|\mathrm{Nm}_{\mathbf{L}/\mathbf{K}}^{(j)}(\eta_i)|)|$$

$$= \left|\det \sum_{k=1}^{m} \ln|\eta_i^{(j,k)}|\right| \leqslant \sum_{(j_1, \ldots, j_m)} \prod_{i=1}^{q} \sum_{k=1}^{m} |\ln|\eta_i^{(j_i,k)}||$$

$$= \sum_{(j_1, \ldots, j_m)} \sum_{k=1}^{m} |\ln|\eta_1^{(j_1,k)}|| \ldots |\ln|\eta_q^{(j_m,k)}||,$$

where $(j_1, \ldots, j_m)$ runs through all possible permutations of the numbers $i, \ldots, m$. Since for any $i = 1, \ldots, r$

$$\max_{(j,k)} |\ln|\eta_i^{(j,k)}|| \geqslant c_{27} > 0,$$

where c_{27} depends only on l (Lemma 4.1), it follows that (37) gives

$$|\ln|\eta_1^{(j_1,k)}|| \ldots |\ln|\eta_q^{(j_m,k)}|| < c_{28}R$$

for any set $(j_1, \ldots, j_m)$. This proves (36).

§5. Proof of Theorem 1

Along with the lemmas of the preceding section, we shall need the following

LEMMA 5.1. *Let* **G** *be an algebraic number field of finite degree over* **Q**, *let* $\mathfrak{a}_1, \ldots, \mathfrak{a}_n$ *be integral ideals in* **G**, *let* $[\mathfrak{a}_1, \ldots, \mathfrak{a}_n]$ *be their least common multiple, and let* $(\mathfrak{a}_i, \mathfrak{a}_j)$ *be the greatest common divisor of the ideals* $\mathfrak{a}_i$ *and* $\mathfrak{a}_j$, $i, j = 1, \ldots, n$.

Then

$$\prod_{i=1}^{n} \mathfrak{a}_i \text{ divides } [\mathfrak{a}_1, \ldots, \mathfrak{a}_n] \prod_{1 \leqslant i < j \leqslant n} (\mathfrak{a}_i, \mathfrak{a}_j).$$

For the proof, see [4], Lemma 2.

We now proceed to analyze equation (1), where we assume that the conditions of Theorem 1 are fulfilled. We may obviously assume that A does not contain prime divisors to powers greater than $m - 1$, i.e., that A is mth power free.

Suppose that the rational integers x and y satisfy (1). In the field $\mathbf{K} = \mathbf{Q}(\alpha)$ the ideal $(x - \alpha)$ can be represented in the form $(x - \alpha) = \mathfrak{a}\mathfrak{b}^m$, where $\mathfrak{a}$ and $\mathfrak{b}$ are integral ideals and $\mathfrak{a}$ is mth power free. Since

$$f(x) = (x - \alpha)\left(f'(\alpha) + \frac{f''(\alpha)}{2}(x - \alpha) + \ldots\right),$$

it follows that, if $\mathfrak{p}$ is a prime ideal of $\mathbf{K}$ and $(f'(\alpha), \mathfrak{p}) = 1$, we have

$$\operatorname{ord}_{\mathfrak{p}} f(x) = \operatorname{ord}_{\mathfrak{p}} (x - \alpha).$$

Then

$$\operatorname{ord}_{\mathfrak{p}} A \equiv \operatorname{ord}_{\mathfrak{p}} \mathfrak{a} \pmod m$$

and hence $\operatorname{ord}_{\mathfrak{p}} \mathfrak{a} \leqslant \operatorname{ord}_{\mathfrak{p}} A$. But if $\mathfrak{p} \mid f'(\alpha)$, then in any case $\operatorname{ord}_{\mathfrak{p}} \mathfrak{a} \leqslant m - 1$. Thus,

$$\mathfrak{a} \mid A\,(f'(\alpha))^{m-1},$$

and, since $f'(\alpha)$ divides R_α, we have $\mathfrak{a} \mid AR_\alpha^{m-1}$. For the conjugate ideals $\mathfrak{a}^{(i)}$ we also have $\mathfrak{a}^{(i)} \mid AR_\alpha^{m-1}$. Consequently, the least common multiple $[\mathfrak{a}^{(1)}, \ldots, \mathfrak{a}^{(s)}]$ divides AR_α^{m-1}.

Further, $(\mathfrak{a}^{(i)}, \mathfrak{a}^{(j)})$ divides $(x - \alpha^{(i)}, x - \alpha^{(j)})$, which in turn divides $\alpha^{(i)} - \alpha^{(j)}$,

$$\prod_{1 \leqslant i < j \leqslant s} (\mathfrak{a}^{(i)}, \mathfrak{a}^{(j)}) \mid D_\alpha.$$

By Lemma 5.1 we find that

$$\mathfrak{a}^{(1)} \ldots \mathfrak{a}^{(s)} \mid AR_\alpha^{m-1} D_\alpha.$$

Hence,

$$\operatorname{Nm}(\mathfrak{a}) \leqslant |AR_\alpha^{m-1} D_\alpha|. \tag{38}$$

Let $\mathfrak{b}'$ be an integral ideal of $\mathbf{K}$ in the inverse ideal class to the class of $\mathfrak{b}$ and with norm not exceeding $|D_{\mathbf{K}}|^{1/2}$. Then $\mathfrak{b}\mathfrak{b}'$ is a principal ideal (ϵ'), and $x - \alpha = \gamma'(\xi')^m$, where $\gamma' = \mathfrak{a}(\mathfrak{b}')^{-m}$, $g\gamma'$ is an integer and $g = \operatorname{Nm}(\mathfrak{b}')^m \leqslant |D_{\mathbf{K}}|^{m/2}$. By (38),

$$|\operatorname{Nm}(\gamma')| \leqslant \operatorname{Nm}(\mathfrak{a}) \leqslant |AR_\alpha^{m-1} D_\alpha|. \tag{39}$$

By Lemma 4.3 applied to the field $\mathbf{K}$ and the number γ', $\mathbf{K}$ has a unit θ for which

$$\overline{|\gamma'\theta|} < |\operatorname{Nm}(\gamma')|^{1/s} e^{c_{29} R_{\mathbf{K}}}, \tag{40}$$

where c_{29} can be expressed in terms of s. θ can be represented in the form $\theta = \theta_1 \theta_2^m$, where $\theta_1, \theta_2 \in U_{\mathbf{K}}$ and θ_1 is made up of basis units of $U_{\mathbf{K}}$ to nonnegative integer powers no greater than $m-1$. Setting

$$\gamma=\gamma'\theta\theta_1^{-1}, \qquad \xi=\xi'\theta_2^{-1},$$

we find that

$$x-\alpha=\gamma\xi^m, \tag{41}$$

ξ and $g\gamma$ are integers, and it follows from (40), Lemma 4.1 and (25) that

$$\overline{|\gamma|} \leqslant \overline{|\gamma'\theta|}\,\overline{|\theta_1^{-1}|} < |\operatorname{Nm}(\gamma')|^{1/s} e^{c_{30}R_{\mathbf{K}}}.$$

Since

$$|R_\alpha| < c_{31}H^{s+n-1}, \qquad |D_\alpha| < c_{32}H^{2s-2}$$

and by Lemma 4.5

$$R_{\mathbf{K}} < c_{33}|D_{\mathbf{K}}|^{1/2}(\ln|D_{\mathbf{K}}|)^{s-1},$$

it follows, if we take (39) into account, that

$$\overline{|\gamma|} < c_{34}|AH^{(m-1)(s+n-1)+2s-2}|^{1/s}\exp\{c_{35}|D_{\mathbf{K}}|^{1/2}(\ln|D_{\mathbf{K}}|)^{s-1}\} \tag{42}$$

($c_{30}, \ldots, c_{35}$ can be expressed in terms of s and n).

We similarly obtain, for the other root α_1 of $f(x)$,

$$x-\alpha_1=\gamma_1\xi_1^m, \tag{43}$$

where γ_1 and ξ_1 belong to the field $\mathbf{K}_1 = \mathbf{Q}_1(\alpha_1)$, ξ_1 is an integer, $g_1\gamma_1$ is an integer, g_1 is a natural number not exceeding $|D_{\mathbf{K}_1}|^{m/2}$ and γ_1 satisfies (42) with s replaced by s_1 and $D_{\mathbf{K}}$ replaced by $D_{\mathbf{K}_1}$.

From (41) and (43) we obtain

$$\alpha_1-\alpha=\gamma\xi^m-\gamma_1\xi_1^m. \tag{44}$$

Now let

$$\mathbf{L}=\mathbf{Q}(\alpha, \alpha_1, \sqrt[m]{\gamma}, \sqrt[m]{\gamma_1}, \zeta),$$

where ζ is a primitive mth root of 1. Equation (44) shows that the numbers

$$\beta_i=\xi\sqrt[m]{\gamma}-\zeta_i\xi_1\sqrt[m]{\gamma_1} \qquad (i=1, 2, \ldots, m; \quad \zeta_i=\zeta^i) \tag{45}$$

lie in $\mathbf{L}$ and have absolute norms not exceeding $c_{36}H^{n^2m\varphi(m)}$ in modulus. The numbers β_i are integers, since $\gamma\xi^m$ and $\gamma_1\xi_1^m$ are integers, and hence so are $\xi\sqrt[m]{\gamma}$ and $\zeta_i\xi_1\sqrt[m]{\gamma_1}$. By Lemma 4.3, the group of units $U_{\mathbf{L}}$ has units $\epsilon_1, \epsilon_2, \epsilon_3$ for which the numbers λ_i defined by $\beta_i = \epsilon_i\lambda_i$, $i = 1, 2, 3$, satisfy the restriction

$$\max(\overline{|\lambda_1|}, \overline{|\lambda_2|}, \overline{|\lambda_3|}) < (c_{36}H^{n^2m\varphi(m)})^{1/l}e^{c_{37}R}, \tag{46}$$

where $R = R_{\mathbf{L}}$ and c_{37} can be expressed in terms of $l = [\mathbf{L} : \mathbf{Q}]$.

Eliminating $\xi\sqrt[m]{\gamma}$, $\xi_1\sqrt[m]{\gamma_1}$ from (45), we obtain, in particular,

$$(\zeta_3-\zeta_2)\beta_1+(\zeta_1-\zeta_3)\beta_2+(\zeta_2-\zeta_1)\beta_3=0,$$

which gives an equation of the form (34) for $\delta_1 = \epsilon_1/\epsilon_3$ and $\delta_2 = \epsilon_2/\epsilon_3$:

$$(\zeta_3-\zeta_2)\lambda_1\delta_1+(\zeta_1-\zeta_3)\lambda_2\delta_2=(\zeta_1-\zeta_2)\lambda_3.$$

From (46) we find that the heights of the numbers

$$(\zeta_3 - \zeta_2)\lambda_1, \qquad (\zeta_1 - \zeta_3)\lambda_2, \qquad (\zeta_1 - \zeta_2)\lambda_3$$

have the following upper bound:

$$T = \exp\{c_{38}(R + \ln H)\},$$

where c_{38} can be expressed in terms of n and m. By Lemma 4.4, we find

$$\max(\overline{|\delta_1|}, \overline{|\delta_2|}) < \exp\{c_{39}R^{3+\varepsilon}(R + \ln H)^{1+\varepsilon}\}, \tag{47}$$

where c_{39} can be expressed in terms of n, m and ϵ and $\epsilon > 0$ is arbitrary.

We set $\mu_i = \delta_i\lambda_i$, $i = 1, 2$. Then (46) and (47) imply an inequality of the form (47):

$$\max(\overline{|\mu_1|}, \overline{|\mu_2|}) < \exp\{c_{40}R^{3+\varepsilon}(R + \ln H)^{1+\varepsilon}\}. \tag{48}$$

Returning to (45), we obtain

$$\xi_1 = \frac{(\mu_2 - \mu_1)\,\varepsilon_3}{(\zeta_1 - \zeta_2)\sqrt[m]{\gamma_1}}, \qquad \xi = \frac{(\mu_1\zeta_1^{-1} - \mu_2\zeta_2^{-1})\,\varepsilon_3}{(\zeta_1^{-1} - \zeta_2^{-1})\sqrt[m]{\gamma}}. \tag{49}$$

Now (44) gives

$$\alpha_1 - \alpha = \left[\left(\frac{\mu_1\zeta_1^{-1} - \mu_2\zeta_2^{-1}}{\zeta_1^{-1} - \zeta_2^{-1}}\right)^m - \left(\frac{\mu_2 - \mu_1}{\zeta_1 - \zeta_2}\right)^m\right]\varepsilon_3^m.$$

Since $\alpha \neq \alpha_1$, we obtain an equation for ϵ_3 which, if we use (48), gives an estimate for $\overline{|\epsilon_3|}$ having the form of the right side of (48) with another c_{40}. Then from (49), (42) and (48) we obtain

$$\overline{|\xi|} < |A|^{c_{41}} \exp\{c_{42}R^{3+\varepsilon}(R + \ln H)^{1+\varepsilon} + c_{43}|D_{\mathbf{K}}|^{1/2}(\ln|D_{\mathbf{K}}|)^{s-1}\}, \tag{50}$$

where c_{41} can be expressed in terms of n and m, c_{42} in terms of n, m and ϵ, and c_{43} in terms of n. Now (41) and (42) give an inequality of the form (50) for $|x|$.

In order to complete the proof of the theorem it remains to estimate the size of the regulator R. By Lemma 4.5, to do this it suffices to estimate the discriminant $D = D_{\mathbf{L}}$.

We consider $\mathbf{L}$ as the field obtained from $\mathbf{Q}$ by successively adjoining the integers α, α_1, $g\sqrt[m]{\gamma}$, $g_1\sqrt[m]{\gamma_1}$ and ζ. We estimate the discriminant D as the norm of the different of $\mathbf{L}$, which, by the multiplicative property of the differents of relative fields, is a divisor of the product Δ of the differents of the numbers which we successively adjoin:

$$\Delta = f'_\alpha(\alpha) f'_{\alpha_1}(\alpha_1)\, m\,(g\sqrt[m]{\gamma})^{m-1}\, m\,(g_1\sqrt[m]{\gamma_1})^{m-1} P'_m(\zeta),$$

where $P_m(x)$ is the mth cyclotomic polynomial. We find

$$\mathrm{Nm}_{\mathbf{L}/\mathbf{Q}}(\Delta) = D_\alpha^{l/s} D_{\alpha_1}^{l/s_1} m^{2l} (gg_1)^{l(m-1)} (\mathrm{Nm}_{\mathbf{K}/\mathbf{Q}}(\gamma))^{(m-1)l/st} (\mathrm{Nm}_{\mathbf{K}_1/\mathbf{Q}}(\gamma_1))^{(m-1)l/s_1t_1} D_\zeta^{l/\varphi(m)},$$

where we denote

$$t=[\mathbf{K}(\sqrt[m]{\gamma}):\mathbf{K}], \qquad t_1=[\mathbf{K}_1(\sqrt[m]{\gamma_1}):\mathbf{K}_1].$$

Since

$$g \leqslant |D_{\mathbf{K}}|^{m/2}, \qquad g_1 \leqslant |D_{\mathbf{K}_1}|^{m/2},$$

and the numbers $D_{\mathbf{K}}$ and $D_{\mathbf{K}_1}$ are divisors of D_α and D_{α_1}, respectively, it follows that

$$g \leqslant |D_\alpha|^{m/2}, \qquad g_1 \leqslant |D_{\alpha_1}|^{m/2}.$$

Inequality (39) shows that

$$|\mathrm{Nm}_{\mathbf{K}/\mathbf{Q}}(\gamma)| \leqslant |AR_\alpha^{m-1}D_\alpha|,$$

and the analogous inequality holds for the norm of γ_1. Hence,

$$\mathrm{Nm}(\Delta)| < c_{44}|A|^{(l/st+l/s_1t_1)(m-1)}|D_\alpha|^{l/s+lm(m-1)/2+(m-1)l/st}|R_\alpha|^{l(m-1)^2/st}$$
$$\times |D_{\alpha_1}|^{l/s_1+lm(m-1)/2+(m-1)l/s_1t_1}|R_{\alpha_1}|^{l(m-1^2/s_1t_1}.$$

Since $l/s \leqslant dm^2\varphi(m)$, and $l/st \leqslant dm\varphi(m)$, and the analogous inequalities hold with s and t replaced by s_1 and t_1, it follows from the inequality

$$|D_{\mathbf{L}}| \leqslant |\mathrm{Nm}_{\mathbf{L}/\mathbf{Q}}(\Delta)|$$

and Lemma 4.5 that

$$R < c_{45}|A|^{m(m-1)\varphi(m)(d+d_1)/2}B_1(\ln|A|+\ln B_1)^{l-1},$$

where we denote

$$B_1 = |D_\alpha D_{\alpha_1}|^{ss_1m^3(m-1)\varphi(m)/4}|D_\alpha^d D_{\alpha_1}^{d_1}|^{m(m-1)\varphi(m)/2}|R_\alpha^d R_{\alpha_1}^{d_1}|^{m(m-1)^2\varphi(m)/2}.$$

Since an estimate of the form (50) holds for $|x|$, we obtain

$$|x| < \exp\{c_{46}|A|^{2m(m-1)\varphi(m)(d+d_1)+\varepsilon}B_1^{4+\varepsilon}(\ln H)^{1+\varepsilon}\},$$

which coincides with (13).

§6. Corollaries of Theorem 1

Inequality (13) makes it possible to estimate from below the mth power free divisor of any value of a polynomial at an integer point.

For $f(x)$ an integral polynomial with at least two simple roots, we set

$$r(f) = \min(d+d_1),$$

where d and d_1 are defined in Theorem 1 and the minimum is taken over all pairs of simple roots of $f(x)$. If $f(x)$ does not have two simple roots, we set $r(f) = 0$. If n is the degree of $f(x)$, then clearly d and d_1 do not exceed $n - 1$, and equality occurs, for example, for polynomials whose Galois group is the symmetric group. Hence we always have

$$r(f) \leqslant 2(n-1).$$

On the other hand, if all the roots of the polynomial are rational, or there is at least one pair of roots such that one can be expressed rationally in terms of the other, then

$d = d_1 = 1$ and $r(f) = 2$. In particular, $r(f) = 2$ if $f(x)$ is a normal polynomial, i.e., if $f(x)$ is irreducible and its root generates a normal field ($n \geqslant 2$).

Theorem 1 shows that, if $f(x)$ has at least two simple roots, then for any integers $m \geqslant 3$ and $q \neq 0$ with $f(q) \neq 0$

$$A_m[f(q)] > c_{47}(\ln|q|)^{(2m(m-1)\varphi(m)r(f))^{-1}-\varepsilon}, \tag{51}$$

where $c_{47} > 0$ depends only on the coefficients of $f(x)$ and its degree, and on m and ϵ ($\epsilon > 0$ is arbitrary).

The following lemma makes it possible to strengthen (51) if $f(x)$ has several irreducible factors.

LEMMA 6.1. *Let $f_1(x), \ldots, f_s(x)$ be integral polynomials no two of which have roots in common, and let*

$$f(x) = f_1(x) \ldots f_s(x). \tag{52}$$

Then for any integers $m \geqslant 2$ and q, with $f(q) \neq 0$,

$$A_m[f(q)] \geqslant P^{-s(m-1)} A_m[f_1(q)] \ldots A_m[f_s(q)],$$

where

$$P = \prod_{1 \leqslant i < j \leqslant s} R(f_i, f_j),$$

and $R(f_i, f_j)$ is the resultant of the polynomials $f_i(x)$ and $f_j(x)$, $i, j = 1, \ldots, s$.

The proof is almost word for word the same as that of Lemma 7 in [4], in which $m = 2$.

Thus, if $f(x)$ has the form (52), it follows by Lemma 6.1 and (51) that

$$A_m[f(q)] > c_{48}(\ln|q|)^{(2m(m-1)\varphi(m)r)^{-1}-\varepsilon}, \tag{53}$$

where $r^{-1} = \Sigma_1^s r^{-1}(f_i)$.

As an example we consider the polynomial $f(x) = x^n - 1$. Then

$$x^n - 1 = \prod_{d \mid n} P_d(x),$$

where $P_d(x)$ is the dth cyclotomic polynomial. Since $P_d(x)$ is a normal polynomial of degree $\varphi(d)$, it follows that $r(P_d) = 2$ for $d \geqslant 3$. Consequently, if $2 \nmid n$, then for the polynomial $x^n - 1$

$$r^{-1} = \sum_{d \mid n} r^{-1}(P_d) = (\tau(n) - 1)/2,$$

and (53) gives

$$A_m[q^n - 1] > c_{49}(\ln|q|)^{\frac{\tau(n)-1}{4m(m-1)\varphi(m)}-\varepsilon}. \tag{54}$$

But if $2 \mid n$, then

$$r^{-1} = r^{-1}(x^2 - 1) + \sum_{\substack{d \mid n \\ d \geqslant 3}} r^{-1}(P_d) = (\tau(n) - 1)/2,$$

and we again obtain (54).

We shall use the special case of (54) for $m = 3$ in the proof of Theorem 2. We make this case into a separate lemma.

LEMMA 6.2. *Let $n > 1$ and $q > 1$ be natural numbers. Then*

$$A_3[q^n - 1] > c_{50}(\ln|q|)^{(\tau(n)-1)/48-\epsilon},$$

where $c_{50} > 0$ can be expressed explicitly in terms of n and ϵ, $\epsilon > 0$ being arbitrary.

The analogous assertion in the case $m = 2$ was obtained in [4].

LEMMA 6.3. *Let $k > 1$ and $q > 1$ be natural numbers. Then*

$$A_2[q^{2k} - 1] > c_{51}(\ln|q|)^{\tau(2k)/24-\delta(k)-\epsilon},$$

where $\delta(k)$ is defined by (12), *$c_{51} > 0$ can be expressed in terms of k and ϵ, and $\epsilon > 0$ is arbitrary.*

For the proof, see [4], p. 106. This lemma is used to prove Theorems 3 and 4. Finally, to prove Theorems 5–7 we need an inequality of the form (53) in the case $m = 2$ which was considered in [4]. It follows from inequality (8) and Lemma 6.1.

§7. Proofs of Theorems 2–4

Let $D = d^3 - 1$, where $d > 0$ is an integer, and let $\alpha = \sqrt[3]{D}$. Then $\eta = d - \alpha$ is a unit of the field $\mathbf{K} = \mathbf{Q}(\alpha)$, and

$$1 < \lceil\eta\rceil < d + |\sqrt[3]{D}| < 2d.$$

Consequently, the regulator $R_{\mathbf{K}} < \ln(2d)$. In addition, the discriminant $D_{\mathbf{K}}$ satisfies

$$|D_{\mathbf{K}}| > A_3[D],$$

since $D_{\mathbf{K}}$ is divisible by $A_3[D]$ (see [21], §25).

If we now take $d = q^k$ in accordance with Theorem 2, we obtain

$$R_{\mathbf{K}} < \ln(2q^k) < 2k\ln q \qquad (q \geqslant 2),$$
$$|D_{\mathbf{K}}| > A_3[q^{3k} - 1] > c_{53}(\ln q)^{(\tau(3k)-1)/48-\epsilon},$$

(the last inequality holds by Lemma 6.2). This proves (14), and applying the Siegel-Brauer theorem gives (15).

Similar arguments will also prove Theorems 3 and 4.

Setting $D = d^4 - 1$ and $\alpha = \sqrt[4]{D}$, we note that the numbers $\eta_1 = d - \alpha$ and $\eta_2 = d^2 - \alpha^2$ are units of the field $\mathbf{K} = \mathbf{Q}(\alpha)$. The regulator of this system of units is equal to

$$R[\eta_1\eta_2] = \pm\begin{vmatrix} \ln|d-\alpha| & \ln|d^2-\alpha^2| \\ \ln|d+\alpha| & \ln|d^2-\alpha^2| \end{vmatrix},$$

from which we find that

$$R[\eta_1, \eta_2] \sim 20(\ln d)^2 \qquad (d \to \infty),$$

if we take into account that

$$\alpha = d\,(1-d^{-4})^{1/4} = d\left(1-\frac{1}{4}d^{-4}+O(d^{-8})\right),$$

$$\alpha^2 = d^2\left(1-\frac{1}{2}d^{-4}+O(d^{-8})\right).$$

Hence $R_{\mathbf{K}} < 21(\ln d)^2$ for d sufficiently large.

The discriminant of **K** is divisible by the square of the discriminant of the quadratic field $\mathbf{Q}(\sqrt{d^4-1})$. Consequently,

$$|D_{\mathbf{K}}| \geqslant A_2^2[d^4-1].$$

If we now take $d = q^k$, by Lemma 6.3 we obtain

$$|D_{\mathbf{K}}| > c_{54}\,(\ln q)^{\tau(4k)/12-\delta_2(k)-\varepsilon},$$

where $\delta_2(k)$ is defined in Theorem 3. This proves (16). Since $R_{\mathbf{K}} < 21(k\ln q)^2$ for q sufficiently large, it follows by the Siegel-Brauer theorem that (17) holds.

Finally, in the case $D = d^6 - 1$ we set $\alpha = \sqrt[6]{D}$. Then $\eta_i = d^i - \alpha^i$, $i = 1, 2, 3$, are units of the field $\mathbf{K} = \mathbf{Q}(\alpha)$. The regulator $R\,[\eta_1, \eta_2, \eta_3]$ of this system of units is equal to

$$\pm\begin{vmatrix} \ln|d-\alpha| & \ln|d^2-\alpha^2| & \ln|d^3-\alpha^3| \\ \ln|d+\alpha| & \ln|d^2-\alpha^2| & \ln|d^3+\alpha^3| \\ \ln|d-\zeta\alpha| & \ln|d^2-\zeta^2\alpha^2| & \ln|d^3-\zeta^3\alpha^3| \end{vmatrix},$$

where $\zeta = e^{\pi i/3}$. To compute the determinant we add the second row to the first and subtract from the last. We find that

$$R\,[\eta_1, \eta_2, \eta_3] = \pm\ln|d^3+\alpha^3|\left(\ln|d^2-\alpha^2|\cdot\ln\left|\frac{d^2-\zeta^2\alpha^2}{d^2-\alpha^2}\right| - 2\ln|d^2-\alpha^2|\cdot\ln\left|\frac{d-\zeta\alpha}{d+\alpha}\right|\right).$$

If we take into account that

$$\alpha^2 = d^2\left(1-\frac{1}{3}d^{-6}+O(d^{-12})\right),$$

we see that as $d \longrightarrow \infty$

$$R\,[\eta_1, \eta_2, \eta_3] \sim \pm 3\ln d\left(-4\ln d\cdot\ln\left|\frac{1-\zeta^2}{2}\right| - 24(\ln d)^2 + 6\ln d\cdot\ln\left|\frac{1-\zeta}{2}\right|\right) \sim 72(\ln d)^3.$$

The discriminant $D_{\mathbf{K}}$ is divisible by the cube of the discriminant of the quadratic field $\mathbf{Q}(\sqrt{d^6-1}$. Consequently,

$$|D_{\mathbf{K}}| \geqslant A_2^3[d^6-1].$$

Setting $d = q^k$, by Lemma 6.3 we obtain

$$|D_{\mathbf{K}}| > c_{55}\,(\ln q)^{\tau(6k)/8-7/16-\varepsilon},$$

which gives (18), and, since $R_{\mathbf{K}} < 73(k\ln q)^3$ for q sufficiently large, it follows that the Siegel-Brauer theorem gives (19).

§8. Compositions of polynomials

We now proceed to the proofs of Theorems 5–7. We shall need some auxiliary facts about compositions of integral polynomials.

The following lemma was communicated to the author by A. Schinzel.

LEMMA 8.1. *Let $f(x)$ and $g(x)$ be polynomials with complex coefficients, where $f(x)$ has no multiple roots, and $\deg(f) \geqslant 2$. Then the square-free divisor of the polynomial $f * g(x)$ has degree no less than*

$$\deg f(x) \deg g(x) - 2 \deg g(x) + 2.$$

PROOF. Let $f * g(x) = u(x)v^2(x)$. Then

$$\frac{d}{dx} f * g(x) = (f' * g(x))\, g'(x) = v(x)\, w(x),$$

where $w(x) = u'(x)v(x) + 2v(x)v'(x)$. Since $(f(x), f'(x)) = 1$, we have

$$(f * g(x), \quad f' * g(x)) = 1$$

and then $v(x) | g'(x)$. Consequently, $\deg v(x) \leqslant \deg g(x) - 1$. Finally,

$$\deg f * g(x) = \deg f(x) \deg g(x) = \deg v(x) + 2 \deg v(x)$$
$$\leqslant \deg u(x) + 2 \deg g(x) - 2,$$

which gives the assertion in the lemma.

LEMMA 8.2. *Let $f(x)$ be a nonconstant integral polynomial with no multiple roots. Then there exist integral polynomials $g(x)$, $p_1(x)$ and $f_1(x)$, which can be effectively determined from $f(x)$, such that*

$$f * g(x) = p_1(x) f_1(x),$$

where $p_1(x)$ is a normal polynomial of degree no less than 3, $(p_1(x), f_1(x)) = 1$ and $\deg p_1(x) = \deg g(x)$.

PROOF. Suppose that α is a root of $f(x)$, $\mathbf{F}$ is a normal field of degree $k \geqslant 3$ containing α, and θ is an integral primitive element of $\mathbf{F}$. Since the numbers θ, $\theta^2, \ldots, \theta^k$ are linearly independent over $\mathbf{Q}$, we have

$$\alpha = r_1\theta + r_2\theta^2 + \ldots + r_k\theta^k, \qquad r_i \in \mathbf{Q} \quad (i = 1, 2, \ldots, k).$$

This implies a relation for the traces:

$$S(\alpha\theta^j) = r_1 S(\theta^{1+j}) + \ldots + r_k S(\theta^{k+j}) \qquad (j = 0, 1, \ldots, k-1).$$

The determinant

$$d = \det\left(S(\theta^{i+j})\right)_{\substack{i=1, 2, \ldots, k \\ j=0, 1, \ldots, k-1}}$$

differs by a rational integer factor $\mathrm{Nm}(\theta)$ from the discriminant $D(\theta)$ of θ, and hence is a nonzero rational integer. Thus, the r_i are integral linear combinations of the numbers $S(\alpha\theta^j)d^{-1}$, $j = 0, 1, \ldots, k-1$. If a is the leading coefficient of $f(x)$, then all of the adr_i are integers. Consequently, the polynomial $r(x) = \Sigma_1^k r_j x^j$ has rational coefficients whose denominators divide ad, and $r_0(x) = r(adx)$ is an integral polynomial.

Let $p(x)$ be the minimal polynomial for θ. Since $f(r(\theta)) = f(\alpha) = 0$, it follows that the polynomial $f * r(x)$ is divisible by the irreducible polynomial $p(x)$, and the integral polynomial $f_0(x) = f * r_0(x)$ is divisible by $p_1(x) = p(adx)$. If $p_1(x)$ occurs in $f_0(x)$ to the first power, we set $g(x) = r_0(x)$. Otherwise we take

$$g(x) = r_0(x) \pm p_1(x),$$

where the sign is chosen so that $g(x)$ has degree k. Then

$$\begin{aligned} f * (r_0(x) \pm p_1(x)) &\equiv f * r_0(x) \pm (f' * r_0(x)) p_1(x) \\ &\equiv \pm (f' * r_0(x)) p_1(x) \not\equiv 0 \pmod{p_1^2(x)}, \end{aligned}$$

since $(f' * r_0(x), p_1(x)) = 1$.

It is clear that $p_1(x)$ is a normal polynomial, and that the degrees of $p_1(x)$ and $g(x)$ coincide.

Finally, we note that the construction of $g(x)$ and $p_1(x)$ can be made effective. In fact, if the degree of α is no less than 3, then we obtain the field **F** by successively adjoining the roots of the minimal polynomial for α, and, replacing the successive extensions by a single extension, we find a primitive element of **F**. If the degree of α is 2 or 1, the construction of **F** is trivial.

LEMMA 8.3. *Let $f(x)$ be a nonconstant integral polynomial with no multiple roots, and let $r > 0$ be an arbitrary integer. Then there exist effectively determinable integral polynomials*

$$g, g_1, \ldots, g_{r-1}; \quad p_1, p_2, \ldots, p_r; \quad f_r, q_r,$$

such that $p_1, \ldots, p_r$ are normal polynomials of degree no less than 3, and, setting

$$\begin{aligned} &G = g * g_1 * \ldots * g_{r-1}, \\ &P_i = p_i * g_i * g_{i+1} * \ldots * g_{r-1} \quad (i = 1, 2, \ldots, r-1), \\ &P_r = p_r, \end{aligned}$$

we have

$$\begin{aligned} &f * G = P_1 P_2 \ldots P_r f_r q_r^2, \\ &(P_i, P_j) = 1, \quad (P_i, f_r) = 1 \quad (i, j, = 1, 2, \ldots, r;\ i \neq j). \end{aligned} \tag{55}$$

PROOF. By Lemma 8.2, there exist polynomials g, p_1 and f_1 for which $f * g = p_1 f_1$, where p_1 is a normal polynomial of degree $k \geqslant 3$, $(p_1, f_1) = 1$, and the degrees of g and p_1 coincide. By Lemma 8.1, the square-free divisor of $f * g$ has degree no less than $nk - 2k + 2$, where n is the degree of $f(x)$. If $n \geqslant 3$, then the square-free part $f_{10}(x)$ of the polynomial $f_1(x)$ has degree no less than $(n-3)k + 2 \geqslant 2$. Applying Lemma 8.2 to $f_{10}(x)$, we find polynomials g_1, p_2 and f_2 for which $f_{10} * g_1 = p_2 f_2$, where p_2 is a normal polynomial of degree no less than 3, $(p_2, f_2) = 1$ and $\deg g_1 = \deg p_2$. Then

$$f * g * g_1 = (p_1 * g_1)(f_{10} * g_1) q_2^2 = (p_1 * g_1) p_2 f_2 q_2^2.$$

Since $(p_1, f_1) = 1$, we have $(p_1, f_{10}) = 1$, and then $(p_1 * g_1, f_{10} * g_1) = 1$.

Consequently,

$$(p_1 * g_1,\ p_2) = 1, \qquad (p_1 * g_1,\ f_2) = 1.$$

If $n = 1$ or 2, we first make the substitution $x \longrightarrow x^3 + l$, where $f(l) \neq 0$, l is an integer, and we consider in the same way as before $f(x^3 + l)$ in place of $f(x)$.

We then take out the square-free part $f_{20}(x)$ of $f_2(x)$, apply Lemma 8.2 to $f_{20}(x)$, and so on. Since this argument can be continued indefinitely, we obtain the assertion in the lemma.

Combining Lemmas 6.1 and 8.3, from (9) we obtain a basic fact which is needed to prove Theorems 5–7.

LEMMA 8.4. *Let $f(x)$ be a nonconstant integral polynomial with no multiple roots, and let $B > 0$ be arbitrary. Then there exists an integral polynomial $G(x)$, which can be effectively determined from $f(x)$ and B, such that for any $q \neq 0$ with $f * G(q) \neq 0$*

$$A_2[f * G(q)] > c_{56} (\ln |q|)^B, \tag{56}$$

where $c_{56} > 0$ can be expressed in terms of B and the coefficients of $f(x)$ and $G(x)$.

PROOF. We take an integer $r > 24B$ and apply Lemma 8.3. From (55) and Lemma 6.1 we obtain

$$A_2[f * G(q)] \geqslant c_{57} A_2[P_1(q)] \ldots A_2[P_r(q)],$$

where c_{57} is the reciprocal of the product of the resultants of the polynomials P_i and P_j, P_i and f_r $(i, j = 1, \ldots, r;\ i \neq j)$. Since

$$P_i(q) = p_i(q_i), \qquad q_i = g_i * \ldots * g_{r-1}(q) \qquad (i = 1, 2, \ldots, r-1),$$

it follows that (9) gives (56).

Proceeding to the proof of Theorem 5, we conclude from (20) that $e(x)f(x)$ is not a perfect square. In fact, for q an integer, $a(q)\sqrt{e(q)} \pm b(q)\sqrt{f(q)}$ is a unit in the ring $\mathbf{Z}[\sqrt{e(q)}, \sqrt{f(q)}]$; the square of this unit is a rational unit if $e(x)f(x)$ is a perfect square. Then, for any q, $a(q)\sqrt{e(q)} \pm b(q)\sqrt{f(q)}$ takes one of the four values ± 1, $\pm i$. This is impossible, since $f(x)$ is not a constant.

Applying Lemma 8.4 to the square-free part of $e(x)f(x)$, we obtain an inequality for the discriminant, and then, by the Siegel-Brauer theorem, also for the class number of the field

$$\mathbf{Q}\left(\sqrt{[e * G(q)]\,[f * G(q)]}\right).$$

The cases $t = 4$, 6 in Theorem 6 are handled as in §7, except that we use Lemma 8.4 instead of Lemma 6.3. The case $t = 3$ is considered in a similar way, except that we must alter Lemmas 8.1, 8.3 and 8.4 so as to apply to cube-free divisors of polynomials. Lemma 8.2 and inequality (51) for $m = 3$ permit us to do this, repeating the above arguments almost word for word.

Finally, the proof of Theorem 7 requires the following additional facts.

LEMMA 8.5. *Let $\alpha_1, \ldots, \alpha_{n-1}$ be complex numbers with*

$$|\alpha_i - \alpha_j| > 2 \qquad (i, j = 1, 2, \ldots, n-1;\ i \neq j),$$

$$f(x, y) = (x - \alpha_1) \ldots (x - \alpha_{n-1})(x - y) \pm 1.$$

Then the discriminant $D(y) = D[f(x|y)]$ *has no multiple roots.*

For the proof, see [3], §4.

LEMMA 8.6. *Let* $a_1, \ldots, a_{n-1}$ *be distinct integers, and let the integer* $N > \max|a_i|$. *Then the regulator of the field* $\mathbf{K}_N$ *generated by a root of the polynomial*

$$f_N(x) = (x - a_1) \ldots (x - a_{n-1})(x - N) \pm 1,$$

does not exceed $c_{58}(\ln N)^{n-1}$, *where* c_{58} *depends only on* $a_1, \ldots, a_{n-1}$ *and* n.

The proof is given in [13], but we include it here for completeness.

Let θ_N be a root of $f_N(x)$. Then $\theta_N - a_1, \ldots, \theta_N - a_{n-1}$ are units of $\mathbf{K}_N$. If $\theta_N^{(i)}$, $i = 1, \ldots, n$, are all the roots of $f_N(x)$, it is easy to see that they can be numbered in such a way that as $N \longrightarrow \infty$

$$\theta_N^{(i)} \to a_i \qquad (i = 1, 2, \ldots, n-1).$$

Consequently, for $j \neq i$

$$|\theta_N^{(i)} - a_j| \to |a_i - a_j| \neq 0 \qquad (N \to \infty),$$

and from the equation

$$(\theta_N^{(i)} - a_1) \ldots (\theta_N^{(i)} - a_i) \ldots (\theta_N^{(i)} - a_{n-1})(\theta_N^{(i)} - N) = \mp 1$$

we note that

$$|\theta_N^{(i)} - a_i| \sim \frac{1}{N \prod\limits_{\substack{j=1 \\ j \neq i}}^{n-1} |a_j - a_i|} \qquad (i = 1, 2, \ldots, n-1).$$

Thus, as $N \longrightarrow \infty$

$$\ln|\theta_N^{(i)} - a_j| \sim \begin{cases} \ln|a_i - a_j| & (i \neq j), \\ -\ln N & (i = j). \end{cases}$$

Consequently,

$$\det(\ln|\theta_N^{(i)} - a_j|)_{i, j = 1, 2, \ldots, n-1} \sim (-\ln N)^{n-1}.$$

This proves the lemma.

We now apply Lemma 8.4 to the discriminant $D(N)$ of the polynomial $f_N(x)$; we can do this because of Lemma 8.5. We find an integral polynomial $G(x)$ such that

$$A_2[D * G(q)] > c_{59}(\ln|q|)^B, \tag{57}$$

where c_{59} depends on $a_1, \ldots, a_{n-1}$, n and B. The discriminant of $\mathbf{K}_N$ is divisible by the square-free part of $D(N)$; hence the discriminant of the field is estimated by the right side of (57). Lemma 8.6 and the Siegel-Brauer theorem now give the inequality for the class number.

§9. Conclusion

Inequality (13) can be given a simpler form if we know that $f(x)$ has no multiple roots. In fact, since D_α and R_α divide D_f, we find that $B \leqslant |D_f|^S$, where

$$S = 2ss_1m^3(m-1)\varphi(m) + (d+d_1)2m(m-1)\varphi(m)$$
$$+ (d+d_1)2m(m-1)^2\varphi(m) = 2m^2(m-1)\varphi(m)(ss_1m+d+d_1).$$

Thus

$$\max(|x|, |y|) < \exp\{c_5|A|^{2m(m-1)\varphi(m)(d+d_1)+\varepsilon}|D_f|^{S+\varepsilon}(\ln H)^{1+\varepsilon}\}.$$

Theorem 7 can be further generalized, where we replace the numbers $a_1, \ldots, a_{n-1}$ by polynomials $a_1(y), \ldots, a_n(y)$ with integral coefficients for which

$$\max_{1\leqslant i\leqslant n-1} \deg a_i(y) < \deg a_n(y).$$

Setting

$$f(x, y) = (x - a_1(y)) \ldots (x - a_n(y)) \pm 1,$$

we require that the polynomial $D(y) = D[f(x|y)]$ not be plus or minus a perfect square in the ring of integral polynomials. Then for any $B > 2(n-1)$ there exists an integral polynomial $g(x)$ for which the field generated by a root of the irreducible polynomial

$$(x - a_1 * g(q)) \ldots (x - a_n * g(q)) \pm 1,$$

has discriminant and class number satisfying the inequalities of Theorem 7.

In a similar way we can consider fields with any fixed number of real and complex isomorphisms.

The structure of the method in [2]–[4] and in §5 of this paper shows that the basic bound for solutions of equation (1) is given in terms of the regulator of a certain auxiliary field (the field **L** in §5), and the estimates we use for the regulator have the most important influence on the resulting bound for the solutions. Lemma 4.6 shows that it is useful to carry out the arguments in the minimal field or to imbed in fields whose regulator has a better estimate than the regulator of the minimal field. In this way we obtain a strengthening of (8) and (13), about which we hope to communicate later.

BIBLIOGRAPHY

1. V. G. Sprindžuk, *On an estimate for solutions of Thue's equation,* Izv. Akad. Nauk SSSR Ser. Mat. **36** (1972), 712–741; English transl. in Math. USSR Izv. **6** (1972).

2. ———, *Square-free divisors of polynomials, and class numbers of algebraic number fields,* Acta Arith. **24** (1973), 143–149. (Russian)

3. ———, *Algebraic number fields with large class number,* Izv. Akad. Nauk SSSR Ser. Mat. **38** (1974), 971–982; English transl. in Math. USSR Izv. **8** (1974).

4. ———, *A hyperelliptic Diophantine equation, and class numbers,* Acta Arith. **30** (1976/77), 95–108. (Russian)

5. L. J. Mordell, *Diophantine equations,* Academic Press, New York and London, 1969.

6. ———, *The diophantine equation* $y^2 - k = x^3$, Proc. London Math. Soc. (2) **13** (1913/14), 60–80.

7. ———, *On the integer solutions of the equation* $ey^2 = ax^3 + bx^2 + cx + d$, Proc. London Math. Soc. (2) **21** (1922/23), 415–419.

8. C. L. Siegel, *The integer solutions of the equation* $y^2 = ax^n + bx^{n-1} + \cdots + k$ (Extract from a letter to Prof. L. J. Mordell [published under the pseudonym X]), J. London Math. Soc. **1** (1926), 66–68; reprinted in *Carl Ludwig Siegel Gesammelte Abhandlungen,* Band I, Springer-Verlag, Berlin and New York, 1966, pp. 207–208.

9. A. Baker, *Contributions to the theory of Diophantine equations.* I, Philos. Trans. Roy Soc. London Ser. A **263** (1967/68), 173–191.

10. ———, *Contributions to the theory of Diophantine equations.* II, Philos. Trans. Roy Soc. London Ser. A **263** (1967/68), 193–208.

11. ———, *The Diophantine equation* $y^2 = ax^3 + bx^2 + cx + d$, J. London Math. Soc. **43** (1968), 1–9.

12. ———, *Bounds for the solutions of the hyperelliptic equation,* Proc. Cambridge Philos. Soc. **65** (1969), 439–444.

13. N. C. Ankeny, R. Brauer and S. Chowla, *A note on the class-numbers of algebraic number fields,* Amer. J. Math. **78** (1956), 51–61.

14. Carl Ludwig Siegel, *Über die Classenzahl quadratischer Zahlkörper,* Acta Arith. **1** (1935/36), 83–86; reprinted in his *Gesammelte Abhandlungen,* Band I, Springer-Verlag, Berlin and New York, 1966, pp. 406–409.

15. Richard Brauer, *On the zeta-functions of algebraic number fields,* Amer. J. Math. **69** (1947), 243–250.

16. H. M. Stark, *Some effective cases of the Brauer-Siegel theorem,* Invent. Math. **23** (1974), 135–152.

17. Carl Ludwig Siegel, *Abschätzung von Einheiten,* Nachr. Akad. Wiss. Göttingen Math.-Phys. Kl. II **1969**, 71–86.

18. P. E. Blanksby and H. L. Montgomery, *Algebraic integers near the unit circle,* Acta Arith. **18** (1971), 355–369.

19. H. M. Stark, *Further advances in the theory of linear forms in logarithms,* Diophantine Approximation and Its Applications (Proc. Conf., Washington, D. C., 1972), Academic Press, New York, 1973, pp. 255–293.

20. Edmund Landau, *Abschätzungen von Charaktersummen, Einheiten und Klassenzahlen,* Nachr. Königl. Ges. Wiss. Göttingen **1918**, 79–97.

21. B. N. Delone and D. K. Faddeev, *The theory of irrationalities of the third degree,* Trudy Mat. Inst. Steklov. **11** (1940); English transl., Amer. Math. Soc., Providence, R. I., 1964.

Translated by N. KOBLITZ

Труды Мат. Инст. Стеклов.
143 (1977)

Proceedings of the Steklov
Institute of Mathematics
1980, Issue 1

ON LOWER ESTIMATES OF INCOMPLETE CHARACTER SUMS OF POLYNOMIALS

UDC 511

S. A. STEPANOV

ABSTRACT. The author proves the existence of square-free mod p polynomials $f(x)$ of degree of order $N/\log p$ for which all of the values $f(1), f(2), \ldots, f(N)$ are quadratic residues mod p.

Bibliography: 2 titles.

A. A. Karacuba [1] proved the existence of an infinite set of primes p and square-free mod p polynomials $f(x)$ of the form $ax^n + b$ with integral coefficients, where $4(p-1)/\log p \leqslant n \leqslant 8(p-1)/\log p$, for which

$$\sum_{x=1}^{p}\left(\frac{f(x)}{p}\right)=p. \tag{1}$$

Later a somewhat more general result was formulated in [2]. It asserted, for all primes $p \geqslant 11$, the existence of square-free mod p polynomials $f(x)$ of degree $n > (2p \log 2)/\log p + 2$ for which (1) holds.

In this present note, using only Dirichlet's "pigeon-hole principle," we prove a result, stronger than that of [2], pertaining not only to complete sums of Legendre symbols, but also to sums over an incomplete residue system.

THEOREM 1. *Suppose $p > 3$ is a prime and N is any natural number not exceeding p. There exist square-free* mod p *polynomials $f(x)$ and $g(x)$ of degree $2n$ $(p \geqslant n \geqslant ((N+1)\log 2)/\log p + 1)$ with integral coefficients such that*

$$\sum_{x=1}^{N}\left(\frac{f(x)}{p}\right)=N \quad \text{and} \quad \sum_{x=1}^{N}\left(\frac{g(x)}{p}\right)=-N.$$

PROOF. Let $f_1(x), \ldots, f_l(x)$ be all standardized irreducible polynomials of degree $n > 1$ with coefficients in the field of residues mod p. We have

$$l=\frac{1}{n}\sum_{d\mid n}\mu(d)\,p^{\frac{n}{d}}$$

and, therefore, $l > p^n/2n$. Put

$$\lambda_i(x)=\left(\frac{f_i(x)}{p}\right)$$

and, for $i = 1, \ldots, l$, consider the sets

$$\lambda_i=(\lambda_i(1),\ \lambda_i(2), \ldots, \lambda_i(N)).$$

1980 *Mathematics Subject Classification.* Primary 10A15, 10G15.

Since $n > 1$, it follows that $f_i(x) \neq 0$ for all $i = 1, \ldots, l$ and $x = 1, \ldots, N$. Consequently, the $\lambda_i(x)$ assume only the two values $+1$ and -1. Now there are precisely 2^N distinct sets $(y_1, \ldots, y_N)$ in which the y_i assume only two values. By hypothesis, $p \geqslant n \geqslant ((N+1)\log 2)/\log p + 1$; hence $l > 2^N$. Therefore, at least two of the sets λ_i must be identical. If λ_s and λ_t are two identical sets, we put $f(x) = f_s(x)f_t(x)$. Since the polynomials $f_s(x)$ and $f_t(x)$ are distinct and irreducible, $f(x)$ is not a perfect square. Moreover,

$$\left(\frac{f(x)}{p}\right) = 1$$

for all $x = 1, \ldots, N$; hence

$$\sum_{x=1}^{N} \left(\frac{f(x)}{p}\right) = N.$$

If we put $g(x) = af(x)$, where a is a quadratic nonresidue mod p, we obtain

$$\sum_{x=1}^{N} \left(\frac{g(x)}{p}\right) = -N.$$

The theorem is proved.

We now consider the case of second-degree polynomials.

THEOREM 2. *Suppose $p > 7$ is a prime and N is any natural number such that*

$$N < \frac{\log p - \log(2\log p)}{\log 2}.$$

There exist square-free mod p *polynomials $f(x) = (x+s)(x+t)$ and $g(x) = a(x+s)(x+t)$, where a, s and t are integers, such that*

$$\sum_{x=1}^{N} \left(\frac{f(x)}{p}\right) = N \quad \text{and} \quad \sum_{x=1}^{N} \left(\frac{g(x)}{p}\right) = -N.$$

PROOF. Consider all standardized polynomials $x, x+1, \ldots, x+p-1$ of first degree with coefficients in the field of residues mod p. Put

$$\lambda_i(x) = \left(\frac{x+i}{p}\right)$$

and consider the sets

$$\lambda_i = (\lambda_i(1), \lambda_i(2), \ldots, \lambda_i(N)), \qquad i = 0, 1, \ldots, p-1.$$

Each of these sets is composed from the numbers ± 1 and possibly one zero. Consequently, there are at most $N2^N$ distinct sets among them. By hypothesis,

$$N < \frac{\log p - \log(2\log p)}{\log 2};$$

hence $p > N2^N$. Therefore at least two of the sets λ_i are identical. Suppose λ_s and λ_t are such sets. Put $f(x) = (x+s)(x+t)$. Clearly, the polynomial $f(x)$ is not a perfect square. Since $x+s$ and $x+t$ are distinct, they cannot have a common root. Since λ_s and λ_t are identical, it follows that

$$\left(\frac{f(x)}{p}\right) = 1$$

for all $x = 1, \ldots, N$. Thus,

$$\sum_{x=1}^{N} \left(\frac{f(x)}{p}\right) = N.$$

Now put $g(x) = af(x)$, where a is a quadratic nonresidue mod p. Then

$$\sum_{x=1}^{N} \left(\frac{g(x)}{p}\right) = -N$$

and Theorem 2 is proved.

BIBLIOGRAPHY

1. A. A. Karacuba, *Lower bounds for character sums of polynomials,* Mat. Zametki **14** (1973), 67–72; English transl. in Math. Notes **14** (1973).

2. D. A. Mit'kin, *Lower bounds for sums of Legendre symbols and for trigonometric sums,* Uspehi Mat. Nauk **30** (1975), no. 5 (185), 214. (Russian)

Translated by G. A. KANDALL

Труды Мат. Инст. Стеклов.
143 (1977)

Proceedings of the Steklov
Institute of Mathematics
1980, Issue 1

BINARY FUNCTIONS ON ORDERED SETS (INVERSION THEOREMS)

UDC 519.1

B. S. STEČKIN

ABSTRACT. By exploiting the possibility of constructing the incidence algebra without appeal to axioms of transitivity or antisymmetry, it is shown that the Möbius principle of inversion, established earlier for partially ordered sets, can be extended to a larger class of ordered sets. A method of computing the Möbius function of such sets by enumerating chains of a special form is proposed.

Bibliography: 7 titles.

§1. Ordered sets

Suppose P is a set; its elements will be denoted by letters such as p, p_i, x, y, $z, \ldots$. Let $\leqslant$ denote a binary relation defined on P; we shall consider those binary relations which satisfy the axiom of reflexivity

1. $x \leqslant x \quad \forall x \in P$

and possibly some of the following axioms:

2. If $x \leqslant y$ and $y \leqslant x$, then $x = y \quad \forall x, y \in P$ (antisymmetry).
3. If $x \leqslant y$ and $y \leqslant z$, then $x \leqslant z \quad \forall x, y, z \in P$ (transitivity).
4. $\forall x, y \in P$, either $x \leqslant y$ or $y \leqslant x$ (linearity).

An arbitrary set P with a binary relation $\leqslant$ of this type will be called an ordered set and denoted by $(P, \leqslant)$, while the binary relation itself will be called an order. A list of possible types of orders and their names is as follows (the names in parentheses will not be used here):

1—reflexive order or simply order;
1, 2—weak order;
1, 2, 3—partial order (order);
1, 2, 3, 4—total (linear, complete, full) order;
1, 3—quasiorder;
1, 4—comparable order;
1, 2, 4—strictly comparable order;
1, 3, 4—transitively comparable order.

We shall say that an order is *proper* if it does not satisfy any further axioms (of these four) apart from the ones for which it is named; in particular, a total order is always proper. It is easy to invent examples of properly ordered sets for each type of order mentioned above. Special names have been attached to certain proper orders:

1980 *Mathematics Subject Classification*. Primary 06A10.

for example, a set satisfying only axioms 1, 2 and 4 is sometimes called a tournament, whereas one satisfying axioms 1, 2, 3 and 4 is called a transitive tournament.

For the sake of conciseness, we shall occasionally employ a formal notation such as $(P, \leqslant) = P(1, 2, \ldots)$ instead of naming the order; in this case, we mean that P is weakly ordered, whereas the notation $P(1, 2)$ would mean that P is properly weakly ordered.

A *zero* of the set $(P, \leqslant)$, denoted by 0_P, is an element in P for which $0_P \leqslant x$ $\forall x \in P$; a *unit* is an element $1_P \in P$ for which $1_P \geqslant x \quad \forall x \in P$.

A *chain* in an ordered set $(P, \leqslant)$ is a sequence $C = \{p_i\}$ of its elements in which every subsequent element is less than the preceding one, i.e. $p_i > p_{i+1}$ $\forall i$. An ordered set may possess *closed chains*, i.e. chains of the type $p_1 > p_2 > \cdots > p_i > p_1$ $(i \geqslant 2)$. Closed chains are naturally called *cycles*. We shall say that a chain is a *chain of type f*, or an *f-chain*, if it contains either a zero or a unit (as an ordered set). By the *length* of a finite chain C we shall understand the number $d(C) = |C| - 1$.(1) It is clear that if $(P, \leqslant) = P(1, 2, \ldots)$, then every cycle contains at least three elements, whereas for $(P, \leqslant) = P(1, 2, 3, \ldots)$ every chain is totally ordered and is not a cycle. An *interval* (*segment*) $[x, y]$ in a set $(P, \leqslant)$ is the subset $[x, y] = \{z \in P: x \leqslant z \leqslant y\}$. An interval is called *closed* if $x, y \in [x, y]$ and *nonclosed* in the opposite case. It is clear that an interval $[x, y]$ is closed if and only if $x \leqslant y$, so that in $P(1, 3, \ldots)$ every interval is either empty or closed. An ordered set is called *locally finite* if every one of its intervals is finite. *Acyclicity* shall mean the absence of cycles; *local acyclicity* means the acyclicity of all closed intervals, while *f-acyclicity* means the absence of f-chains with cycles (see [1]–[5]).

§2. Binary functions on $(P, \leqslant)$

We shall consider only real-valued functions; the set of such functions of two variables, defined on P^2, will be denoted by $A(P)$. If a function $f \in A(P)$ is associated with the binary relation $\leqslant$ introduced in P, it will itself be called *binary*. The simplest examples of binary functions are the Kronecker delta

$$\delta(x, y) = \chi_{\{x=y\}} \qquad \forall x, y \in P,$$

the order function

$$\zeta(x, y) = \chi_{\{x \leqslant y\}} \qquad \forall x, y \in P,$$

also called the zeta function of the set $(P, \leqslant)$, and the strict order function

$$\eta(x, y) = \zeta(x, y) - \delta(x, y)$$

($\chi_{\{\ \}}$ denotes the indicator of the occurrence contained in the braces).

If $(P, \leqslant)$ is an ordered set, the set $AI(P) = A_\zeta(P, \leqslant)$ of all binary functions on $(P, \leqslant)$ is defined as the subset of those functions in $A(P)$ which admit a representation

$$f(x, y) = f(x, y)\,\zeta(x, y) \qquad \forall x, y \in P, \tag{I}$$

(1) The count of elements in a chain is carried out with regard for multiplicity.

i.e. the function f may assume nonzero values only in pairs $x \leqslant y$ or on closed intervals, whereas otherwise its values must be zero. It is easy to see that the functions δ, ζ and η satisfy (I).

We remark that if in the representation (I) the zeta function is replaced by another fixed function $\lambda(x, y)$ for which $\lambda(x, x) = 1 \ \forall x \in P$, i.e. if one considers the set of functions $A_\lambda(P, \leqslant) \subseteq A(P)$ which may assume nonzero values only on those pairs x, y for which $\lambda(x, y) = 1$, then this is equivalent to considering the set $AI(P)$ relative to a certain new order on P; namely, if one defines the new order $\preccurlyeq$ by the rule

$$x \preccurlyeq y \Leftrightarrow \lambda(x, y) = 1,$$

then $A_\lambda(P, \leqslant) = A_\zeta(P, \preccurlyeq)$.

Generally speaking, the restriction imposed by requiring axiom 1 to be satisfied is not essential and its denial does not lead to serious alteration of the following results, although its fulfilment enables one to avoid awkward calculations in certain places. This is related to the fact that there are in reality two binary relations defined on P, the order $\leqslant$ and the equivalence relation of 'equality' (=), while the Kronecker delta, which plays a major role in the following results, is defined as a function of only the second relation (see [1]–[5]).

§3. Operations on functions

We first introduce two simple operations in the set $A(P)$:

Scalar multiplication: if $\alpha \in \mathbf{R}^1$ and $f \in A(P)$, then $\alpha \cdot f = \alpha f(x, y)$.

Addition: if $f, g \in A(P)$, then $f + g = f(x, y) + g(x, y)$.

It is easy to see that $AI(P)$ is closed under these operations. We now introduce in the set $AI(P)$, for locally finite P, the following general operation, whose modifications will be used throughout the paper.

K-convolution: if $f, g \in AI(P)$, then

$$f * g = \sum_{z \in P} f(x, z) K(x, z, y) g(z, y),$$

where $K(x, z, y)$ is a certain function (from P^3 to $\mathbf{R}^1$), called the *kernel* of the convolution. Since $f, g \in AI(P)$, we have

$$f * g(x, y) = \sum_z f(x, z) \zeta(x, z) K(x, z, y) \zeta(z, y) g(z, y)$$
$$= \sum_{x \leqslant z \leqslant y} f(x, z) K(x, z, y) g(z, y);$$

in view of the local finiteness of P, this shows that convolution is well defined. A kernel will be called *correct* if it admits a representation

$$K(x, z, y) = \zeta(x, z) K(x, z, y) \zeta(z, y) \qquad \forall x, y, z \in P;$$

i.e. if K can assume nonzero values only on chains $x \leqslant z \leqslant y$.(2) The convolution

(2) It is clear that correct kernels allow the $*$ operation to be introduced in all of $A(P)$.

operation allows one to introduce the concepts of unit and inverse function; namely a (right) unit function with respect to the K-convolution is a function $e \in AI(P)$ which satisfies

$$f * e(x, y) = f(x, y) \qquad \forall x, y \in P$$

for any function $f \in AI(P)$.

Inverse function: if $f \in AI(P)$, then a (right) inverse function of f is a function $g = f^{-1}$ which satisfies

$$f * g(x, y) = e(x, y) \qquad \forall x, y \in P,$$

where e is some unit function. In particular, a function inverse to the zeta function is called a *Möbius function* and is denoted by $\mu(x, y)$, where the unit function is assumed to be the delta function (see [2]–[5]).

§4. Stable kernels

It is clear that $AI(P)$ is not always closed with respect to the operations of convolution and the taking of inverses. We shall now attempt to select those convolutions for which

a) $AI(P)$ is closed with respect to convolution;

b) the Kronecker delta is a unit function;

c) there exists a Möbius function.

Convolutions (and their kernels) which satisfy conditions a)–c) will be called *stable on P*; if, furthermore, a Möbius function belongs to $AI(P)$, such convolutions (and their kernels) will be called *Möbius-stable on P.*

A criterion for a) to hold is given by

PROPOSITION 1. *The set $AI(P)$ is closed with respect to K-convolution if and only if*

$$K(x, z, y) = K(x, z, y)\zeta(x, y) \qquad \forall x, y, z \in P, \qquad z \in [x, y]. \qquad (K1)$$

PROOF. *Sufficiency.* We have

$$\begin{aligned} f * g(x, y) &= \sum_{x \leqslant z \leqslant y} f(x, z) K(x, z, y) g(z, y) \\ &= \sum_{x \leqslant z \leqslant y} f(x, z) K(x, z, y) \zeta(x, y) g(z, y) \\ &= \zeta(x, y) \sum_{z} f(x, z) K(x, z, y) g(z, y) = \zeta(x, y) f * g(x, y). \end{aligned}$$

Necessity. Assuming that $AI(P)$ is closed, we deduce that if

$$K(x_0, z_0, y_0) \neq K(x_0, z_0, y_0)\zeta(x_0, y_0), \qquad z_0 \in [x_0, y_0],$$

then $K(x_0, z_0, y_0) \neq 0$ and $1 \neq \zeta(x_0, y_0)$, i.e. $x_0 \nleqslant y_0$. But then the functions $f_0(x, y) = \delta(x, x_0)\delta(y, z_0)$ and $g_0(x, y) = \delta(x, z_0)\delta(y, y_0)$ on the one hand belong to $AI(P)$, whereas on the other their convolution is nonzero at (x_0, y_0), since

$$f_0 * g_0(x_0, y_0) = \sum_z \delta(x_0, x_0)\delta(z, z_0) K(x_0, z, y_0), \delta(z, z_0)\delta(y_0, y_0)$$
$$= \sum_z \delta(z, z_0) K(x_0, z, y_0) = K(x_0, z_0, y_0) \neq 0,$$

which contradicts the closure of $AI(P)$, completing the proof.

COROLLARY 1. *The set $AI(P(1, 3, \ldots))$ is closed with respect to any correct convolution.*

In fact, starting from the correctness of the kernel, the transitivity implies that

$$K(x, z, y) = K(x, z, y)\zeta(x, z)\zeta(z, y) \leqslant K(x, z, y)\zeta(x, y);$$

since $1 \geqslant \zeta(x, y)$, we have

$$K(x, z, y) \geqslant K(x, z, y)\zeta(x, y),$$

which means that

$$K(x, z, y) = K(x, z, y)\zeta(x, y).$$

Condition b) imposes quite severe restrictions on the nature of the kernel.

PROPOSITION 2. *The delta function is a unit function with respect to K-convolution with kernel $K(x, z, y)$ if and only if*

$$K(x, y, y) = 1 \qquad \forall x, y \in P, \qquad x \leqslant y. \tag{K2}$$

PROOF. *Sufficiency.* If $f \in AI(P)$, we have

$$f * \delta(x, y) = \sum_z f(x, z) K(x, z, y)\delta(z, y)$$
$$= \zeta(x, y) K(x, y, y) f(x, y) = \zeta(x, y) f(x, y) = f(x, y).$$

Necessity. Assuming that δ is a unit function, we deduce that if $K(x_0, y_0, y_0) \neq 1$ for $x_0 \leqslant y_0$, then $f_0(x, y) = \delta(x, x_0)\delta(y, y_0)$ on the one hand belongs to $AI(P)$, while on the other its convolution with the delta function is not equal to it at the point (x_0, y_0), since

$$f_0 * \delta(x_0, y_0) = \sum_z \delta(x_0, x_0)\delta(z, y_0) K(x_0, z, y_0)\delta(z, y_0)$$
$$= \sum_z \delta(z, y_0) K(x_0, z, y_0) = K(x_0, y_0, y_0) \neq 1 = f_0(x_0, y_0),$$

which contradicts the assumption.

In a completely analogous way one can demonstrate that delta is a left unit if and only if

$$K(x, x, y) = 1 \qquad \forall x, y \in P, \qquad x \leqslant y. \tag{2K}$$

Sufficient conditions for Möbius-stability in the class of weakly ordered sets are given by the

INVERSE FUNCTION LEMMA. *Suppose* $(P, \leqslant) = P(1, 2, \ldots)$ *is a locally finite f-acyclic weakly ordered set and assume that the kernel* $K(x, z, y)$ *satisfies conditions* $(K1)$ *and* $(K2)$, *while*

$$K(x, x, y) \neq 0 \qquad \forall x, y \in P, \qquad x \leqslant y. \tag{$K3$}$$

Then a function $f \in AI(P)$ *has an inverse (relative to the delta function) function* $f^{-1} \in AI(P)$ *if and only if* $f(x, x) \neq 0 \ \forall x \in P$.

The proof repeats almost verbatim that of Lemma 2.2.1 in [3]. Carried out by induction, it immediately follows from the possibility of defining f^{-1} recursively in the form

$$f^{-1}(x, y) = \frac{\delta(x, y)}{f(x, x)} - \sum_{\substack{x < z \leqslant y \\ x \leqslant y}} \frac{f(x, z)\, K(x, z, y)}{f(x, x)\, K(x, x, y)} f^{-1}(z, y);$$

from this it follows in particular that $f^{-1} \in AI(P)$. One should note that the assumptions of antisymmetry and f-acyclicity are essential for the inductive argument.

COROLLARY 2. *If a* $K1$-*kernel is defined on an f-acyclic* $P(1, 2, \ldots)$ *for which delta is a right and left unit, then* $f \in AI(P)$ *has an inverse if and only if* $f(x, x) \neq 0$ $\forall x \in P$.

In particular, the kernel $K(x, z, y) = \zeta(x, y)$ is Möbius-stable on every f-acyclic weakly ordered set. Furthermore, one can prove that ζ may be Möbius-stable even on a $P(1, 2, \ldots)$.

It is clear that in general the existence of a Möbius function is equivalent to solvability of the system of equations

$$\zeta * \mu(x, y) = \delta(x, y), \qquad x, y \in P,$$

for the unknown μ. In the case of a finite P, a precise criterion for the solvability of such a system can be expressed in terms of minors.

§5. New kernels on $P(1, 2, 3, \ldots)$

Earlier authors (see [2]–[5]) in general considered only the unit kernel $K \equiv 1$ on partially ordered sets; this kernel, of course, is Möbius-stable on $P(1, 2, 3, \ldots)$. Effective extensions of the set of kernels were only made for sets $P(1, 2, 3, \ldots)$ of a special kind (see, for example, [6]). The precise conditions for stability obtained here permit one to introduce new kernels without diminishing the generality of $P(1, 2, 3, \ldots)$. We shall now consider several of these.

Let $K(x, z, y) = |[z, y]|$, the cardinality of the interval $[z, y]$. It is clear that conditions $(K1)$, $(K2)$ and $(K3)$ are fulfilled for this kernel, so that it is Möbius-stable.

The Möbius function itself is defined by

$$\mu(x, y) = \delta(x, y) - \sum_{x<z\leqslant y} \frac{|[z, y]|}{|[x, y]|} \mu(z, y). \tag{1}$$

In particular, if $P(1, 2, 3, \ldots) = (\mathscr{P}(S_n), \subseteq)$ is the partially ordered set of all subsets of a set $S_n = \{a_1, \ldots, a_n\}$, ordered by inclusion, then

$$|[z, y]| = 2^{|y|-|z|}\zeta(z, y) \quad \text{and} \quad \mu(x, y) = \left(-\frac{1}{2}\right)^{|y|-|x|}\zeta(x, y). \tag{2}$$

Let $r(x, y)$ denote the length of the largest chain with a 'lower' end x and an 'upper' end y. Then the following functions may also serve as examples of Möbius-stable kernels on a partially ordered set:

$$K(x, z, y) = r(z, y) + \zeta(z, y); \tag{3}$$

$$K(x, z, y) = q^{r(z,y)}, \qquad q \in \mathbf{N}; \tag{4}$$

$$K(x, z, y) = \binom{|[x, y]|}{|[z, y]| - 1}; \tag{5}$$

$$K(x, z, y) = \binom{|[x, y]|}{|[x, z]| \cdot |[z, y]|}. \tag{6}$$

The enumeration and classification of stable (Möbius-stable) kernels permits one, in particular, to count and classify ordinary combinatorial identities (and their inverses).

§6. The inversion principle

Suppose that a K-convolution with respect to a kernel $K(x, z, y)$ is defined on the set of all binary functions $AI(P)$.

THEOREM 1. *Suppose $(P, \leqslant) = P(1, \ldots)$ is a locally finite reflexively ordered set with a zero 0_P; suppose that functions $f, g, \lambda, \kappa \in A(P)$ are defined on P^2 and satisfy*

$$g(y, x) = \sum_{z\leqslant y}{}_z f(z, x) K(z, y, x) \qquad \forall x, y \in P, \tag{7}$$

$$\zeta * \zeta\lambda(x, y) = \delta(x, y)\varkappa(x, y) \qquad \forall x, y \in P. \tag{8}$$

Then

$$f(x, x)\varkappa(x, x) = \sum_{y\leqslant x} g(y, x)\lambda(y, x). \tag{9}$$

PROOF. Since $0_P \in P$, the local finiteness of $(P, \leqslant)$ shows that all of the following sums are well defined:

$$\sum_{y\leqslant x^y}{}_y g(y, x)\lambda(y, x) = \sum_{y\leqslant x^y}{}_y \Big(\sum_{z\leqslant y^z}{}_z f(z, x) K(z, y, x)\Big)\lambda(y, x)$$

$$= \sum_{y\leqslant x^y}{}_y \sum_z f(z, x)\zeta(z, y) K(z, y, x)\lambda(y, x) =$$

$$= \sum_z f(z, x) \sum_y \zeta(z, y) K(z, y, x) \zeta(y, x) \lambda(y, x)$$

$$= \sum_z f(z, x) (\zeta * \zeta\lambda(z, x))$$

$$= \sum_z f(z, x) \delta(z, x) \varkappa(z, x) = f(x, x) \varkappa(x, x),$$

as required.

It is clear that the presence of a zero element can be replaced by the finiteness of sums of type (7) and (9).

If $(P, \leqslant) = P(1, 2, \ldots)$, then $\kappa(x, x) = K(x, x, x)\lambda(x, x)$, and if δ is a unit function with respect to K-convolution, then $\kappa(x, x) = \lambda(x, x)$.

If $\kappa \equiv 1$, then $\zeta\lambda = \mu$, whereas if $\lambda \in AI(P)$, then $\zeta\lambda = \lambda = \mu$. If K is a Möbius-stable kernel, then for $\kappa \equiv 1$ we can always choose a $\lambda \in AI(P)$ which satisfies (8); in this case, the requirement of Möbius-stability may be replaced by the requirement of correctness and stability of the kernel K. One should also note that in case $\kappa \equiv 1$ and $(P, \leqslant) = P(1, 2, \ldots)$ there exist $P(1, 2)$ for which (8) is fulfilled for $K \equiv 1$ and 'ordinary' μ:

$$\mu(x, y) = \begin{cases} 1, & x = y, \\ -\sum_{x < z \leqslant y} \mu(z, y), & x < y, \\ 0, & x \nleqslant y. \end{cases} \tag{10}$$

An example of such a $P(1, 2)$ without cycles is the set shown in the diagram (where the presence of an arrow is equivalent to strict inequality $>$). This example is clearly not transitive, while a simple check shows that (8) is satisfied. If one adds an arrow from a to b, one obtains a set with cycles which still satisfies (8).

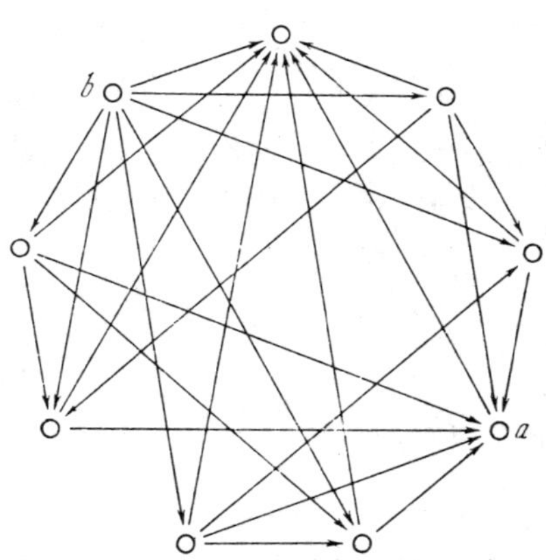

Finally, if we let $(P, \leqslant) = P(1, 2, 3, \ldots)$, $K \equiv 1$, $\kappa = 1$ and $f(z, x) = f(z)$, then $\lambda = \mu \in AI(P)$ and Theorem 1 is the familiar Möbius inversion principle for partially ordered sets (see [2]–[5]).

§7. Calculation of μ on $P(1, 2, \ldots)$

In the following $(P, \leqslant) = P(1, 2, \ldots)$ will denote a locally finite f-acyclic weakly ordered set; we also define on the set of all binary functions $AI(P)$ convolution with respect to the kernel $K(x, z, y) = \zeta(x, y)$. Then the set $AI(P)$, together with the operations of addition, scalar multiplication and ζ-convolution, forms an algebra with unit element δ.

Let $\tau_0(x, k, y)$ denote the number of chains of length k with initial element y and zero element x in the set $(P, \leqslant)$. A method for calculating the values of a Möbius function, which does not rely on recursion, is given by

THEOREM 2. *Suppose* $(P, \leqslant) = P(1, 2, \ldots)$ *is a locally finite f-acyclic weakly ordered set; then*

$$\mu(x, y) = \sum_{k \geqslant 0} (-1)^k \tau_0(x, k, y) \qquad \forall x, y \in P. \tag{11}$$

PROOF. We introduce the notation

$$f^{(n)} = \underbrace{(\ldots((f * f) * f) * \ldots * f) * f}_{n},$$

where $f^{(0)} = \delta$. Then, since $AI(P)$ is an algebra, it follows immediately that

$$\mu = \zeta^{-1} = (\delta + \eta)^{-1} = \delta - \eta^{(1)} + \eta^{(2)} - \eta^{(3)} + \ldots,$$

i.e.

$$\mu(x, y) = \sum_{k \geqslant 0} (-1)^k \eta^{(k)}(x, y) \qquad \forall x, y \in P. \tag{12}$$

Since it is easily verified that

$$\eta^{(k)}(x, y) = \tau_0(x, k, y) \qquad (k \geqslant 0),$$

we obtain the desired result.

It should be remarked that (11) gives the values only of a left Möbius function; the completely analogous formula for a right Möbius function has the form

$$\mu(x, y) = \sum_{k \geqslant 0} (-1)^k \tau_1(x, k, y), \tag{11'}$$

where $\tau_1(x, k, y)$ denotes the number of chains (in P) of length k with largest element y and terminal element x.

In particular, if $P = P(1, 2, 3, \ldots)$, then $\tau_0 = \tau_1$, and formulas (11) and (11′) coincide and represent P. Hall's theorem (see [1], [4], [5] or [7]). Furthermore, even if the ζ-kernel is Möbius-stable on a locally finite $P(1, 2, \ldots)$ (which is not necessarily f-acyclic, as was remarked above), then the requirement of f-acyclicity in Theorem 2 is essential, since the presence of a cycle in an f-chain would imply an infinite number of summands in (11) and, in particular, would result in equalities of the type

$$\frac{1}{2} = \sum_{k=0}^{\infty} (-1)^k. \tag{13}$$

§8. Remarks and generalizations

Of special interest is the study of bikernels (kernels assuming only the values 0 and 1), since they can always be considered as a restriction of the number of summands of a convolution.

Also of interest are decomposable kernels, i.e. kernels having a decomposition

$$K(x, z, y) = K'(x, z)\, Q(x, z, y)\, K''(z, y),$$

since the identities of the type

$$f \underset{K}{*} g = fK' \underset{Q}{*} K''g$$

generated by them are useful for the study of corresponding algebras of functions.

Finally, by analogy with a binary relation one can consider a two-place relation on sets X and Y (defined either as a subset of $X \times Y$ or as a multivalued function from X to Y), which becomes a binary relation in the case $X = Y$. A special interest is presented by the cases when in addition to the multivalued functions between X and Y there exist binary relations on X and Y themselves.

BIBLIOGRAPHY

1. Garrett Birkhoff, *Lattice theory*, 2nd rev. ed., Amer. Math. Soc., Providence, R.I., 1948.

2. K. A. Rybnikov, *Introduction to combinatorial analysis*, Izdat. Moskov. Univ., Moscow, 1972. (Russian)

3. Marshall Hall, Jr., *Combinatorial theory*, Blaisdell, Waltham, Mass., 1967.

4. Gian-Carlo Rota, *On the foundations of combinatorial theory*. I, Z. Wahrscheinlichkeitstheorie und Verw. Gebiete 2 (1964), 340–368.

5. Robin Wilson, *The Möbius function in combinatorial mathematics*, Combinatorial Mathematics and its Applications (Proc. Conf., Oxford, 1969), Academic Press, New York, 1971, pp. 315–333.

6. Ingrid Popa Fotino, *Generalized convolution ring of arithmetic functions*, Pacific J. Math. 61 (1975), 103–116.

7. P. Hall, *A contribution to the theory of groups of prime-power order*, Proc. London Math. Soc. (2) 36 (1933/34), 29–95.

Translated by G. MAXWELL

Труды Мат. Инст. Стеклов.
143 (1977)

Proceedings of the Steklov
Institute of Mathematics
1980, Issue 1

ESTIMATE OF A COMPLETE RATIONAL TRIGONOMETRIC SUM

UDC 511

S. B. STEČKIN

ABSTRACT. Suppose n and q are natural numbers, $n \geqslant 2$, a_ν $(\nu = 1, \ldots, n)$ are integers, $(a_1, \ldots, a_n, q) = 1$, and

$$f(x) = \sum_{\nu=1}^{n} a_\nu x^\nu, \qquad S(f, q) = \sum_{k=1}^{q} e^{2\pi i f(k)/q}.$$

It is shown that $|S(f, q)| \leqslant \overline{B}_n q^{1-1/n}$, where $\overline{B}_n = \exp\{n + o(n/\ln n)\}$.

Bibliography: 24 items.

Introduction

In this paper we use the following notation: $\mathbf{N}$ is the set of all natural numbers, $\mathbf{P}$ the set of all primes, $\mathbf{R}$ the field of all real numbers, $\mathbf{Z}$ the ring of all integers, $\mathbf{Z}_+$ the set of all nonnegative integers, $\mathbf{F}_p = \mathbf{Z}/p\mathbf{Z}$ $(p \in \mathbf{P})$ the residue field modulo p, $\mathbf{Z}[x]$ the ring of polynomials with coefficients in $\mathbf{Z}$, and $\mathbf{F}_p[x]$ the ring of polynomials with coefficients in $\mathbf{F}_p$; also, as usual, $\exp(y) = e^y$ and $e(y) = e^{2\pi i y}$ $(y \in \mathbf{R})$. Throughout this paper, $n, q, s \in \mathbf{N}$, $p \in \mathbf{P}$, and $C_1, \ldots, C_{29}$ are absolute positive constants.

Suppose $f \in \mathbf{Z}[x]$, $\deg f = n \geqslant 1$. By a *complete rational trigonometric sum* we mean a sum of the form

$$S(f, q) = \sum_{k=1}^{q} e(f(k)/q). \tag{0.1}$$

The sums $S(f, q)$ have important applications in number theory (see [1]–[4]) and have been studied by many authors. If $f(x) = \Sigma_0^n a_\nu x^\nu$ and $h = (a_1, \ldots, a_n, q) > 1$, then

$$S(f, q) = h e(a_0/q) S((f - a_0)/h, q/h). \tag{0.2}$$

Therefore, it suffices to investigate $S(f, q)$ for the case

$$a_0 = 0, \qquad (a_1, \ldots, a_n, q) = 1. \tag{0.3}$$

If $f \in \mathbf{Z}[x]$, $\deg f = n \geqslant 1$ and the conditions (0.3) are satisfied, we write $f \in K_n(q)$.

The case $n = 1$ is trivial: $S(f, q) = 0$ $(f \in K_1(q), q > 1)$. The sums $S(f, q)$ with $n = 2$ were thoroughly investigated by Gauss [5]. In particular, it follows from his results that

$$|S(f, q)| \leqslant \sqrt{2q} \qquad (f \in K_2(q)). \tag{0.4}$$

Therefore, in the sequel we will assume, unless it is stated otherwise, that $n \geqslant 3$.

1980 *Mathematics Subject Classification.* Primary 10G10.

The sums $S(f, q)$ with arbitrary $n \in \mathbf{N}$ were studied by Hardy and Littlewood in connection with Waring's problem (see [6]). Using Weyl's method [7] for estimating trigonometric sums, they showed that if $(a_n, q) = 1$, then for any $\epsilon > 0$

$$|S(f, q)| \leqslant A_1(n, \varepsilon)\, q^{1-2^{1-n}+\varepsilon} \tag{0.5}$$

(see also [8], Theorem 265). Moreover, they showed that for the Gaussian sums

$$S_n(a, q) = \sum_{k=1}^{q} e(ak^n/q), \qquad (a, q) = 1 \tag{0.6}$$

(which had already been encountered in the theory of cyclotomy) it was possible to obtain the following estimates:

$$|S_n(a, p)| \leqslant (d-1)\sqrt{p}, \qquad d = (n, p-1), \tag{0.7}$$

$$|S_n(a, q)| \leqslant A_2(n)\, q^{1-\frac{1}{n}}. \tag{0.8}$$

These estimates are attainable, in the sense of order, with respect to p and q, respectively. More precisely (see the Introduction to [1], Russian pp. 6–7),

$$\overline{\lim_{p\to\infty}} \sup_{\substack{a \\ (a,p)=1}} |S_n(a, p)|/\sqrt{p} \geqslant \sqrt{n-1} \tag{0.9}$$

and (see the Introduction to [1], Russian pp. 7–8)

$$S_n(a, p^n) = p^{n-1} = p^{n\left(1-\frac{1}{n}\right)}, \qquad (n, p) = 1; \tag{0.10}$$

hence

$$\overline{\lim_{q\to\infty}} \sup_{\substack{a \\ (a,q)=1}} |S_n(a, q)|/q^{1-\frac{1}{n}} \geqslant 1. \tag{0.11}$$

More important is the problem of estimating the sums $S(f, q)$ depending on q and n *uniformly with respect to the coefficients of the polynomial* $f \in K_n(q)$, i.e. the problem of estimating

$$H_n(q) = \sup_{f \in K_n(q)} |S(f, q)|. \tag{0.12}$$

This problem was first considered in 1924 by Kamke [9] in connection with the problem of the distribution of fractional parts of polynomials and Waring's problem for polynomials. He proved that if $f \in K_n(q)$, then (0.5) holds.

The behavior of $H_n(q)$ essentially depends on the arithmetic properties of q. The most difficult problem is that of estimating $H_n(q)$ when $q = p$, a prime. In 1932 Mordell [10] proved that

$$H_n(p) \leqslant A_3(n)\, p^{1-\frac{1}{n}}, \tag{0.13}$$

and Davenport [11] improved this result somewhat (see also [3], §12). Hua investigated the problem of estimating $H_n(q)$ in connection with the mean value theorem for trigonometric sums and Tarry's problem. Using (0.13), Hua [12] (see also [1], Chapter 3) established in 1940 that for any $\epsilon > 0$

$$H_n(q) \leqslant A_4(n, \varepsilon)\, q^{1-\frac{1}{n}+\varepsilon}. \tag{0.14}$$

In 1948 Weil [13], [14] obtained the estimate

$$H_n(p) \leqslant (n-1)\sqrt{p} \qquad (p > n \geqslant 2), \tag{0.15}$$

which, in view of (0.9), is attainable, in the sense of order, with respect to p.([1]) Using (0.15), Nečaev [16] proved that

$$H_n(q) \leqslant A_5(n) q^{1-\frac{1}{n}}, \tag{0.16}$$

which, in view of (0.11), is attainable with respect to q.

Karacuba [17] considered the problem of estimating $H_n(q)$ from below.

Let

$$B_n = \sup_{q \in \mathbf{N}} H_n(q)/q^{1-\frac{1}{n}}. \tag{0.17}$$

This present paper is devoted to investigating the behavior of B_n. In view of (0.11), $B_n \geqslant 1$. In 1953 Nečaev [16] proved that $B_n \leqslant \exp(2^n)$ $(n \geqslant 12)$, and in 1975 he showed [18] that

$$B_n \leqslant \exp(5n^2/\ln n) \qquad (n \geqslant 3). \tag{0.18}$$

Here we establish the estimate

$$B_n \leqslant \exp\{n + O(n/\ln n)\} \qquad (n \to \infty), \tag{0.19}$$

from which it follows, in particular, that

$$B_n \leqslant \exp(C_1 n). \tag{0.20}$$

In other words, we have this

THEOREM. *Suppose $n \geqslant 2$, $q \in \mathbf{N}$ and $f \in K_n(q)$. Then*

$$|S(f, q)| \leqslant \bar{B}_n q^{1-\frac{1}{n}}, \tag{0.21}$$

where

$$\bar{B}_n = \exp(n + O(n/\ln n)). \tag{0.22}$$

Note that if we use Weil's estimate (0.15) for $S(f, p)$, the asymptotic quantity in the exponent of (0.19) cannot be improved.

To prove (0.19) we require estimates of the sums $S(f, p^s)$ $(f \in K_n(p))$ which are uniform with respect to $s \in \mathbf{N}$. The history of this auxiliary problem is as follows. Hua [12] (see also [1], Chapter III, Lemma 6) proved in 1940 that

$$|S(f, p^s)| \leqslant n^{2n} p^{s\left(1-\frac{1}{n}\right)}, \tag{0.23}$$

and later (see [4], Lemma 1.6) he obtained the estimate

$$|S(f, p^s)| \leqslant n^3 p^{s\left(1-\frac{1}{n}\right)}.$$

Improving the results of [16] and [19], Nečaev showed that for all $s \in \mathbf{N}$

$$|S(f, p^s)| \leqslant p^{s\left(1-\frac{1}{n}\right)} \quad \text{for} \quad p \geqslant (n-1)^{\frac{2n}{n-2}}, \tag{0.24}$$

and obtained several new estimates of $S(f, p^s)$ for smaller p. In particular, it follows from these estimates that

([1]) Recently, Stepanov [15] gave a new, and elementary, proof of (0.15).

$$|S(f, p^s)| \leqslant n^2 p^{s\left(1-\frac{1}{n}\right)} \qquad (\forall p, \ \forall s \in \mathbf{N}).$$

To prove (0.19) we will essentially use (0.24) and a strengthening of Nečaev's estimate for $p \leqslant (n-1)^{2n/(n-2)}$. Here we obtain the following result (Theorem 1): for all $n \geqslant 3$, $p \in \mathbf{P}$, and $s \in \mathbf{N}$

$$|S(f, p^s)| \leqslant C_2 p^{s\left(1-\frac{1}{n}\right)}. \tag{0.25}$$

Our main investigative tools will be properties of multiple roots of polynomials in the ring $\mathbf{F}_p[x]$ (Lemma 2) and the partition formula (Lemma 3), which reduces the estimation of $S(f, p^s)$ to estimations of analogous sums with smaller exponents s.

As an application of (0.21) we obtain the following estimate for the number $\rho(f, m)$ of distinct solutions of the congruence

$$f(x) \equiv 0 \pmod{m}, \tag{0.26}$$

where $f \in \mathbf{Z}[x]$, $\deg f = n \geqslant 2$, $m \in \mathbf{N}$, $m \geqslant 3$ and $(a_1, \ldots, a_n, m) = 1$:

$$\rho(f, m) \leqslant m^{1-\frac{1}{n}} \exp\left\{C_3 n \left(\frac{(\ln m)^{1/n}}{\ln \ln m} + 1\right)\right\}. \tag{0.27}$$

The author would like to thank S. A. Stepanov for a tutorial on the theory of polynomials over finite fields.

§1. Auxiliary results

1.1. Let $\mathbf{A}$ be a commutative ring with unity and $\mathbf{A}[x]$ the ring of polynomials in one variable with coefficients in $\mathbf{A}$. For any $l \in \mathbf{Z}_+$ there is defined in $\mathbf{A}[x]$ a differentiation operation of order l: if $f(x) = \Sigma_0^n a_\nu x^\nu$, $a_\nu \in \mathbf{A}$, then

$$f^{[l]}(x) = \sum_{\nu=l}^{n} \binom{\nu}{l} a_\nu x^{\nu-l} \qquad (l \in \mathbf{Z}_+).$$

Clearly, $f^{[l]} \in \mathbf{A}[x]$, and if the polynomial ring is $\mathbf{Z}[x]$, then $f^{[l]} = f^{(l)}/l!$. Therefore, instead of $f^{[1]}$ we will write f'.

LEMMA 1. *For an element $a \in \mathbf{A}$ to be a root of multiplicity $k \in \mathbf{N}$ of a polynomial $f \in \mathbf{A}[x]$, $f \neq 0$, it is necessary and sufficient that*

$$f^{[l]}(a) = 0 \quad (l = 0, \ldots, k-1), \qquad f^{[k]}(a) \neq 0. \tag{1.1}$$

See [20]–[22] and [15]. We will give a simple proof.

Suppose $a \in \mathbf{A}$ is a root of multiplicity k of the polynomial f, i.e.

$$f(x) = (x-a)^k g(x), \qquad \text{where } g \in \mathbf{A}[x], \qquad g(a) \neq 0. \tag{1.2}$$

According to Leibniz,

$$f^{[l]}(x) = \sum_{\lambda=0}^{l} \{(x-a)^k\}^{[\lambda]} g^{[l-\lambda]}(x) = \sum_{\lambda=0}^{l} \binom{l}{\lambda} (x-a)^{k-\lambda} g^{[l-\lambda]}(x) \ (l = 0, 1, \ldots, k);$$

hence $f^{[l]}(a) = 0$ $(l = 0, \ldots, k-1)$ and $f^{[k]}(a) = g(a) \neq 0$.

Conversely, if (1.1) holds, then, by Taylor's formula,

$$f(x) = \sum_{\nu=0}^{n} f^{[\nu]}(a)(x-a)^\nu = \sum_{\nu=k}^{n} f^{[\nu]}(a)(x-a)^\nu =$$

$$=(x-a)^k \sum_{\nu=k}^{n} f^{[\nu]}(a)(x-a)^{\nu-k}=(x-a)^k g(x),$$

where $g \in \mathbf{A}[x]$ and $g(a) = f^{[k]}(a) \neq 0$, i.e. a is a root of multiplicity k of the polynomial f.

1.2. Suppose $f \in \mathbf{Z}[x]$, $p \in \mathbf{P}$, and $\varphi = \varphi_p$ is the reduction modulo p of the ring $\mathbf{Z}[x]$ to the ring $\mathbf{F}_p[x]$. Put $\bar{f} = \varphi(f)$. A number $a \in \mathbf{Z}$ is called a root of multiplicity $k \in \mathbf{N}$ of the congruence

$$f(x) \equiv 0 \pmod p, \tag{1.3}$$

if $\bar{f} \neq 0$ and $\bar{a} = \varphi(a)$ is a root of multiplicity k of the polynomial $\bar{f} \in \mathbf{F}_p[x]$. This definition is equivalent to saying that

$$f(x)=(x-a)^k g_1(x)+pg_2(x), \tag{1.4}$$

where $g_1, g_2 \in \mathbf{Z}[x]$ and $g_1(a) \neq 0$. Roots $a_1, a_2 \in \mathbf{Z}$ of (1.3) are considered distinct if $\bar{a}_1 \neq \bar{a}_2$. We denote by $N_p(f)$ the number of roots of (1.3), taking their multiplicities into account. In other words, if $a_1, \ldots, a_r$ ($r \in \mathbf{Z}_+$) are all of the distinct roots of (1.3) and $k_1, \ldots, k_r$ their respective multiplicites, then we put

$$N_p(f)=\sum_{\nu=1}^{r} k_\nu. \tag{1.5}$$

It is well known that we always have

$$0 \leqslant N_p(f) \leqslant \deg \bar{f} \leqslant \deg f. \tag{1.6}$$

1.3. Suppose $f \in \mathbf{Z}[x]$, $f \neq 0$, and $p \in \mathbf{P}$. Let $u = u_p(f)$ denote the largest power of p dividing all coefficients of f. Clearly, if we put $F(x) = p^{-u}f(x)$, then $F \in \mathbf{Z}[x]$ and $\bar{F} \neq 0$. Suppose, in addition, that $\deg f \geqslant 1$, so that $f' \neq 0$. Put

$$t(f)=t_p(f)=u_p(f'), \qquad g(x)=p^{-t(f)}f'(x), \tag{1.7}$$

$$J(f)=J_p(f)=N_p(g), \tag{1.8}$$

$$h_a(x)=f(a+px)-f(a) \qquad (a \in \mathbf{Z}), \tag{1.9}$$

$$m(a)=u_p(h_a), \qquad f_a(x)=p^{-m(a)}h_a(x). \tag{1.10}$$

We will need several properties of these characteristics and polynomials.

LEMMA 2. *Suppose $n \geqslant 2$ and $f \in K_n(p)$. Then the following assertions hold*:

1) $$0 \leqslant t_p(f) \leqslant t=\left[\frac{\ln n}{\ln p}\right]; \tag{1.11}$$

in particular, $t_p(f) = 0$ for $p > n$.

2) $$0 \leqslant J_p(f) \leqslant n-1. \tag{1.12}$$

3) $$1 \leqslant m(a) \leqslant n \quad (a \in \mathbf{Z}). \tag{1.13}$$

4) *If b is a root of multiplicity k of the congruence*

$$g(x) \equiv 0 \pmod p, \tag{1.14}$$

then

$$0 \leqslant J_p(f_b) \leqslant k, \tag{1.15}$$

$$2 \leqslant m(b) \leqslant k+t_p(f)+1. \tag{1.16}$$

Except for the inequality

$$m(b) \leqslant k + t(f) + 1, \tag{1.17}$$

all of these properties are known (see [4] and [18]).

Let us prove (1.17). By Taylor's formula,

$$h_b(x) = f(b + px) - f(b) = \sum_{\nu=1}^{n} p^\nu f^{[\nu]}(b) x^\nu.$$

Since $k \leqslant \deg g \leqslant n - 1$, the coefficient of x^{k+1} in this formula is

$$p^{k+1} f^{[k+1]}(b) = p^{k+t(f)+1} \frac{g^{[k]}(b)}{k+1},$$

where, by Lemma 1, $g^{[k]}(b) \not\equiv 0 \pmod p$. Therefore, if $p^u \| p^{k+1} f^{[k+1]}(b)$, we have $u \leqslant k + t(f) + 1$, which implies (1.17).

The notation $t = [\ln n/\ln p]$ will be preserved throughout this paper.

1.4. Let $b_1, \ldots, b_r$ $(r \in \mathbf{Z}_+)$ be all of the distinct roots of congruence (1.14), $k_1, \ldots, k_r$ their respective multiplicities, and put

$$m_\nu = m(b_\nu), \qquad \varphi_\nu(x) = f_{b_\nu}(x) \qquad (\nu = 1, \ldots, r). \tag{1.18}$$

In view of the definition of $f_a(x)$ and Lemma 2, we have

$$2 \leqslant m_\nu \leqslant n, \qquad m_\nu \leqslant k_\nu + t_p(f) + 1, \tag{1.19}$$

$$\varphi_\nu \in K_n(p), \qquad J_p(\varphi_\nu) \leqslant k_\nu \qquad (\nu + 1, \ldots, r), \tag{1.20}$$

$$0 \leqslant \sum_{\nu=1}^{r} k_\nu = J_p(f) \leqslant n - 1. \tag{1.21}$$

LEMMA 3 (PARTITION FORMULA). *Suppose $f \in K_n(p)$, $s \in \mathbf{N}$ and*

$$s \geqslant 2t_p(f) + 2. \tag{1.22}$$

Then

$$|S(f, p^s)| \leqslant \sum_{\nu=1}^{r} \left| \sum_{x=1}^{p^{s-1}} e(p^{m_\nu - s} \varphi_\nu(x)) \right|; \tag{1.23}$$

in particular, if $J_p(f) = 0$ (i.e. if (1.14) has no solutions), then $S(f, p^s) = 0$.

See [18], Lemmas 5 and 6, or [4], Chapter I. Note that

$$\left| \sum_{x=1}^{p^{s-1}} e(p^{m_\nu - s} \varphi_\nu(x)) \right| = \begin{cases} p^{m_\nu - 1} |S(\varphi_\nu, p^{s - m_\nu})| & \text{if } m_\nu < s, \\ p^{s-1} & \text{if } m_\nu \geqslant s. \end{cases} \tag{1.24}$$

Thus,

$$|S(f, p^s)| \leqslant \sum_{\nu=1}^{r} p^{m_\nu - 1} |S(\varphi_\nu, p^{s - m_\nu})|, \tag{1.25}$$

if $m_\nu < s$ $(\nu = 1, \ldots, r)$, i.e. in this case $S(f, p^s)$ is estimated in terms of sums of the same form, but with smaller exponents s.

LEMMA 4. *Suppose $w \in \mathbf{R}$, $w \geqslant 2t + 2$, and for any polynomial $f \in K_n(p)$*

$$|S(f, p^s)| \leqslant J(f) \gamma_n(p) p^{s\left(1 - \frac{1}{n}\right)} \qquad \text{for } w \leqslant s < n + w. \tag{1.26}$$

Then for any $f \in K_n(p)$

$$|S(f, p^s)| \leqslant J(f)\gamma_n(p)p^{s\left(1-\frac{1}{n}\right)} \quad \textit{for all} \quad s \geqslant w. \tag{1.27}$$

PROOF. We proceed by induction on s. Suppose (1.27) holds for $w \leqslant s < n + w_1$, where $w_1 \geqslant w$. We will show that it is then true for $n + w_1 \leqslant s < n + w_1 + 1$. Since $s \geqslant n + w_1 \geqslant w \geqslant 2t + 2$, the partition formula is applicable. It shows that

$$|S(f, p^s)| \leqslant \sum_{\nu=1}^{r} p^{m_\nu - 1}|S(\varphi_\nu, p^{s-m_\nu})| \qquad (\varphi_\nu \in K_n(p)),$$

where, in view of (1.19),

$$w \leqslant w_1 \leqslant s - n \leqslant s - m_\nu \leqslant s - 2 < n + w_1 + 1 - 2 < n + w_1.$$

Therefore, (1.26) is applicable, and we obtain

$$|S(f, p^s)| \leqslant \sum_{\nu=1}^{r} p^{m_\nu\left(1-\frac{1}{n}\right)} J(\varphi_\nu)\gamma_n(p)p^{(s-m_\nu)\left(1-\frac{1}{n}\right)}$$

$$\leqslant \sum_{\nu=1}^{r} J(\varphi_\nu)\gamma_n(p)p^{s\left(1-\frac{1}{n}\right)} \leqslant J(f)\gamma_n(p)p^{s\left(1-\frac{1}{n}\right)};$$

the lemma is proved.

§2. Estimates of $S(f, p^s)$

LEMMA 5. *Suppose* $f \in K_n(p)$, $s \in \mathbf{N}$ *and* $k \in \mathbf{N}$. *Put*

$$q_1 = 0, \qquad q_k = p^{k-1} + \ldots + p \qquad (k > 1). \tag{2.1}$$

Then

$$|S(f, p^s)| \leqslant J(f)p^{s-k} \qquad \textit{for} \quad s \geqslant q_k + (k+1)t + 2, \tag{2.2}$$

$$|S(f, p^s)| \leqslant \frac{J(f)}{s - q_k - (k+2)t - 2} p^{s-1} \tag{2.3}$$

$$\textit{for} \quad q_k + (k+2)t + 2 < s \leqslant q_{k+1} + (k+2)t + 2. \tag{2.4}$$

PROOF. By the conditions of the lemma, $s \geqslant 2t + 2 \geqslant 2t(f) + 2$. Hence, according to the partition formula,

$$|S(f, p^s)| \leqslant \sum_{\nu=1}^{r} \left| \sum_{x=1}^{p^{s-1}} e(p^{m_\nu - s}\varphi_\nu(x)) \right|. \tag{2.5}$$

Estimating the inner sums in (2.5) trivially, we obtain

$$|S(f, p^s)| \leqslant rp^{s-1} \leqslant J(f)p^{s-1} \qquad (s \geqslant 2t(f) + 2), \tag{2.6}$$

which proves (2.2) for $k = 1$.

We now proceed by induction on k. Assume that (2.2) holds for some $k \in \mathbf{N}$; we will show that (2.3) then holds for this k, and (2.2) for $k + 1$. Suppose $s > q_k + (k + 2)t + 2$. In (2.5) we divide the summation over ν into two parts: a sum Σ_1 in which $k_\nu \geqslant s - q_k - (k + 2)t - 2$, and a sum Σ_2 in which $k_\nu \leqslant s - q_k - (k + 2)t - 3$. Put

$$S_h = \sum\nolimits_h \left| \sum_{x=1}^{p^{s-1}} e(p^{m_\nu - s}\varphi_\nu(x)) \right|, \qquad J_h = \sum\nolimits_h k_\nu \quad (h = 1, 2). \tag{2.7}$$

Then

$$|S(f,\ p^s)|\leqslant S_1+S_2, \tag{2.8}$$

$$J_1\geqslant 0,\qquad J_2\geqslant 0,\qquad J_1+J_2=\sum_{\nu=1}^{r}k_\nu=J=J(f). \tag{2.9}$$

Let us estimate S_1. Let Q be the number of terms in Σ_1. Since in Σ_1 we have $k_\nu \geqslant s-q_k-(k+2)t-2$, it follows that

$$Q(s-q_k-(k+2)t-2)\leqslant \textstyle\sum_1 k_\nu=J_1\leqslant J.$$

Hence, by a trivial estimate,

$$S_1\leqslant Qp^{s-1}\leqslant \frac{J_1}{s-q_k-(k+2)t-2}p^{s-1}. \tag{2.10}$$

Let us estimate S_2. In Σ_2 we have $k_\nu \leqslant s-q_k-(k+2)t-3$. Consequently, in view of (1.19) and the definition of t,

$$m_\nu\leqslant k_\nu+t+1\leqslant s-q_k-(k+1)t-2,\ \ s-m_\nu\geqslant q_k+(k+1)t+2\ \ (k\in \mathbf{N}).$$

Therefore,

$$S_2=\textstyle\sum_2 p^{m_\nu-1}|S(\varphi_\nu,\ p^{s-m_\nu})|,$$

and (2.2) is applicable to each term on the right. In view of (1.20) and (2.9),

$$S_2\leqslant \textstyle\sum_2 J(\varphi_\nu)p^{s-k-1}\leqslant \sum_2 k_\nu p^{s-k-1}=J_2p^{s-k-1}=(J-J_1)p^{s-k-1}. \tag{2.11}$$

Combining (2.8), (2.10), and (2.11), we obtain

$$|S(f,\ p^s)|\leqslant \frac{J_1}{s-q_k-(k+2)t-2}p^{s-1}+(J-J_1)p^{s-k-1}$$

$$=J_1\left(\frac{1}{s-q_k-(k+2)t-2}-\frac{1}{p^k}\right)p^{s-1}+Jp^{s-k-1}.$$

But $0\leqslant J_1\leqslant J$. Therefore, if $s-q_k-(k+2)t-2\leqslant p^k$, i.e.

$$s\leqslant q_k+p^k+(k+2)t+2=q_{k+1}+(k+2)t+2,$$

then

$$|S(f,\ p^s)|\leqslant J\left(\frac{1}{s-q_k-(k+2)t-2}-\frac{1}{p^k}\right)p^{s-1}+Jp^{s-k-1}$$

$$=\frac{J(f)}{s-q_k-(k+2)t-2}p^{s-1},$$

and if $s-q_k-(k+2)t-2\geqslant p^k$, i.e. $s\geqslant q_{k+1}+(k+2)t+2$, then

$$|S(f,\ p^s)|\leqslant J(f)p^{s-k-1},$$

and the lemma is proved.

Lemma 6. *Suppose $n\geqslant 3$, $f\in K_n(p)$, $s\in \mathbf{N}$ and $s\geqslant 2t+2$. Then*

$$|S(f,\ p^s)|\leqslant J(f)p^{\frac{3t+3}{n}-1}p^{s\left(1-\frac{1}{n}\right)}\quad \textit{if}\quad p\leqslant (n-1)^{\frac{n}{n-2}}, \tag{2.12}$$

$$|S(f,\ p^s)|\leqslant \frac{J(f)}{n-1}p^{\frac{1}{n}}p^{s\left(1-\frac{1}{n}\right)}\quad \textit{if}\quad p\geqslant (n-1)^{\frac{n}{n-2}}. \tag{2.13}$$

PROOF. First note that

$$n<(n-1)^{\frac{n}{n-2}} \qquad (n\geqslant 3). \tag{2.14}$$

The case $n = 3$ can be verified directly, and if $n \geqslant 4$, then

$$\ln\frac{n}{n-1}<\frac{1}{n-1}<\frac{2}{n-2}<\frac{2}{n-2}\ln(n-1),$$

which is equivalent to (2.14).

We will show that (2.12) and (2.13) are valid when

$$2t+2\leqslant s<n+2t+2, \tag{2.15}$$

i.e. when $2t + 2 \leqslant s \leqslant n + 2t + 1$. We use the previous lemma with $k = 1$ and $k = 2$.

Let us consider several different cases.

If $2t + 2 \leqslant s \leqslant 3t + 2$, then, according to (2.2) with $k = 1$,

$$|S(f,\ p^s)|\leqslant J(f)\,p^{\frac{s}{n}-1}p^{s\left(1-\frac{1}{n}\right)}\leqslant J(f)\,p^{\frac{3t+2}{n}-1}p^{s\left(1-\frac{1}{n}\right)}. \tag{2.16}$$

Suppose $3t + 3 \leqslant s \leqslant n + 2t + 1$. Inequality (2.3) with $k = 1$ can be written in the form

$$|S(f,\ p^s)|\leqslant J(f)\,\frac{p^{\frac{s}{n}-1}}{s-3t-2}\,p^{s\left(1-\frac{1}{n}\right)} \qquad (3t+3\leqslant s\leqslant p+3t+2). \tag{2.17}$$

Note that the function $\psi(s) = p^{s/n-1}(s - 3t - 2)^{-1}$ is convex and therefore its maximum on an interval $[\alpha, \beta]$ $(\alpha \geqslant 3t + 3)$ is attained at an end point.

Suppose $p \geqslant n - t - 1$, i.e. $n + 2t + 1 \leqslant p + 3t + 2$. Then the interval $3t + 3 \leqslant s \leqslant n + 2t + 1$ falls within the scope of (2.17). Therefore,

$$|S(f,\ p^s)|\leqslant J(f)\max\left\{p^{\frac{3t+3}{n}-1},\ \frac{p^{\frac{2t+1}{n}}}{n-t-1}\right\}p^{s\left(1-\frac{1}{n}\right)}$$
$$(3t+3\leqslant s\leqslant n+2t+1).$$

In view of (2.16), we obtain

$$|S(f,\ p^s)|\leqslant J(f)\max\left\{p^{\frac{3t+3}{n}-1},\ \frac{p^{\frac{2t+1}{n}}}{n-t-1}\right\}p^{s\left(1-\frac{1}{n}\right)} \tag{2.18}$$

for

$$p\geqslant n-t-1, \qquad 2t+2\leqslant s\leqslant n+2t+1. \tag{2.19}$$

Now suppose $p \leqslant n - t - 2$, i.e. $p + 3t + 2 < n + 2t + 1$. We apply (2.17) on the interval $3t + 3 \leqslant s \leqslant p + 3t + 2$, and we use (2.3) with $k = 2$ on the remaining part $p + 3t + 2 \leqslant s \leqslant n + 2t + 1$. We obtain

$$|S(f,\ p^s)|\leqslant J(f)\max\left\{p^{\frac{3t+3}{n}-1},\ p^{\frac{p+3t+2}{n}-2}\right\}p^{s\left(1-\frac{1}{n}\right)}\ (3t+3\leqslant s\leqslant p+3t+2),$$

$$|S(f,\ p^s)|\leqslant J(f)\,p^{s-2}=J(f)\,p^{\frac{s}{n}-2}p^{s\left(1-\frac{1}{n}\right)}\leqslant J(f)\,p^{\frac{n+2t+1}{n}-2}p^{s\left(1-\frac{1}{n}\right)}$$
$$=J(f)\,p^{\frac{2t+1}{n}-1}p^{s\left(1-\frac{1}{n}\right)} \qquad (p+3t+2\leqslant s\leqslant n+2t+1).$$

Therefore, on the interval $3t + 3 \leqslant s \leqslant n + 2t + 1$

$$|S(f,\ p^s)| \leqslant J(f) \max\left\{p^{\frac{3t+3}{n}-1},\ p^{\frac{2t+1}{n}-1}\right\} p^{s\left(1-\frac{1}{n}\right)},$$

i.e.

$$|S(f,\ p^s)| \leqslant J(f)\, p^{\frac{3t+3}{n}-1} p^{s\left(1-\frac{1}{n}\right)} \qquad (3t+3 \leqslant s \leqslant n+2t+1).$$

In view of (2.16), we obtain

$$|S(f,\ p^s)| \leqslant J(f)\, p^{\frac{3t+3}{n}-1} p^{s\left(1-\frac{1}{n}\right)} \tag{2.20}$$

for

$$p \leqslant n-t-2, \qquad 2t+2 \leqslant s \leqslant n+2t+1. \tag{2.21}$$

Let us examine estimate (2.18). Suppose $n - t - 1 \leqslant p \leqslant n$. We will show that then

$$\max\left\{p^{\frac{3t+3}{n}-1},\ \frac{p^{\frac{2t+1}{n}}}{n-t-1}\right\} = p^{\frac{3t+3}{n}-1}. \tag{2.22}$$

This assertion is equivalent to the inequality

$$p^{\frac{n-t-2}{n}} \leqslant n-t-1 \qquad (n-t-1 \leqslant p \leqslant n). \tag{2.23}$$

If $n = 3$, the possibilities are $p = 2$, $t = 1$ or $p = 3$, $t = 1$, and if $n = 4$, the possibilities are $p = 2$, $t = 2$ or $p = 3$, $t = 1$. In all of these cases (2.23) holds. Therefore, suppose $n \geqslant 5$. It is easy to see that $n > \ln n/\ln 2 + 2$ for $n \geqslant 5$; hence

$$n > t+2 \qquad (n \geqslant 5). \tag{2.24}$$

Therefore, when $n \geqslant 5$, (2.23) is equivalent to the inequality

$$p \leqslant (n-t-1)^{\frac{n}{n-t-2}} \qquad (n-t-1 \leqslant p \leqslant n),$$

and hence it is enough to show that

$$n \leqslant (n-t-1)^{\frac{n}{n-t-2}} \qquad (n-t-1 \leqslant p \leqslant n,\ n \geqslant 5).$$

Put $\chi(\eta) = y^{1/(\eta-1)}$ $(\eta > 1)$. We have

$$\frac{\chi'}{\chi} = (\eta-1)^{-2}\left(\frac{\eta-1}{\eta} - \ln \eta\right) = (\eta-1)^{-2}\left(\frac{\eta-1}{\eta} - \int_1^{\eta} \frac{du}{u}\right) < 0 \qquad (\eta > 1),$$

i.e. the function $\chi(\eta)$ decreases for $\eta > 1$. Therefore, in view of (2.14) and (2.24),

$$n < (n-1)^{\frac{n}{n-2}} < (n-t-1)^{\frac{n}{n-t-2}} \qquad (n \geqslant 5),$$

and (2.23) is proved. It follows from (2.18) and (2.22) that

$$|S(f,\ p^s)| \leqslant J(f)\, p^{\frac{3t+3}{n}-1} p^{s\left(1-\frac{1}{n}\right)} \tag{2.25}$$

for $n - t - 1 \leqslant p \leqslant n$ and $2t + 2 \leqslant s \leqslant n + 2t + 1$. If we combine (2.20) and (2.25), we see that (2.25) holds for all $p \leqslant n$ and $2t + 2 \leqslant s \leqslant n + 2t + 1$.

Now suppose $p > n$. Then $t = 0$, and (2.18) assumes the form

$$|S(f, p^s)| \leqslant J(f) \max\left\{p^{\frac{3}{n}-1}, \frac{p^{\frac{1}{n}}}{n-1}\right\} p^{s\left(1-\frac{1}{n}\right)}$$

$$(p > n, \quad 2t+2 \leqslant s \leqslant n+2t+1). \tag{2.26}$$

Here, max is equal to $p^{3/n-1}$ when $p \leqslant (n-1)^{n/(n-2)}$ and is equal to $p^{1/n}(n-1)^{-1}$ when $p \geqslant (n-1)^{n/(n-2)}$. Combining (2.25) and (2.26), we see that when $2t + 2 \leqslant s < n + 2t + 2$ we have

$$|S(f, p^s)| \leqslant J(f)\, p^{\frac{3t+3}{n}-1} p^{s\left(1-\frac{1}{n}\right)} \quad \left(p \leqslant (n-1)^{\frac{n}{n-2}}\right)$$

and

$$|S(f, p^s)| \leqslant J(f) \frac{p^{\frac{1}{n}}}{n-1} p^{s\left(1-\frac{1}{n}\right)} \quad \left(p \geqslant (n-1)^{\frac{n}{n-2}}\right).$$

Finally, applying Lemma 4, we see that these inequalities hold for all $s \geqslant 2t + 2$, and the lemma is completely proved.

We can now obtain estimates of $S(f, p^s)$ valid for all $s \in \mathbf{N}$.

LEMMA 7. *Suppose $n \geqslant 3$, $f \in K_n(p)$ and $s \in \mathbf{N}$. Then*

$$|S(f, p^s)| \leqslant \max\left\{p^{\frac{2t+1}{n}}, J(f)\, p^{\frac{3t+3}{n}-1}\right\} p^{s\left(1-\frac{1}{n}\right)} \quad \textit{if} \quad p \leqslant (n-1)^{\frac{n}{n-2}}, \tag{2.27}$$

$$|S(f, p^s)| \leqslant p^{\frac{1}{n}} p^{s\left(1-\frac{1}{n}\right)} \quad \textit{if} \quad p \geqslant (n-1)^{\frac{n}{n-2}}. \tag{2.28}$$

In particular,

$$|S(f, p^s)| \leqslant (n-1)\, p^{\frac{3t+3}{n}-1} p^{s\left(1-\frac{1}{n}\right)} \leqslant (n-1)\, n^{\frac{3}{n}} p^{\frac{3}{n}-1} p^{s\left(1-\frac{1}{n}\right)} \quad \textit{if} \quad p \leqslant n, \tag{2.29}$$

$$|S(f, p^s)| \leqslant (n-1)\, p^{\frac{3}{n}-1} p^{s\left(1-\frac{1}{n}\right)} \quad \textit{if} \quad n < p \leqslant (n-1)^{\frac{n}{n-2}}. \tag{2.30}$$

PROOF. When $1 \leqslant s \leqslant 2t + 1$, a trivial estimate yields

$$|S(f, p^s)| \leqslant p^s = p^{\frac{s}{n}} p^{s\left(1-\frac{1}{n}\right)} \leqslant p^{\frac{2t+1}{n}} p^{s\left(1-\frac{1}{n}\right)} \quad (1 \leqslant s \leqslant 2t+1).$$

In view of Lemma 6 and the fact that we always have $J(f) \leqslant n - 1$, we obtain (2.27) and (2.28).

Also, for $p \leqslant (n-1)^{n/(n-2)}$ we have

$$\max\left\{p^{\frac{2t+1}{n}}, J(f)\, p^{\frac{3t+3}{n}-1}\right\} \leqslant \max\left\{p^{\frac{2t+1}{n}}, (n-1)\, p^{\frac{3t+s}{n}-1}\right\}$$

$$= (n-1)\, p^{\frac{3t+3}{n}-1} \leqslant \begin{cases} (n-1)\, n^{\frac{3}{n}} p^{\frac{3}{n}-1} & (p \leqslant n), \\ (n-1)\, p^{\frac{3}{n}-1} & \left(n < p \leqslant (n-1)^{\frac{n}{n-2}}\right), \end{cases}$$

from which (2.29) and (2.30) follow, and the lemma is proved.

We will use (2.28) when $p \leqslant (n-1)^{2n/(n-2)}$, and the stronger estimate (0.21) when the reverse inequality holds.

Note that in the case $(n-1)^{n/(n-2)} \leqslant p \leqslant (n-1)^{2n/(n-2)}$ Nečaev ([18], Lemma 9), using (0.15), obtained the estimate

$$|S(f,\ p^s)| \leqslant (n-1)\, p^{\frac{1}{n}-\frac{1}{2}} p^{s\left(1-\frac{1}{n}\right)}; \tag{2.31}$$

it will be used in §3 to prove Theorem 1.

§3. Uniform estimate of $S(f, p^s)$

It follows easily from the results of §2 that

$$|S(f,\ p^s)| \leqslant A_6(n)\, p^{s\left(1-\frac{1}{n}\right)} \qquad (f \in K_n(p),\ s \in \mathbf{N}),$$

where $A_6(n) = O(n)$, uniformly with respect to p and s. Here we will prove that, in fact, $A_6(n) \leqslant C_4$.

This result will not be used in proving (0.19).

As usual when obtaining uniform estimates, the domain of variation of the parameters must be divided into many parts, each with its own estimates. We sharpen Lemma 6 for small p.

LEMMA 8. *Suppose $f \in K_n(p)$, $w = n/\ln p$, $s \in \mathbf{N}$ and*

$$n \geqslant C_5, \qquad p \leqslant n, \qquad s > w. \tag{3.1}$$

Then

$$|S(f,\ p^s)| \leqslant C_6 J(f)\, n^{-1} p^{s\left(1-\frac{1}{n}\right)}. \tag{3.2}$$

PROOF. 1. We will show that there exists a constant C_5 such that for all

$$n \geqslant C_5, \qquad p \leqslant n, \qquad w < s \leqslant n + w \tag{3.3}$$

we have

$$|S(f,\ p^s)| \leqslant C_7 J(f)\, n^{-1} p^{s-1}. \tag{3.4}$$

Since $t \leqslant \ln n/\ln p$ and $p \leqslant n$, it follows that $2(3t + 2) \leqslant w$ $(n \geqslant C_8)$. Suppose $n \geqslant C_8$. We determine $k \in \mathbf{N}$ from the conditions

$$2(q_k + (k+2)t + 2) < s \leqslant 2(q_{k+1} + (k+3)t + 2) \tag{3.5}$$

and consider two cases. Suppose

$$2(q_k + (k+2)t + 2) < s \leqslant q_{k+1} + (k+2)t + 2. \tag{3.6}$$

Then (2.3) is applicable, and since, in view of (3.6),

$$(s - q_k - (k+2)t - 2)^{-1} \leqslant 2s^{-1},$$

we have

$$|S(f,\ p^s)| \leqslant \frac{J(f)}{s - q_k - (k+2)t - 2}\, p^{s-1} \leqslant 2J(f)\, s^{-1} p^{s-1}. \tag{3.7}$$

Now suppose

$$q_{k+1} + (k+2)t + 2 < s < 2(q_{k+1} + (k+3)t + 2). \tag{3.8}$$

Since $p^k \leqslant q_{k+1} \leqslant 2p^k$ $(k \in \mathbf{N})$, it follows that

$$p^k \leqslant q_{k+1} < s \leqslant n + w \leqslant C_9 n, \qquad k \leqslant C_{10} \frac{\ln n}{\ln p},$$

and hence

$$(k+3)t+2 \leqslant C_{11}\frac{\ln^2 n}{\ln p} \leqslant \frac{w}{4} \leqslant \frac{s}{4} \qquad (n \geqslant C_{12}). \tag{3.9}$$

Therefore, using (3.8),

$$s \leqslant 2(q_{k+1}+(k+3)t+2) \leqslant 4p^k + \frac{s}{2} \qquad (n \geqslant C_{12}),$$

i.e. $s \leqslant 8p^k$ $(n \geqslant C_{12})$.

In view of (3.8), estimate (2.2) is applicable (with $k+1$ in place of k), and we obtain

$$|S(f, p^s)| \leqslant J(f)\, p^{s-k-1} \leqslant 8J(f)\, s^{-1}p^{s-1}. \tag{3.10}$$

Estimates (3.7) and (3.10) establish (3.4) with $C_6 = 8$ and $C_5 = \max(C_8, C_{12})$.

2. We will now show that (3.2) holds when $n \geqslant C_5$ and $w < s \leqslant n + w$. By what has been proved,

$$|S(f, p^s)| \leqslant C_5 J(f)\, s^{-1}p^{s-1} = C_6 J(f)\, s^{-1}p^{\frac{s}{n}-1}p^{s\left(1-\frac{1}{n}\right)} \qquad (w < s \leqslant n+w).$$

But $s^{-1}p^{s/n-1}$ is a convex function of s. Therefore, since $w = n/\ln p$, we have

$$\max_{w \leqslant s \leqslant n+w} s^{-1}p^{\frac{s}{n}-1} = \max\left\{w^{-1}p^{\frac{w}{n}-1},\ (w+n)^{-1}p^{\frac{w+n}{n}-1}\right\}$$
$$= \max\{n^{-1}\ln p \cdot e \cdot p^{-1},\ n^{-1}e\} \leqslant C_{12}n^{-1}. \tag{3.11}$$

Therefore,

$$|S(f, p^s)| \leqslant C_6C_{13}J(f)\, n^{-1}p^{s\left(1-\frac{1}{n}\right)} = C_6 J(f)\, n^{-1}p^{s\left(1-\frac{1}{n}\right)}$$
$$(n \geqslant C_5,\ w < s \leqslant n+w). \tag{3.12}$$

Finally, applying Lemma 4, we see that (3.12) holds for all $s > w$, and the lemma is proved.

THEOREM 1. *Suppose $n \geqslant 3$, $f \in K_n(p)$ and $s \in \mathbf{N}$. Then*

$$|S(f, p^s)| \leqslant C_2 p^{s\left(1-\frac{1}{n}\right)}. \tag{3.13}$$

PROOF. We consider two cases.

1) $p \leqslant n$. If $3 \leqslant n \leqslant C_5$, then, by Lemma 7,

$$|S(f, p^s)| \leqslant (n-1)\, n^{\frac{3}{n}}p^{\frac{3}{n}-1}p^{s\left(1-\frac{1}{n}\right)} \leqslant C_{14}p^{s\left(1-\frac{1}{n}\right)} \qquad (3 \leqslant n \leqslant C_3).$$

Suppose $n \geqslant C_5$. If $s \leqslant w = n/\ln p$, then, by a trivial estimate,

$$|S(f, p^s)| \leqslant p^s \leqslant p^{\frac{w}{n}}p^{s\left(1-\frac{1}{n}\right)} \leqslant ep^{s\left(1-\frac{1}{n}\right)} \qquad (s \leqslant w).$$

If $s > w$, then, by the previous lemma,

$$|S(f, p^s)| \leqslant C_6 J(f)\, n^{-1}p^{s\left(1-\frac{1}{n}\right)} \leqslant C_6 p^{s\left(1-\frac{1}{n}\right)} \qquad (s > w).$$

It follows from these estimates that

$$|S(f, p^s)| \leqslant C_{15}p^{s\left(1-\frac{1}{n}\right)} \qquad (n \geqslant 3,\ p \leqslant n). \tag{3.14}$$

2) $p > n$. If $n < p \leqslant (n-1)^{n/(n-2)}$, then, by (2.30),

$$|S(f, p^s)| \leqslant (n-1)\, p^{\frac{3}{n}-1} p^{s\left(1-\frac{1}{n}\right)} \leqslant p^{\frac{3}{n}} p^{s\left(1-\frac{1}{n}\right)} \leqslant (n-1)^{\frac{3}{n-2}} p^{s\left(1-\frac{1}{n}\right)}.$$

If $(n-1)^{n/(n-2)} < p \leqslant (n-1)^{2n/(n-2)}$, then, by (2.28),

$$|S(f, p^s)| \leqslant p^{\frac{1}{n}} p^{s\left(1-\frac{1}{n}\right)} \leqslant (n-1)^{\frac{2}{n-2}} p^{s\left(1-\frac{1}{n}\right)}.$$

If $p \geqslant (n-1)^{2n/(n-2)}$, then, by (0.21),

$$|S(f, p^s)| \leqslant p^{s\left(1-\frac{1}{n}\right)}.$$

Therefore,

$$|S(f, p^s)| \leqslant (n-1)^{\frac{3}{n-2}} p^{s\left(1-\frac{1}{n}\right)} \qquad (p > n); \tag{3.15}$$

in particular,

$$|S(f, p^s)| \leqslant C_{16} p^{s\left(1-\frac{1}{n}\right)} \qquad (n \geqslant 3,\ p > n). \tag{3.16}$$

In view of (3.14) and (3.16), we see that for all $n \geqslant 3$, $f \in K_n(p)$, and $s \in \mathbf{N}$

$$|S(f, p^s)| \leqslant C_2 p^{s\left(1-\frac{1}{n}\right)},$$

where $C_2 = \max(C_{15}, C_{16})$, and the theorem is proved.

It follows from (0.10) and (3.15) that

$$\sup_{s \in \mathbf{N}} H_n(p^s)/p^{s\left(1-\frac{1}{n}\right)} = 1 + O(n^{-1} \ln n) \tag{3.17}$$

uniformly with respect to $p > n$.

§4. Estimate of $S(f, q)$

LEMMA 9 (MULTIPLICATIVE PROPERTY). *Suppose $f \in K_n(q)$, where*

$$q = d_1 \dots d_l, \qquad d_j \in \mathbf{N}, \qquad (d_i, d_j) = 1 \qquad (i \neq j). \tag{4.1}$$

Then there exist polynomials $\varphi_j \in K_n(d_j)$ $(j = 1, \dots, l)$ such that

$$S(f, q) = \prod_{j=1}^{l} S(\varphi_j, d_j). \tag{4.2}$$

See [1], Chapter III, Lemma 1.

Put, as is customary,

$$\pi(x) = \sum_{p \leqslant x} 1, \qquad \vartheta(x) = \sum_{p \leqslant x} \ln p \qquad (x \geqslant 1). \tag{4.3}$$

We will need the following classical estimates for these functions:

$$\pi(x) = O\left(\frac{x}{\ln x}\right), \qquad \vartheta(x) = x + O\left(\frac{x}{\ln x}\right), \tag{4.4}$$

$$\pi(x) \ln x - \vartheta(x) = \int_1^x \pi(u) \frac{du}{u} = O\left(\frac{x}{\ln x}\right) \qquad (x \geqslant 2). \tag{4.5}$$

THEOREM 2. *Suppose $n \geqslant 3$ and $f \in K_n(q)$. Then*

$$|S(f, q)| \leqslant \bar{B}_n q^{1-\frac{1}{n}}, \tag{4.6}$$

where

$$\bar{B}_n = \exp(n + O(n/\ln n)). \tag{4.7}$$

PROOF. According to Lemma 9, if q has canonical decomposition

$$q = \prod_{j=1}^{l} p_j^{s_j} \qquad (p_j \in \mathbf{P},\ s_j \in \mathbf{N}),$$

then

$$S(f, q) = \prod_{j=1}^{l} S(\varphi_j, p_j^{s_j}),$$

where $\varphi_j \in K_n(p_j)$ $(j = 1, \ldots, l)$. Using Lemma 7 and (0.21) and putting $X_n = (n-1)^{n/(n-2)}$, we obtain

$$\begin{aligned} q^{\frac{1}{n}-1} |S(f, q)| &= \prod_{p_j \mid q} p_j^{s_j\left(\frac{1}{n}-1\right)} |S(\varphi_j, p_j^{s_j})| \\ &\leqslant \prod_{p \leqslant n} (n-1) n^{\frac{3}{n}} p^{\frac{3}{n}-1} \prod_{n < p \leqslant X_n} (n-1) p^{\frac{3}{n}-1} \prod_{X_n < p \leqslant X_n^2} p^{\frac{1}{n}} \\ &= \prod_{p \leqslant X_n^2} p^{\frac{1}{n}} \prod_{p \leqslant X_n} (n-1) p^{\frac{2}{n}-1} \prod_{p \leqslant n} n^{\frac{3}{n}} \\ &= \exp\left\{\frac{1}{n} \vartheta(X_n^2) + \ln(n-1) \cdot \pi(X_n) + \left(\frac{2}{n} - 1\right) \vartheta(X_n) + \frac{3 \ln n}{n} \pi(n)\right\} \\ &= \exp\left\{\frac{1}{n} \vartheta(X_n^2) + \left(1 - \frac{2}{n}\right)(\pi(X_n) \ln X_n - \vartheta(X_n)) + \frac{3 \ln n}{n} \pi(n)\right\}. \end{aligned}$$

Thus, we can put

$$\bar{B}_n = \exp(F(n)), \tag{4.8}$$

where

$$F(n) = \frac{1}{n} \vartheta(X_n^2) + \left(1 - \frac{2}{n}\right)(\pi(X_n) \ln X_n - \vartheta(X_n)) + \frac{3 \ln n}{n} \pi(n). \tag{4.9}$$

Let us estimate $F(n)$. We have

$$n < X_n = (n-1)^{\frac{n}{n-2}} = n + O(\ln n), \qquad X_n^2 = n^2 + O(n \ln n);$$

hence, in view of (4.4) and (4.5),

$$\begin{aligned} F(n) &= \frac{1}{n}\left(X_n^2 + O\left(\frac{X_n^2}{\ln X_n^2}\right)\right) + O\left(\frac{X_n}{\ln X_n}\right) + O(1) \\ &= \frac{1}{n}(n^2 + O(n \ln n)) + O\left(\frac{n}{\ln n}\right) + O\left(\frac{n}{\ln n}\right) + O(1) = n + O(n/\ln n). \end{aligned}$$

Therefore, $\bar{B}_n = \exp(n + O(n/\ln n))$, and the theorem is proved.

COROLLARY. *Suppose $n \geqslant 2$ and $f \in K_n(q)$. Then*

$$|S(f, q)| \leqslant e^{C_1 n} q^{1-\frac{1}{n}}. \tag{4.10}$$

If $p \leqslant (n-1)^2$, nontrivial estimates for $S(f, p)$ valid for arbitrary polynomials $f \in K_n(p)$ are not known. Putting $q = \Pi_{p \leqslant (n-1)^2} p$ and using the multiplicative property of $S(f, q)$ we obtain for $S(f, q)$ only the trivial estimate

$$|S(f, q)| \leqslant q = \Big\{\prod_{p \leqslant (n-1)^2} p\Big\}^{\frac{1}{n}} q^{1-\frac{1}{n}}. \tag{4.11}$$

In view of (4.4), the factor of $q^{1-1/n}$ is equal to

$$\Big\{\prod_{p \leqslant (n-1)^2} p\Big\}^{\frac{1}{n}} = \exp\Big\{\frac{1}{n}\vartheta((n-1)^2)\Big\} = \exp(n + o(n)),$$

i.e. it has the form of the right-hand side of (4.7). Thus, it is impossible to improve Theorem 2 without first improving the estimates for $S(f, p)$.

§5. On the number of solutions of the congruence $f(x) \equiv 0 \pmod m$

Suppose $f \in \mathbf{Z}[x]$, $\deg f = n \geqslant 2$, $m \in \mathbf{N}$, $m \geqslant 2$ and $(a_1, \ldots, a_n, m) = 1$. Let $\rho(f, m)$ denote the number of distinct solutions of the congruence

$$f(x) \equiv 0 \pmod m. \tag{5.1}$$

It is known that $\rho(f, p) \leqslant n$; hence

$$\rho(f, p^s) \leqslant np^{s-1} \qquad (s \in \mathbf{N}). \tag{5.2}$$

When $1 \leqslant s \leqslant n$, this estimate is attainable, in the sense of order, with respect to p, since the congruence $\bar{f}(x) = x^n \equiv 0 \pmod{p^s}$, where $1 \leqslant s \leqslant n$, has p^{s-1} solutions. In particular,

$$\rho(\bar{f}, m) = m^{1-\frac{1}{n}} \tag{5.3}$$

for all m of the form p^n.

Kamke [9] proved that for any $m \in \mathbf{N}$

$$\rho(f, m) \leqslant d^{n-1}(m)\, m^{1-\frac{1}{n}}, \tag{5.4}$$

where $d(m)$ is the number of divisors of m. Using known estimates for $d(m)$, it follows from (5.4) that

$$\rho(f, m) \leqslant m^{1-\frac{1}{n}} \exp\Big\{C_{17} n \frac{\ln m}{\ln\ln m}\Big\} \qquad (m \geqslant 3). \tag{5.5}$$

For polynomials with nonzero discriminant, other estimates of $\rho(f, m)$ were obtained by Øre and Nagell (see [23], Chapter III).

As an application of the estimates established in §4 for $S(f, q)$ we will prove here still another estimate for $\rho(f, m)$, incomparable with (5.4), but stronger than (5.5).

We will need two simple lemmas pertaining to classical functions of number theory.

LEMMA 10. *Suppose $x > 1$ and $0 \leqslant \alpha < 1$. Then*

$$\sum_{p \leqslant x} p^{-\alpha} \leqslant C_{18}\Big\{\frac{x^{1-\alpha}}{(1-\alpha)\ln x} + \ln\frac{1}{1-\alpha}\Big\}. \tag{5.6}$$

PROOF. It suffices to consider $x \geqslant 2$. Since $\pi(u) \leqslant C_{19} u/\ln u$ $(u \geqslant 2)$, we have

$$\sum_{p\leqslant x} p^{-\alpha}=\int_{2-0}^{x} u^{-\alpha}d\pi(u)=x^{-\alpha}\pi(x)+\alpha\int_{2}^{x}\pi(u)u^{-\alpha-1}du$$

$$\leqslant C_{19}\left\{\frac{x^{1-\alpha}}{\ln x}+\int_{2}^{x}\frac{du}{u^{\alpha}\ln u}\right\}\quad (x\geqslant 2). \tag{5.7}$$

Let us estimate the integral

$$I(x)=\int_{2}^{x} u^{-\alpha}(\ln u)^{-1}\,du.$$

We put $B=\exp(2/(1-\alpha))$ and consider two cases.

Suppose $2\leqslant x\leqslant B$. Then $u^{1-\alpha}\leqslant e^2$ $(2\leqslant u\leqslant x)$ and

$$I(x)\leqslant e^2\int_{2}^{x} u^{-1}(\ln u)^{-1}\,du=e^2(\ln\ln x-\ln\ln 2)$$

$$\leqslant e^2\left(\ln\frac{2}{1-\alpha}+\ln\frac{1}{\ln 2}\right)\leqslant C_{20}\ln\frac{2}{1-\alpha}\qquad (2\leqslant x\leqslant B). \tag{5.8}$$

Now suppose $x\geqslant B$. In view of (5.8),

$$I(x)=\int_2^B+\int_B^x\leqslant C_{20}\ln\frac{2}{1-\alpha}+\int_B^x u^{-\alpha}(\ln u)^{-1}\,du.$$

But, by the definition of B,

$$\int_B^x u^{-\alpha}(\ln u)^{-1}\,du=\frac{u^{1-\alpha}}{(1-\alpha)\ln u}\bigg|_B^x+\frac{1}{1-\alpha}\int_B^x u^{-\alpha}(\ln u)^{-2}\,du$$

$$\leqslant\frac{1}{1-\alpha}\left\{\frac{x^{1-\alpha}}{\ln x}+\frac{1}{\ln B}\int_B^x u^{-\alpha}(\ln u)^{-1}\,du\right\}\leqslant\frac{x^{1-\alpha}}{(1-\alpha)\ln x}+\frac{1}{2}\int_B^x u^{-\alpha}(\ln u)^{-1}\,du.$$

Hence

$$\int_B^x u^{-\alpha}(\ln u)^{-1}\,du\leqslant\frac{2x^{1-\alpha}}{(1-\alpha)\ln x}\qquad (x\geqslant B)$$

and

$$I(x)\leqslant\frac{2x^{1-\alpha}}{(1-\alpha)\ln x}+C_{20}\ln\frac{2}{1-\alpha}\qquad (x\geqslant B). \tag{5.9}$$

Combining (5.7)–(5.9), we obtain

$$\sum_{p\leqslant x} p^{-\alpha}\leqslant C_{21}\left\{\frac{x^{1-\alpha}}{(1-\alpha)\ln x}+\ln\frac{2}{1-\alpha}\right\}\qquad (x\geqslant 2). \tag{5.10}$$

Finally, if $0\leqslant\alpha\leqslant 1/2$, then the first term in (5.10) absorbs the second, and if $1/2<\alpha<1$, then

$$\ln\frac{2}{1-\alpha}\leqslant C_{22}\ln\frac{1}{1-\alpha}.$$

Hence (5.6) holds for all $x\geqslant 2$, and the lemma is proved.

LEMMA 11. *Suppose $m\geqslant 3$ and $1/2\leqslant\alpha<1$. Then*

$$\sigma_{-\alpha}(m)=\sum_{d\mid m} d^{-\alpha}\leqslant\exp\left\{C_{23}\left(\frac{(\ln m)^{1-\alpha}}{(1-\alpha)\ln\ln m}+\ln\frac{1}{1-\alpha}\right)\right\}. \tag{5.11}$$

PROOF. Let the canonical decomposition of m be $m = \Pi_1^{\omega} p_{k_s}^{a_s}$. Since $\ln(1 - p^{-\alpha}) \leqslant C_{24}p^{-\alpha}$ $(1/2 \leqslant \alpha < 1)$, we have

$$\sigma_{-\alpha}(m) = \prod_{s=1}^{\omega} \frac{1 - p_{k_s}^{-\alpha(a_s+1)}}{1 - p_{k_s}^{-\alpha}} < \prod_{p \mid m} (1 - p^{-\alpha})^{-1} = \exp\left\{\sum_{p \mid m} \ln(1 - p^{-\alpha})^{-1}\right\}$$
$$\leqslant \exp\left\{C_{24} \sum_{p \mid m} p^{-\alpha}\right\} \qquad (1/2 \leqslant \alpha < 1). \tag{5.12}$$

Also,

$$m = \prod_{s=1}^{\omega} p_{k_s}^{a_s} \geqslant \prod_{s=1}^{\omega} p_{k_s} \geqslant \prod_{p \leqslant p_\omega} p,$$

and hence

$$\ln m \geqslant \sum_{p \leqslant p_\omega} \ln p = \vartheta(p_\omega) \geqslant C_{25} p_\omega, \qquad p_\omega \leqslant C_{26} \ln m,$$

where $C_{26} > 1$. Using the previous lemma, we deduce that

$$\sum_{p \mid m} p^{-\alpha} \leqslant \sum_{p \leqslant p_\omega} p^{-\alpha} \leqslant \sum_{p \leqslant C_{26} \ln m} p^{-\alpha} \leqslant C_{18} \left\{\frac{(C_{26} \ln m)^{1-\alpha}}{(1-\alpha) \ln \ln m} + \ln \frac{1}{1-\alpha}\right\}$$
$$\leqslant C_{27} \left\{\frac{(\ln m)^{1-\alpha}}{(1-\alpha) \ln \ln m} + \ln \frac{1}{1-\alpha}\right\} \qquad (0 \leqslant \alpha < 1). \tag{5.13}$$

Combining this with (5.12), we obtain (5.11), and the lemma is proved.

THEOREM 3. *Suppose*

$$f \in \mathbf{Z}[x], \quad \deg f = n \geqslant 2, \qquad m \geqslant 3, \qquad (a_1, \ldots, a_n, m) = 1. \tag{5.14}$$

Then

$$\rho(f, m) \leqslant m^{1-\frac{1}{n}} \exp\left\{C_3 n \left(\frac{(\ln m)^{\frac{1}{n}}}{\ln \ln m} + 1\right)\right\}. \tag{5.15}$$

PROOF. We have (see [24], Chapter IV, Problem 1)

$$\rho(f, m) = \frac{1}{m} \sum_{a=1}^{m} \sum_{k=1}^{m} e\left(\frac{a}{m} f(k)\right) = \frac{1}{m} \sum_{d \mid m} \sum_{a(d)}^{*} \sum_{k=1}^{m} e\left(\frac{a}{d} f(k)\right),$$

where * denotes summation over a reduced residue system mod d. Using the notation established in the Introduction, we obtain

$$\rho(f, m) \leqslant \frac{1}{m} \sum_{d \mid m} \varphi(d) \frac{m}{d} H_n(d) = \sum_{d \mid m} \frac{\varphi(d)}{d} H_n(d), \tag{5.16}$$

where $\varphi(d)$ is Euler's function. But $H_n(d) \leqslant B_n d^{1-1/n}$ and $\varphi(d) \leqslant d$. Therefore

$$\rho(f, m) \leqslant B_n \sum_{d \mid m} d^{1-\frac{1}{n}} = B_n m^{1-\frac{1}{n}} \sigma_{\frac{1}{n}-1}(m). \tag{5.17}$$

Also, according to the corollary of Theorem 2, $B_n \leqslant \exp(C_1 n)$. Hence, using Lemma 11,

$$\rho(f,\ m)\leqslant m^{1-\frac{1}{n}}\exp\left\{C_1 n+C_{23}\left(n\frac{(\ln m)^{\frac{1}{n}}}{\ln\ln m}+\ln n\right)\right\}$$

$$\leqslant m^{1-\frac{1}{n}}\exp\left\{C_3 n\left(\frac{(\ln m)^{\frac{1}{n}}}{\ln\ln m}+1\right)\right\},$$

and the theorem is proved.

Note that the function

$$\psi(m)=\frac{(\ln m)^{\frac{1}{n}}}{\ln\ln m}+1$$

is such that $\psi(m)\leqslant C_{28}$ for $3\leqslant m\leqslant \exp\{(n\ln n)^n\}$, and hence

$$\rho(f,\ m)\leqslant e^{C_{29}n}m^{1-\frac{1}{n}}\qquad (3\leqslant m\leqslant \exp\{(n\ln n)^n\}). \tag{5.18}$$

The possibility that this estimate holds for all m is not excluded.

BIBLIOGRAPHY

1. I. M. Vinogradov, *Trigonometrical sums in number theory,* rev. ed., "Nauka", Moscow, 1971; English transl., Statistical Publishing Soc., Calcutta, 1975.

2. Hua Loo-keng, *Additive theory of prime numbers,* Trudy Mat. Inst. Steklov. **22** (1947); English transl. of rev. Chinese ed., Amer. Math. Soc., Providence, R.I., 1965.

3. ———, *Die Abschätzung von Exponentialsummen und ihre Anwendung in der Zahlentheorie,* Enzyklopädie der math. Wissenschaften, Bände 1, 2, Heft 13, Teil 1, Teubner, Leipzig, 1959.

4. ———, *Additive Primzahltheorie,* Teubner, Leipzig, 1959. [German transl. of rev. Chinese ed. of [2].]

5. Karl Friedrich Gauss, *Summatio quarundam serierum singularium,* Comment. Soc. Reg. Sci. Gottingen. Recentiores 1 (1808/11); reprinted in his *Werke,* Band II, Königl. Ges. Wiss. Göttingen, 1863, 1876.

6. G. H. Hardy, *Collected papers.* Vol. 1, Clarendon Press, Oxford, 1966.

7. Hermann Weyl, *Über die Gleichverteilung von Zahlen mod. Eins,* Math. Ann. **77** (1915/16), 313–352; reprinted in his *Gesammelte Abhandlungen,* Band I, Springer-Verlag, Berlin and New York, 1968, pp. 563–599.

8. Edmund Landau, *Vorlesungen über Zahlentheorie.* Band I, Hirzel, Leipzig, 1927.

9. E. Kamke, *Zur Arithmetik der Polynome,* Math. Z. **19** (1923/24), 247–264.

10. L. J. Mordell, *On a sum analogous to a Gauss's sum,* Quart. J. Math. Oxford Ser. **3** (1932), 161–167.

11. H. Davenport, *On certain exponential sums,* J. Reine Angew. Math. **169** (1932/33), 158–176.

12. Hua Loo-keng, *On an exponential sum,* J. Chinese Math. Soc. **2** (1940), 301–312.

13. André Weil, *On some exponential sums,* Proc. Nat. Acad. Sci. U.S.A. **34** (1948), 204–207.

14. ———, *Sur les courbes algébriques et les variétés qui s'en déduisent,* Actualités Sci. Indust., no. 1041, Hermann, Paris, 1948.

15. S. A. Stepanov, *On estimating rational trigonometric sums with prime denominator,* Trudy Mat. Inst. Steklov. **112** (1971), 346–371; English transl. in Proc. Steklov Inst. Math. **112** (1971).

16. V. I. Nečaev, *On the representation of natural numbers as a sum of terms of the form* $(x(x+1)\cdots(x+n-1))/n!$, Izv. Akad. Nauk SSSR Ser. Mat. **17** (1953), 485–498. (Russian)

17. A. A. Karacuba, *On estimates of complete trigonometric sums,* Mat. Zametki **1** (1967), 199–208; English transl. in Math. Notes **1** (1967).

18. V. I. Nečaev, *An estimate of a complete rational trigonometric sum,* Mat. Zametki **17** (1975), 839–849; English transl. in Math. Notes **17** (1975).

19. Chen Jing-run, *On the representation of a natural number as sum of terms of the form* $x(x+1)\cdots(x+k-1)/k!$, Acta Math. Sinica **9** (1959), 264–270; English transl. in Chinese Math. Acta **9** (1967).

20. Helmut Hasse, *Theorie der höheren Differentiale in einem algebraischen Funktionenkörper mit vollkommenen Konstantenkörper bei beliebiger Charakteristik*, J. Reine Angew. Math. **175** (1936), 50–54.

21. ———, *Noch eine Begrundung der Theorie der höheren Differentialquotienten in einem algebraischen Funktionenkörper einer Unbestimmten* (Nach einer brieflichen Mitteilung von F. K. Schmidt in Jena), J. Reine Angew. Math. **177** (1937), 215–237.

22. Oswald Teichmüller, *Differentialrechnung bei Charakteristik p*, J. Reine Angew. Math. **175** (1936), 89–99.

23. Trygve Nagell, *Introduction to number theory*, Almqvist & Wiksell, Stockholm; Wiley, New York, 1951.

24. I. M. Vinogradov, *Foundations of the theory of numbers*, 8th ed., "Nauka", Moscow, 1972; English transl. of 6th ed., Pergamon Press, New York, 1955.

Translated by G. A. KANDALL

ABCDEFGHIJ–CM–89876543210